Optimizing Stormwater Treatment Practices

Andrew J. Erickson • Peter T. Weiss
John S. Gulliver

Optimizing Stormwater Treatment Practices

A Handbook of Assessment and Maintenance

Springer

Andrew J. Erickson
St. Anthony Falls Laboratory
University of Minnesota
Minneapolis, MN, USA

Peter T. Weiss
Department of Civil Engineering
Valparaiso University
Valparaiso, IN, USA

John S. Gulliver
St. Anthony Falls Laboratory
University of Minnesota
Minneapolis, MN, USA

Additional material to this book can be downloaded from
http://extras.springer.com

ISBN 978-1-4899-9401-1 ISBN 978-1-4614-4624-8 (eBook)
DOI 10.1007/978-1-4614-4624-8
Springer New York Heidelberg Dordrecht London

Printed on acid-free paper

Cover illustration: Cover Image of Algal Blooms © Stormwater Maintenance, LLC, www.SWMaintenance.com, reprinted with permission

Springer is part of Springer Science+Business Media (www.springer.com)

Preface

This book is intended to be a resource to cost-effectively assess the performance of, and schedule maintenance for, stormwater treatment practices. Maintenance should never occur without an accurate assessment of the operating condition of a stormwater treatment practice. Thus, this book first details how to assess the performance of a stormwater treatment practice. It provides distinct levels of standardized assessment methodology, in increasing cost and difficulty, from which the user can select only those methods that are necessary. It also provides instructions on how to successfully complete an assessment of a stormwater treatment practice, including all required tasks, sample and data analysis, and other items. Finally, the book provides detailed guidance on how to use the information gathered during assessment to select and schedule the most appropriate maintenance actions.

The methods presented in this book will:

- Help users select cost-efficient assessment methods
- Help users develop an assessment program
- Ensure that an assessment program yields meaningful results
- Provide guidelines for reporting results and scheduling maintenance
- Allow for more meaningful comparisons between assessment and maintenance results of different stormwater treatment practices

The intended audience for this book includes engineers, planners, consultants, watershed district personnel, municipal staff, and natural resource managers, among others. Thus, case studies have been included, when possible, to provide practical examples related to the concepts discussed.

The research project that preceded this book and led to the development of much of the material was funded by the Minnesota Pollution Control Agency with C. Bruce Wilson as project manager. The authors would like to thank Bruce and all the staff at the Minnesota Pollution Control Agency for their confidence in us to deliver a quality product. Several partner projects that provided material for the case studies were funded by the Local Road Research Board of Minnesota, Metropolitan Council Environmental Services, Minnehaha Creek Watershed District, Minnesota Department of Transportation, Mississippi Watershed Management Organization, and the US Environmental Protection Agency. The authors wish to thank many individuals and organizations for their contribution to the completion of these projects which led to the development of this book: Brooke C. Asleson, Lawrence A. Baker, William R. Herb, Raymond M. Hozalski, Omid Mohseni, John L. Nieber, Bruce N. Wilson, the University of Wisconsin Extension, Emmons and Olivier Resources, South Washington Watershed District, Ramsey-Washington Metro Watershed District, Sarah M. Stai, Westwood Professional Services, Three Rivers Park District, City of Blaine, and Wenck Associates, Inc. The authors especially thank Brooke C. Asleson, Lawrence A. Baker, William R. Herb and John L. Nieber for contributing material that led to the development of this book. The authors also thank Bob Newport at US EPA Region 5 for his continual support of this effort.

Minneapolis, MN, USA — Andrew J. Erickson
Valparaiso, IN, USA — Peter T. Weiss
Minneapolis, MN, USA — John S. Gulliver

Contents

Introduction

1

Abstract

Countries and organizations around the world are working to reduce stormwater runoff volumes and increase the quality of runoff before it enters receiving water bodies. These efforts have resulted in the development of stormwater treatment practices, designed to retain contaminants such as suspended solids, nutrients, bacteria, metals, and others. The stormwater treatment practices are designed to perform at a certain level of treatment, but, over time, the performance level will decline due to factors such as clogging with sediment, reaching some finite contaminant storage capacity, excessive vegetative growth, and a host of other factors.

Stormwater treatment practices must receive intentional and regular maintenance to perform at predetermined, desired levels of runoff volume reduction, contaminant load reduction, or other primary objective over an extended period of time. In order to perform cost-effective maintenance at optimal time intervals, the practice must be regularly assessed. Assessment, as defined in this book, is the determination of the level of performance of a stormwater treatment practice with regard to the treatment goals of the practice and/or the determination of the state of a stormwater practice. For example, the former may include determining the percent load reduction a practice obtains with regard to suspended solids, while the latter may involve inspecting inlet and outlet structures of a pond for structural integrity and potential blockage from trash and debris. Only after a practice is assessed can optimal maintenance actions be planned and performed.

This chapter presents three levels of assessment, including visual inspection, testing (including capacity and synthetic runoff testing), and monitoring. The levels increase in complexity as presented and should be used selectively in accordance with stormwater management goals.

As society has developed and population has increased, human impact on the Earth has increased substantially. Houses, office complexes, shopping malls, airports, road networks, and scores of other amenities that provide conveniences and increase quality of life also create environmental challenges that cannot be ignored.

A.J. Erickson et al., *Optimizing Stormwater Treatment Practices: A Handbook of Assessment and Maintenance*, DOI 10.1007/978-1-4614-4624-8_1,

These challenges include changes in stormwater runoff quantity and quality as a result of anthropogenic activities.

Rain falling on an open field or prairie may be intercepted by vegetation, infiltrated into the soil, and stored in low-lying areas. Over time this water can be transferred to the atmosphere through evaporation and transpiration (i.e., evapotranspiration). Together, these mechanisms, called abstractions, can significantly reduce the fraction of rain that becomes runoff as it travels across the watershed and into a receiving water body.

If an open field is developed for human occupation, however, heavy construction equipment can compact the soil and reduce its infiltration capacity, vegetation will likely be removed, which will reduce interception and evapotranspiration, and pervious soil will be covered by impervious surfaces such as asphalt, concrete, and buildings. These changes significantly increase the volume of stormwater runoff and the velocity at which it travels across the surface. Furthermore, the composition of stormwater is affected. Rainwater that runs off metal roofs and buildings can acquire dissolved metals such as copper, lead, and zinc (Davis et al. 2001). The application of fertilizers, herbicides, and pesticides can contribute excess nutrients (i.e., phosphorus and nitrogen) and toxic chemicals to the runoff (APHA 1998a, US EPA 1999a). Through tire wear, brake pad erosion, and other mechanisms, vehicular traffic generates metals and solid particles that build up on the road and nearby ground surface and are washed off during a rain event. These changes have a significant negative impact on the quality of surface waters that receive urban stormwater runoff.

1.1 Need for Treatment

Formal action to reduce stormwater pollution in the USA was taken in 1987 with the passage of amendments to the Clean Water Act that required the United States Environmental Protection Agency (US EPA) to address this issue. The US EPA did so with the passage of the National Pollution Discharge Elimination Systems (NPDES) Phase I (1990) and Phase II (2003) requirements. Under this program, most stormwater discharges are considered point sources and require an NPDES permit. Also, under section 303(d) of the Clean Water Act, states are required to develop a list of impaired waters, which are water bodies that do not meet water quality standards. States must also develop Total Maximum Daily Loads (TMDLs) for these impaired waters, which are the maximum daily pollutant load a water body can receive and still meet water quality standards. In order to meet the goals of a TMDL, the pollutant loading to a water body, of which stormwater runoff often contributes a large fraction, will have to be reduced. Thus, to meet this need, technologies (herein called stormwater treatment practices) have been developed for reducing stormwater runoff volume and improving its quality.

The United Nations Millennium Project, which was commissioned in 2002, seeks to develop an action plan to reduce world hunger, poverty, and disease

(UNMP 2005). Millennium Goal 7 seeks to ensure sustainability and has driven countries to address stormwater pollution and reduce stormwater runoff contamination. For example, in order to help achieve Goal 7, China is addressing stormwater pollution in the city of Wuhan as part of a plan to improve living conditions in this highly urban area (ADB 2006).

In the European Union, the Water Framework Directive (WFD), which established regulations to protect and restore surface and ground waters across Europe, was passed in 2000 (European Commission 2008). With regard to surface waters, the WFD lists 33 priority pollutants and sets limits on the concentration of these contaminants in sources discharging to water bodies and in the water bodies themselves. Contaminants addressed by the WFD include phosphorus, suspended solids, and metals, among others. The regulation also calls for all surface waters to meet good ecological status, which means if the water body becomes impaired or threatened, relevant regulations can be made stricter. To meet WFD requirements, member countries must address point and nonpoint source pollution in their river basin management plans and adopt control measures to limit contamination of surface waters.

Australia has also focused efforts on managing urban stormwater runoff, as evidenced by its National Water Quality Management Strategy (ARMC-ANZECC 2000), Urban Stormwater Initiative, Clean Seas programs (Commonwealth of Australia 2002), and other programs. Overall, 12% of rainfall in Australia reaches surface waters as runoff, but in urban areas the amount jumps to 90%. Polluted urban runoff has been recognized as a significant source of water pollution. As a result, Australia is focusing efforts to reduce pollution in urban stormwater runoff and develop water management policies that result in ecological sustainable development.

Because of increased attention to stormwater quality, municipalities, watershed districts, and other organizations around the globe have spent countless resources on the installation, operation, and maintenance of stormwater treatment practices. Maintenance for a stormwater treatment practice ensures the practice is performing as designed and extends the useable life of the practice. Because maintenance is an ongoing task, resources, that are often limited, must be allocated to the maintenance of stormwater treatment practices each year. In order to optimally allocate resources and plan maintenance, the performance of stormwater treatment practices must be assessed. Historically, however, there has been little guidance on assessment and/or maintenance strategies. Therefore, assessment and maintenance are rarely performed in the most cost-effective manner. In order to optimize maintenance, stormwater treatment practices must be regularly assessed.

1.2 Need for Maintenance

In order to keep performing as designed, stormwater treatment practices require regular maintenance. For example, as a detention pond fills with sediment over time, it will approach its storage capacity, and previously settled solids may be resuspended and washed out of the pond. Also, if there is no remaining capacity to store newly settled solids, total suspended solids removal by the pond may decrease or not occur at all. In this case, removal of accumulated sediment will help restore the pond and improve its performance. As another example, Lindsey et al. (1991) found that 53% of the infiltration trenches investigated were not operating as designed, 36% were partially or totally clogged, and 22% exhibited slow infiltration. Detention ponds and infiltration trenches are not unique; any stormwater treatment practice will experience a drop in performance over time and, at some point, will require maintenance if it is to perform at desired levels. This fact generates two important questions that form the foundation of this book: (1) How can one most cost-effectively assess the performance of a stormwater treatment practice and, (2) Based on the assessment results, what kind of maintenance action is warranted, if any?

1.3 Maintenance Challenges and Limitations

Stormwater treatment practice maintenance is the purposeful management of a stormwater treatment practice so as to ensure proper function and extend the useable life by maintaining a desired level of performance and efficiency. Maintenance activities can be broken down into three different categories: routine (regular and relatively frequent), nonroutine (irregular and less frequent), and major (irregular and rare) actions, as shown in Fig. 1.1. The purpose of routine and nonroutine maintenance activities is to prevent or limit the need for major maintenance, and therefore the combination of these activities is called preventative maintenance.

Stormwater treatment practices have a life cycle from their creation (design and construction) through operative stages (functional or not) that is largely dictated by operation and maintenance actions. As maintenance involves a significant amount of resources (personnel, equipment, materials, sediment disposal expense, etc.), a major challenge is how to most cost-effectively budget these resources so that nonroutine and major maintenance activities occur as infrequently as possible without underutilizing resources on excessive and unnecessary routine maintenance. In order to successfully balance limited resources and meet this challenge, assessment must occur on a regular basis.

Accurate assessment results will indicate if maintenance is needed, what level of maintenance activity is warranted, and can be used to help estimate when future maintenance will be required. By identifying the level and timing of maintenance required, assessment prevents unnecessary maintenance activity, which helps to conserve valuable limited resources. Regular assessment also identifies the need for maintenance, so that problems with a stormwater treatment practice can be identified and resolved as soon as possible, before the state of the practice deteriorates further. In this way, assessment can minimize the frequency of nonroutine and major maintenance actions.

The impact of regular assessment resulting in optimal maintenance does have limits, however. The best maintenance action performed at the best possible time will not always restore the practice to previous performance levels. At some time over the

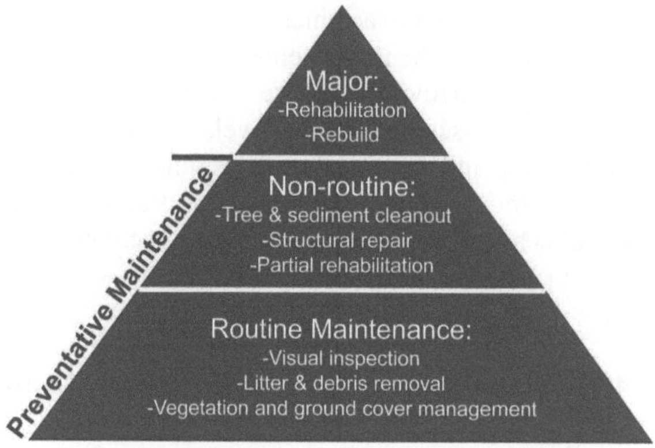

Fig. 1.1 Stormwater treatment practice operation and maintenance pyramid

life of any practice, rehabilitation, replacement, or rebuilding of the practice will be required. For example, a sand filter accumulates relatively large solid particles on its surface, while some smaller particles travel into the filter media before being strained and retained by the filter media. Because the top solid layer tends to more quickly reduce filtration capacity than the solids trapped below the media surface, removing the surface layer of solids from the top of a filter is a maintenance action that is often recommended when the flow rate through the filter drops to unacceptable levels. Although this action can significantly increase the flow rate through the filter, the filtration capacity is also somewhat reduced because of the finer solid particles that have accumulated within the media. Eventually, as the smaller particles trapped within the filter become more and more numerous, the filter media will clog throughout a significant portion of the media depth and removing the top layer of solids will have little effect on the overall capacity of the filter. When this occurs, major maintenance, such as replacing the entire media bed, will be necessary. In this example, no routine or even nonroutine maintenance could have prevented the need for major maintenance.

1.4 Assessment Strategies

For reasons discussed above, before maintenance of any stormwater treatment practice is performed, the practice must be assessed to determine its level of performance with regard to stormwater management goals (e.g., volume reduction, suspended solids removal). The assessment results should be accurate and within an acceptable level of uncertainty. Only after the current level of performance is determined and compared to the original or desired level of performance can a wise decision regarding maintenance be made. If maintenance is not immediately warranted, assessment results can help estimate when maintenance may be required in the future. Thus, assessment must not only be an integral part of any stormwater treatment practice maintenance plan, but it must also be the first requirement and must take place prior to any maintenance activity. Also, because assessment is a part of maintenance, if the implementation of a maintenance plan is to be cost-effective, all assessment activity must be cost-effective. In order to help users perform cost-effective assessment, three levels of assessment strategies, in order of increasing time, cost, and complexity, are presented and described in this book. To best use resources, should choose the lowest level of assessment, or combination of levels, that will achieve the storm water management goals.

1.4.1 Visual Inspection, Testing, and Monitoring

Implementing cost-effective assessment and maintenance is specific to each stormwater treatment practice and each assessment goal. Thus, only a brief overview of the three assessment techniques, is presented in this chapter. More detailed information on assessment techniques, including specifics for practices that utilize sedimentation, filtration, infiltration, and biological processes for stormwater treatment, is presented in subsequent chapters. The three levels of assessment, visual inspection, testing (capacity and synthetic runoff), and monitoring, are presented and briefly discussed below.

1. *Visual inspection*: A rapid assessment procedure that qualitatively evaluates and documents the functionality of a stormwater treatment practice by sight only. The primary purpose of visual inspection is to identify, diagnose, and schedule maintenance for stormwater treatment practices. The results can be used to select and schedule maintenance.
2. *Testing*: Testing of stormwater treatment practices consists of making a series of measurements under conditions *other than* a natural runoff event. Testing is further subdivided into two kinds of testing, capacity testing and synthetic runoff testing.
   (a) *Capacity testing*: An assessment method that uses a series of spatially distributed, relatively rapid, and simple point measurements. Specifically, capacity testing can be used to estimate the surface saturated hydraulic conductivity at specific locations within a practice or the depth of accumulated sediment, which can be related to the sediment removal capacity (remaining sediment storage volume) of an entire stormwater treatment practice. The results can be used to select and schedule maintenance.
   (b) *Synthetic runoff testing*: An assessment method in which a prescribed amount of synthetic stormwater is applied to a stormwater treatment practice under controlled conditions to assess its effectiveness. With measurements such as drain time and mass of pollutant capture, synthetic runoff testing can be used to assess the performance of a stormwater treatment practice for runoff volume reduction (e.g., through infiltration) and pollutant removal efficiency. Results from synthetic runoff testing can also be used to calibrate watershed models for simulation of performance during natural rainfall events.

3. *Monitoring*: An assessment method which measures performance of a practice during natural rainfall or snowmelt events by measuring influent and effluent flow rates, collecting influent and effluent stormwater samples for analysis, and comparing influent and effluent volume, pollutant concentration, or pollutant load. Monitoring is the most comprehensive form of assessment and can assess multiple aspects of stormwater treatment practice performance (e.g., peak flow reduction and pollutant removal). It also requires a significant amount of data to calculate reliable results because the number and range of variables are large. The results from monitoring can be used to describe the runoff and pollutant load characteristics of a watershed and the associated response of stormwater treatment practices.

Developers of an assessment program should consider each of the three levels of assessment based on effort and uncertainty aspects. The lowest assessment level should be considered first, and the next highest level should only be considered when warranted by the goals of the assessment program. By this process, assessment may include any combination of the three assessment levels, but inclusion of all three levels is not mandatory (and often not recommended). Each level of assessment will vary in application based on the stormwater treatment practice and the assessment goals. A summary of the three levels of assessment, including the relative effort, typical elapsed time, advantages, and disadvantages, is given in Table 1.1.

Visual inspection (level 1) and capacity testing (level 2a) do not depend on the size of the stormwater treatment practice and therefore can be applied to any practice. The applicability of synthetic runoff testing (level 2b), however, is dependent on the size of the practice and the available water supply. Monitoring is only limited by the site design and accessibility of the practice.

1.5 Need for This Book

Other stormwater management books and manuals discuss the design of stormwater treatment practices and may sometimes include hypothetical or assumed maintenance schedules. In contrast, this book provides instructions on how to directly measure the level of performance of stormwater treatment practices and bases proposed maintenance schedules on actual performance and historical maintenance efforts and costs. The inspection methods, which are proven in the field and have been implemented successfully, are necessary as regulatory agencies begin requiring measured performance of such devices.

In order to determine the effectiveness of stormwater treatment practices, it is common to monitor the practice during actual rainfall or snowmelt events. This process is time consuming, expensive, and uncertain. It involves waiting for a rainfall event, hoping the depth of rain and duration are adequate for measurement and sampling equipment, and collecting water samples for an unknown duration of time. It is not uncommon to monitor many (~20) rainfall events over 2 years before obtaining sufficient information to minimize uncertainty. The continuous change in discharge, concentrations, and performance means that uncertainties are still great.

Table 1.1 Comparison of the three levels of assessment

	1. Visual inspection	2a. Capacity testing	2b. Synthetic runoff testing	3. Monitoring
Objectives	Determine if stormwater treatment practice is malfunctioning	Determine infiltration or sedimentation capacity and rates	Determine infiltration rates, capacity, and pollutant removal performance	Determine infiltration rates, capacity, and pollutant removal performance
Relative effort	1	10	10–100	400
Typical elapsed time	1 day	1 week	1 week–1 month	14+ months
Advantages	Quick, inexpensive	Less expensive, no equipment left in field	Controlled experiments, more accurate with fewer tests required for statistical significance as compared to monitoring, no equipment left in field	Most comprehensive, assesses stormwater treatment practice within watershed without modeling
Disadvantages	Limited knowledge gained	Limited to infiltration and sedimentation capacity/rates, uncertainties can be substantial	Cannot be used without sufficient water supply, limited scope	Uncertainty in results due to lack of control, equipment left in field

We have developed a three-tiered assessment technique in which each tier increases in complexity and cost. In this technique, monitoring is the highest tier; thus, significant amounts of time and money can be saved if one of the two lower tiers can be used to assess the stormwater treatment practice. This book describes how to determine which tier is appropriate and provides detailed information on how to perform each tier of assessment. The assessment can then be used to schedule maintenance, document performance for regulatory agencies, and perform construction due diligence. This book also documents the maintenance actions and frequency needed to maintain performance once the appropriate assessment techniques have been chosen and implemented, and the cost of maintenance. This book also contains a substantial number of examples and case studies to illustrate the use of the material.

1.6 About This Book

This book, "*Optimizing Stormwater Treatment Practice: A Handbook of Assessment and Maintenance*," is organized into 13 chapters, References, and one appendix. Each chapter is intended to provide guidance and information on stormwater (e.g., stormwater processes), assessment (e.g., water budget measurement), or maintenance. To help the reader find specific information within this book, each chapter and the Appendix are briefly described below.

Chapter 1: Introduction. Chapter 1 describes the need for stormwater treatment and maintenance of stormwater treatment practices; assessment strategies including inspection, testing, and monitoring; and the need for and organization of this book.

Chapter 2: Impacts and Composition of Urban Stormwater. Urban development results in impacts by stormwater on water resources. Chapter 2 describes these impacts, including flow and channel alteration as well as pollutants such as nutrients and metals. Chapter 2 also provides numerical values for typical concentrations of some pollutants of concern in urban stormwater.

Chapter 3: Stormwater Treatment Processes. Urban stormwater runoff can be treated to reduce runoff volume, peak flow, and pollutants. Chapter 3 discusses processes relevant to stormwater treatment, including physical, biological, and chemical processes. Understanding these processes is critical to developing a successful assessment and maintenance program.

Chapter 4: Stormwater Treatment Practices. Chapter 4 provides a detailed description of the most common stormwater treatment practices, including dry and wet ponds, filtration practices, infiltration basins and trenches, biofiltration/bioinfiltration practices (rain gardens), constructed wetlands, and swales.

Chapter 5: Visual Inspection of Stormwater Treatment Practices. The first step in understanding the performance of a stormwater treatment practice is visual inspection. Chapter 5 provides detailed information for visual inspection of stormwater treatment practices including the most common inspection criteria and inspection considerations specific to each stormwater treatment practice.

Chapter 6: Capacity Testing of Stormwater Treatment Practices. Capacity testing is a testing methodology comprising a series of spatially distributed point measurements. Chapter 6 provides details about applying capacity testing to stormwater treatment practices with details specific to each treatment practice and a case study of capacity testing to measure infiltration rate in a bioinfiltration (rain garden) practice.

Chapter 7: Synthetic Runoff Testing of Stormwater Treatment Practices. Synthetic runoff testing is a testing methodology in which synthetic stormwater is applied to a stormwater treatment practice and the response by the practice is measured. Chapter 7 provides details about applying synthetic runoff testing to stormwater treatment practices with details specific to each treatment practice.

Chapter 8: Monitoring of Stormwater Treatment Practices. Monitoring stormwater treatment involves setting up equipment that will measure flow and collect samples during natural rainfall events. Through monitoring, the performance of a stormwater treatment practice can be determined for actual runoff events from the contributing watershed. Chapter 8 provides details about applying monitoring to stormwater treatment practices with details specific to each treatment practice and a case study of monitoring a dry pond.

Chapter 9: Water Budget Measurement. Assessment of stormwater treatment practices requires an understanding and accurate measurement of the water budget. Chapter 9 describes several methods for measuring water budget inflows and outflows, such as open channel flow, conduit flow, infiltration, and rainfall, and provides recommendations for simple, accurate water budget measurement.

Chapter 10: Water Sampling Methods. One possible goal of an assessment program is to determine the pollutant removal efficiency of a stormwater treatment practice. To determine pollutant removal efficiency, pollutant amounts (e.g., mass, concentration) in stormwater runoff must be measured. Chapter 10 discusses methods for measuring pollutant(s) in stormwater runoff, including the number of storm events and samples, sampling methodology (e.g., flow-weighted), and handling, as well as special considerations such as winter sampling in cold climates and automatic sampling of suspended solids.

Chapter 11: Analysis of Water and Soils. Stormwater often contains several pollutants at various concentrations. Determining target pollutants and accurate analytical methods is important in developing a simple and cost-effective assessment and maintenance program. Chapter 11 describes common stormwater analyses and quality assurance/quality control considerations such as bias, precision, and inspection.

Chapter 12: Data Analysis. Once assessment data have been collected, the data must be analyzed to determine the performance of the practice. Chapter 12 describes methods for analyzing assessment data, such as summation of loads and the Event Mean Concentration (EMC) and calculating the corresponding uncertainty.

Chapter 13: Maintenance of Stormwater Treatment Practices. Performance will determine whether a stormwater treatment practice is functioning adequately. For practices that are functioning below desired levels, appropriate maintenance should be performed. Chapter 13 provides guidance for determining what maintenance is required, describes maintenance activities specific to each stormwater treatment practice, and presents actual maintenance frequency, effort, and costs for various kinds of stormwater treatment practices.

Appendix A: Visual Inspection Checklists. Appendix A contains all the checklists for visual inspection of stormwater treatment practices.

References: A complete list of references cited throughout the book is provided.

Impacts and Composition of Urban Stormwater

2

Abstract

If construction or development occurs in a watershed, the area of impervious surfaces such as roads, parking lots, and buildings typically increases, with a corresponding decrease in the area of natural pervious surfaces. The result is an increase in stormwater runoff volumes, peak flow rates, and a degradation of runoff quality. The degradation of runoff quality can be observed in increased concentrations and total mass loads of nutrients and other organics, metals, chlorides, bacteria, viruses, hydrocarbons, and other substances, as well as increases in runoff temperature. The increased loading of these substances to receiving water bodies can be quite detrimental. This chapter discusses the most common contaminants found in urban stormwater runoff, their impacts, and typical concentrations.

2.1 Impacts of Urban Stormwater

The impact of the increase in urban stormwater runoff volumes and pollutant loads is substantial. Urban stormwater is responsible for about 15% of impaired river miles in the USA (US EPA 2000b) and urban stormwater is the leading cause of pollution to fresh and brackish receiving waters (Mallin et al. 2009). Stormwater impacts can be hydrologic, chemical, biological, or physical, but the impacts of greatest concern are biological integrity and habitat alteration due to the loading of sediment, nutrients, metals, chloride, bacteria, high temperature water, oxygen-demanding substances, and hydrocarbons (US EPA 1992). Although the impacts tend to increase as the urbanization within the watershed increases, negative impacts can be significant in watersheds that are less than 10% urbanized (Pitt 2002).

A.J. Erickson et al., *Optimizing Stormwater Treatment Practices: A Handbook of Assessment and Maintenance*, DOI 10.1007/978-1-4614-4624-8_2,

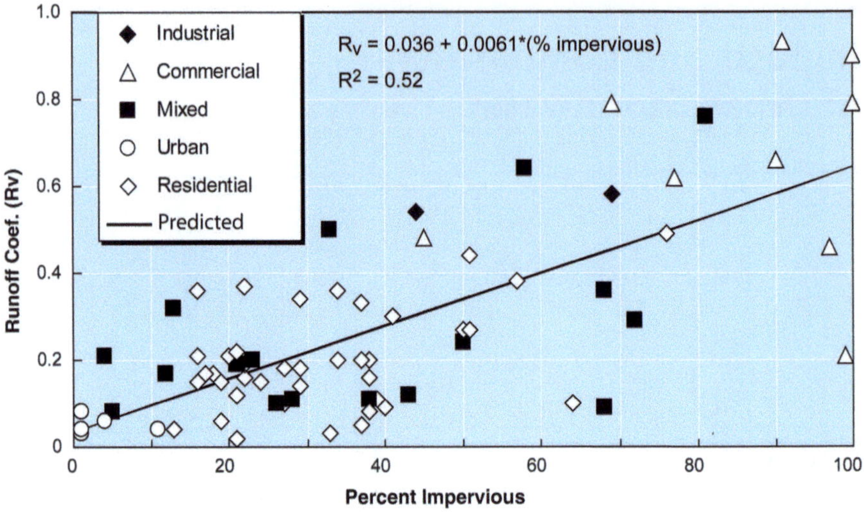

Fig. 2.1 Percent impervious surface versus runoff coefficient for watersheds included in the National Urban Runoff Program (NURP) study (US EPA 1983)

2.1.1 Flow and Channel Alteration

Urbanization, as reflected by increased impervious surface, alters watershed hydrology in several ways. As shown in Fig. 2.1, for sites studied in the US EPA's National Urban Runoff Program (US EPA 1983), one way is an increase in the runoff coefficient (ratio of inches of runoff to inches of rainfall) as the percentage of impervious surface in the watershed increases. Increasing imperviousness also leads to hydrographs with shorter durations and greater peak flows, larger flood flows, and smaller base flows (Paul and Meyer 2001). Some of the effects of altered flow on biota are due to larger peak temperatures, altered sediment discharge, unstable channels, fewer pools, and degraded habitat due to channelization. Evaluations of stream habitats indicate that flow and channel alteration are major contributors to the observed decline in biological integrity often associated with increased imperviousness (Paul and Meyer 2001; Pitt 2002; Booth et al. 2002; and Schueler 2000a).

2.1.2 Nutrients

Nutrients, primarily phosphorus and nitrogen, increase plant growth in streams, reservoirs, and lakes in a process called eutrophication. In many parts of the country, stormwater containing a large concentration of nutrients enters lakes, causing nutrient enrichment, reduced water clarity, and increased presence of undesirable blue-green algae and other plants. Upon decomposition and oxidation of the plant matter, dissolved oxygen in the water body is consumed, and can be reduced to zero or near zero levels.

Because of urban sprawl, residential land is now the dominant land use in 64% of the nation's water supply reservoirs (Robbins et al. 1991). Eutrophication caused by nutrients in stormwater often impairs municipal drinking water supplies. One example is the New Croton Reservoir, which provides daily drinking water to about 900,000 New York City residents. Due to excessive phosphorus loading, the reservoir suffers from algae blooms, low dissolved oxygen, and poor taste. As a result, it is common for the use of this reservoir to be reduced or temporarily suspended in the summer (NYSAGO 2011).

Excessive nutrient loading can also stimulate the growth of undesirable rooted aquatic plants in streams. The US EPA reports that approximately 11% of the nation's assessed stream miles are threatened or impaired due to excess nutrients (US EPA 2000b). With only 26% of the total stream miles assessed, the total number of stream miles that are threatened or impaired is likely significantly higher.

2.1.3 Metals

A large number of potentially toxic substances, including metals, occur in stormwater. Metals of primary concern (based on toxicity and occurrence) are cadmium, copper, zinc, and lead (Jang et al. 2005; Rangsivek and Jekel 2005), with roughly 50% of the metal load in dissolved form (Morrison et al. 1983). Lead concentration in the environment has declined since the 1970s, when lead in gasoline and paint was banned, but there is still substantial degrading lead paint present in the urban environment, making this a continuing concern. Note the smaller lead concentration in the three more recent stormwater studies in Table 2.4, as compared with that in the NURP study (US EPA 1983).

Large concentrations of metals can be lethal, and moderate concentrations can reduce growth, reproduction, and survival in aquatic organisms. Small concentrations of metals also have been documented to alter the behavior and competitive advantage of invertebrates, a result that could change the balance of ecosystems (Clements and Kiffney 2002). Kayhanian et al. (2008) investigated the toxicity of stormwater runoff from urban highway sites near Los Angeles, USA. Results indicated that the toxicity to water fleas and flathead minnows of the most toxic samples was mostly, but not entirely, due to copper and zinc.

Once in an aquatic environment, metals can accumulate in freshwater biofilms to such an extent that the biofilm concentrations are larger than sediment metal concentrations. Fish and invertebrates feed on biofilms, as a result, the metals can be transferred through the food chain (Ancion et al. 2010), and bioaccumulation will continue to occur.

Of the stream miles assessed in the USA as of 2011, approximately 7% have been categorized as threatened or impaired due to metals other than mercury. Mercury, which is a metal more common to runoff from industrial land uses and

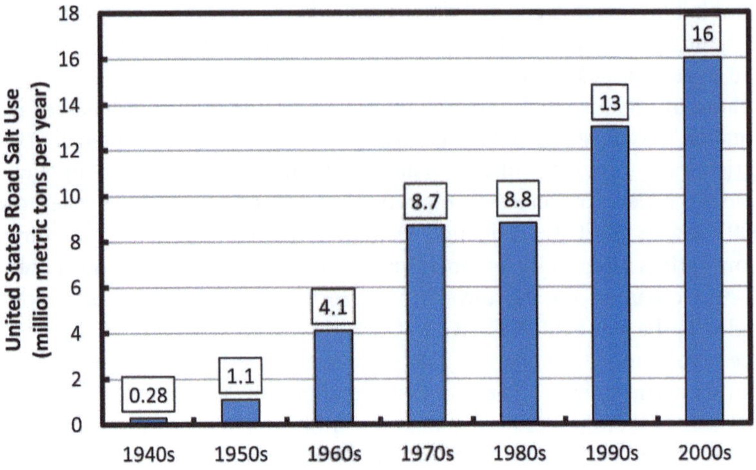

Fig. 2.2 Increase in annual road salt use in the USA (NURP) study (US EPA 1983)

atmospheric deposition, has threatened or impaired approximately 5% of assessed stream miles (US EPA 2011). As more stream miles are assessed, these numbers are likely to increase.

2.1.4 Chloride

Chloride is an emerging urban pollutant as a result of road deicing (Novotny et al. 2009). The chloride concentration in streams has been directly correlated with the percent of impervious surface area (Kaushal et al. 2005) and the quantity of rock salt purchases (Novotny et al. 2008). Furthermore, annual road salt use in the USA has continually increased since the 1940s (Fig. 2.2).

After application on a road surface, salt will typically travel to receiving waters, where it can increase the salt concentration of the water body. Peak chloride concentration in winter runoff has been observed close to sea water (35,000 mg/L) at 11,000 mg/L (Corsi et al. 2010), and peak chloride concentration in urban streams during winter can be several thousand mg/L. At these concentrations, chloride can negatively impact the water body. For example, salt increases the density of water, and highly saline waters can settle to the bottom of lakes and alter lake mixing patterns. This process can extend periods of low oxygen in or near the sediment which, in turn, can cause the release of dissolved phosphorus and metals (Wetzel 1975; p. 224). Increased salt concentration can also negatively impact aquatic life by decreasing biodiversity, increasing mortality rates of tadpoles, and decreasing the overall health of organisms (Novotny and Stefan 2010).

The US EPA's acute and chronic water quality limits for chloride in fresh water are 860 mg/L and 230 mg/L, respectively. Studies have found that these limits are often exceeded in northern metropolitan areas during the winter, and less often

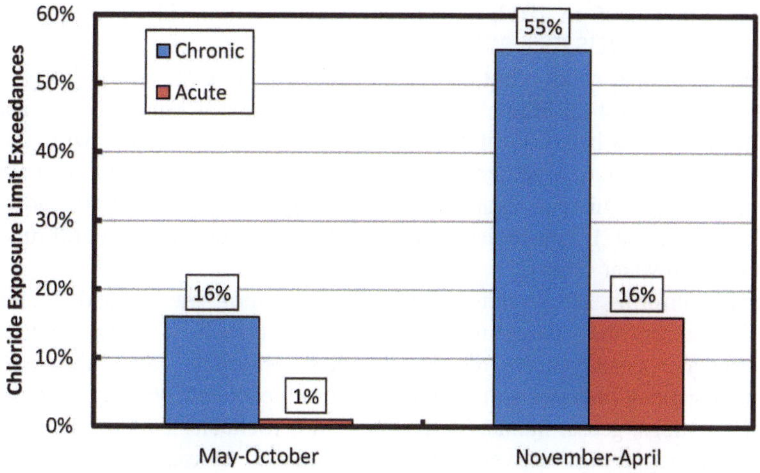

Fig. 2.3 Chloride exposure limit exceedances in summer (May–October) and winter (November–April) for 168 monitoring stations in 13 northern metropolitan areas (Corsi et al. 2010)

during the summer. Exceedance was negligible in all southern monitoring sites (Fig. 2.3).

In the Minneapolis-St. Paul metropolitan region of Minnesota, background chloride concentration in urban lakes was 3–10 mg/L before development, but in 2005, after development, averaged 87 mg/L. Detailed modeling (Novotny and Stefan 2010) has shown that at current road salt application rates, the salt concentration will continue to rise such that some urban lakes will exceed the established chronic standard of 230 mg/L (four-day average) (MPCA 2003) for impairment to aquatic habitat. Clearly, salt concentration in road runoff and surface waters cannot be ignored.

2.1.5 Bacteria and Viruses

The potential for bacterial contamination of water is generally measured by the concentration of fecal coliforms, *Escherichia coli*, or enterococci. Although most fecal coliforms are not pathogenic, they are currently the best established representative surrogate, or indicator, of human pathogens.

Rain and increased runoff increase the presence of microbial pathogens in marine and estuarine waters, an effect that can be a direct health threat to humans and can contaminate shellfish. In fact, urban stormwater is the cause of 40% of shellfish closures in US waters (Mallin et al. 2009). In one study, the number of gastrointestinal diseases per 1,000 swimmers was shown to increase linearly with coliform counts (Durfour 1984). One outcome of elevated coliform levels is beach closings. From 2006 to 2009, 32–43% of beaches nationwide were affected by closings each year (US EPA 2011).

Fecal coliform concentrations are generally largest immediately after rainstorms. A study of Minnehaha Creek in Minnesota (Wenck 2003) reported that fecal coliforms in excess of 2000 CFU/100 mL were found only within 3 days of a rainstorm. Fecal streptococci and *E. coli* were found in 94% and 95.5%, respectively, of municipal separate storm sewer system (MS4) outfalls monitored (Clark and Pitt 2007). This indicates that a large percentage of fecal coliforms are a result of stormwater runoff.

Fecal coliforms are excreted from the bodies of warm-blooded animals. For urban stormwater, sources may include humans (via illicit sewage connections to stormwater conveyances), dogs, cats, geese, raccoons, and other wildlife. Although generation rates (number of coliforms excreted per day) for various organisms (dogs, geese, humans) are well known (Schueler 2000b), there is little information regarding "delivery ratios" (the fraction of excreted coliforms that enters runoff) for urban stormwater.

Potential for groundwater contamination by bacteria and pathogens depends on the soil chemical properties, adsorption capability, the ability of the soil to physically strain the pathogens, and pathogen survival. Bacteria survive longer in low pH (acidic) soils and in soils with large organic content. Bacteria and viruses can move through soil media and may be transported to aquifers by infiltrating stormwater. The transport distance of bacteria seems to be a function of bacteria density and water velocity through the soil (Camesano and Logan 1998; Unice and Logan 2000). Pitt et al. (1996) rate enteroviruses as having high groundwater contamination potential for all surface and subsurface infiltration/injection systems and a variety of other pathogens as having high groundwater contamination potential for subsurface infiltration/injection systems.

Although documented cases of groundwater contamination do exist, bacteria are generally removed by straining at the soil surface and sorption to solid particles. Once removed from the water, the ability of bacteria to survive is a function of factors such as temperature, pH, and presence of metals, among others. Bacteria survival may be between two and three months, but survival for up to 5 years has been documented (Pitt et al. 1999). Although not readily modeled in natural environments, fecal coliforms can also regrow in the environment under warm conditions with a supply of organic matter for food, conditions commonly found in wetlands or stormwater ponds.

As part of the National Urban Runoff Program, fecal coliforms were evaluated at 17 sites for 156 storm events, and based on the results, it was concluded that coliform bacteria in urban runoff may exceed US EPA water quality criteria during and after storm events (US EPA 1999a). There existed a high degree of variability within the data, but land use did not appear to correlate with coliform concentration. During warmer months, concentrations were approximately 20 times larger than cold months.

A study by the National Academy of Sciences (NAS 2000) noted that very large removal rates—on the order of 99% would be needed to reduce coliforms from the levels observed in urban stormwater (15,000–20,000/100 mL) to the EPA's 200/100 mL criterion for recreational water. Their review indicated that bacterial

Table 2.1 Comparison of mean bacterial removal rates achieved by different stormwater treatment practices

	Bacterial removal rate%		
Bacterial indicator	Ponds	Sand filters	Swales
Fecal Coliform	65% (n = 9)	51% (n = 9)	–58% (n = 5)
Fecal Streptococci	73% (n = 4)	58% (n = 7)	N/A
E. coli	51% (n = 2)	N/A	N/A

The number (n) of practices analyzed indicated in parenthesis (data from NAS 2000)
N/A = Information not reported in the source

removal rates in several types of stormwater treatment practices were significantly less than 99% (Table 2.1). Studies of coliform regrowth in stormwater ponds have apparently not been reported in the peer-reviewed literature.

2.1.6 Temperature

Urbanization generally requires removing crops, trees, and native plants from parcels of land and replacing them with roads, parking lots, lawns, and buildings. Along with the impacts previously mentioned, these changes in land use affect riparian shading and heating of runoff in these areas, which results in increases in summertime temperatures of nearby streams. This can significantly impact relatively cool waters, such as trout streams that are fed by groundwater, because increases in the volume and temperature of runoff from impervious surfaces will dilute the colder groundwater, lower the volume of groundwater entering the water body, and reduce coldwater fish habitat.

In most temperate climates, the risk to salmon and trout populations due to increased temperature is of concern. Water temperature affects many areas of fish health, such as migration, disease resistance, growth, and mortality (Sullivan et al. 2000).

The US EPA reports that, of the 935,393 stream miles assessed nationwide, approximately 5% (46,786 miles) are threatened or impaired due to thermal pollution (US EPA 2011). With only 26% of the nation's stream miles assessed, the total length of impaired streams is certain to increase. In a study of 39 trout streams in Wisconsin and Minnesota, stream temperatures increased 0.25°C (0.5°F) per 1% increase in watershed imperviousness (Wang et al. 2003). In Minnesota, no temperature increase is allowed in cold water streams (Class 2A) while warm water streams (Class 2B) are allowed a temperature increase of 3°C (5°F) (MPCA 2003).

The temperature of stormwater runoff is controlled by the initial rainfall temperature and by the heating/cooling processes with the land and other surfaces during runoff. The temperature of land surfaces is controlled by several processes including solar radiation during the daytime, atmospheric long wave radiation, long wave back radiation from the surface, evaporative heat flux, and sensible heat flux. Land surfaces are heated above ambient air temperature primarily by solar radiation. Asphalt and roof surfaces in Minnesota reach daily maximum temperatures that average 50°C (122°F) in July, while concrete reaches an average of 46°C (115°F) in July (Herb et al. 2007a). Maximum roof temperatures in Mississippi and Wisconsin were found to be over 70°C (158°F), with similar temperatures reported for Arizona, Georgia, Oregon, and Texas (Winandy et al. 2004). Asphalt temperatures in Arizona exceeded 71°C (160°F) in June and July with a mean daily maximum of over 68°C (154°F), while 25% of maximum asphalt temperatures in July and August were over 54°C (129°F) (Harrington et al. 1995).

Land surfaces that are warmed by solar radiation will cool prior to and during storm events, with the amount of cooling depending on the surface properties and the amount of cloud cover prior to the onset of rainfall. Pavement has relatively large thermal mass, and therefore cools off more slowly, while asphalt shingle rooftops cool quickly, and typically reach ambient air temperature by the start of a storm (Janke et al. 2009). Storms with a rapid onset of cloud cover and rainfall after bright sun give less time for land surfaces to cool off, which leads to larger land temperatures at the onset of rainfall and larger runoff temperatures.

Overall, pavement produces the largest runoff temperatures, which can approach 30°C (86°F) (Herb et al. 2007a; Herb et al. 2008). Tar-gravel commercial rooftops also have sufficient thermal mass to produce large runoff temperatures (Janke et al. 2009). The largest runoff temperatures are typically observed at the beginning of storm events, when the land surfaces are warmest. Because the amount of heat available to heat surface runoff is finite, land surfaces have more impact on runoff temperatures for smaller storms.

Bare soil can produce thermal pollution if the infiltration capacity is exceeded. Vegetated land surfaces are cooler, due to evaporation and the shading effect of vegetation. Figure 2.4 is an illustration of the simulated average weekly surface temperature of five types of land uses in St. Paul, Minnesota, from April to November of 2004, where asphalt weekly surface temperatures are about 18°C (64°F) warmer than grasslands or vegetated ponds in midsummer months. Soil evaporation keeps bare soil surfaces somewhat cooler, with average July maximum temperatures of about 34°C (93°F), and July average temperatures of about 24°C (75°F). Vegetated, pervious surfaces produce relatively little thermal pollution per unit area, because both runoff rates and runoff temperatures are lower than pavement temperatures. Vegetated surfaces, however, can produce thermal pollution for storms of large volume and dew point temperature (Herb et al. 2007a).

2.1.7 Oxygen-Demanding Substances

Degradation of organic matter in streams utilizes oxygen, often rapidly enough to reduce the dissolved oxygen concentration to an extent that it impairs aquatic life. Unlike point source discharges that cause the most severe oxygen depletion during low-flow conditions, reduced oxygen concentration due to stormwater in urban streams often occurs just after major storms because of the transport of oxygen-demanding substances into streams.

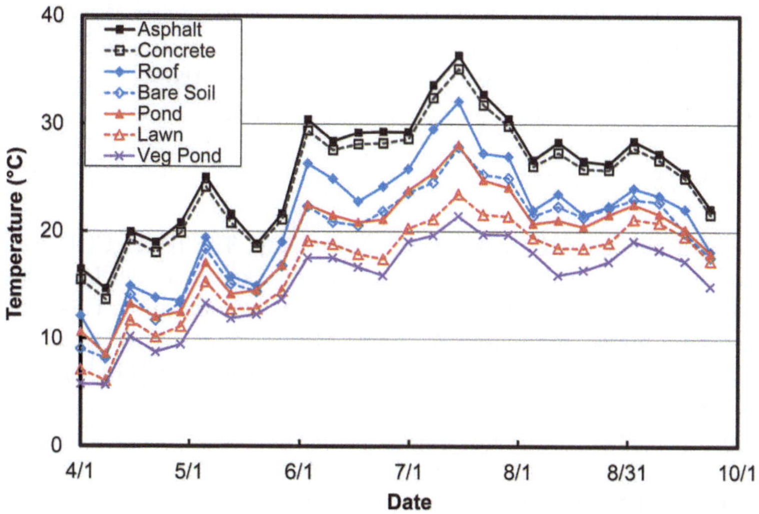

Fig. 2.4 Simulated average weekly surface temperature for five land uses calculated with hourly climate data from St. Paul, MN (2004); the vegetated pond represents a pond with fully covered emergent macrophytes (from Herb et al. 2007a)

The amount of degradable organic matter in water is usually quantified by measuring the amount of oxygen that is consumed as the organic matter decomposes or is oxidized. The biochemical oxygen demand (BOD) test measures the amount of oxygen that bacteria consume, typically over a five-day span (BOD_5), while decomposing organic matter. Chemical oxygen demand (COD) measures the oxygen consumed in oxidizing all organic matter into carbon dioxide and water, not just that portion that can be oxidized by bacteria. For this reason, COD values are larger than BOD values. Because the COD test oxidizes organic matter with a strong chemical, this test can usually be completed in a day or less, while BOD tests take 5 days.

Some sources of oxygen demand, such as animal waste and decaying vegetation, are natural; others, like oils and greases, grass clippings, and pet waste, are anthropogenic (due to human activity). Mallin et al. (2009) found BOD in urban runoff to be directly correlated with the percent of watershed development and the percent impervious surface cover within a watershed.

Maestre and Pitt (2005) report BOD_5 median values of 8.6 mg/L ($n = 3{,}105$) and COD median values of 53 mg/L ($n = 2{,}751$) for approximately 100 municipal separate storm sewer system (MS4) discharge sites throughout the USA, with freeway sites having median values of 8 mg/L ($n = 26$) and 100 mg/L ($n = 67$) for BOD_5 and COD, respectively. Mijangos-Montiel et al. (2010) evaluated contaminant concentrations in runoff from gas stations in Tijuana, Mexico, and compared results to values of contaminant concentrations in runoff from gas stations in urban areas in Washington, DC, USA, and in Genoa, Italy. BOD_5 was not evaluated, but COD values of gas station runoff were 169, 9, and 27 mg/L for

Table 2.2 Estimated pollutant loads from urban watersheds (from Mijangos-Montiel et al. 2010)

Parameter	Skudai, Malaysia	Johor Bahru, Malaysia	Saskatoon, Canada	Dallas-Fort Worth, Texas, USA	Tijuana, Mexico
Drainage area (ha)	3.3	171.4	616	4–65	2–45
COD (kg/ha)	9.0	12	24	3	54

Tijuana, Washington, DC, and Genoa, respectively. The larger COD values in Mexico were thought to be due to the large fraction of older vehicles in Mexico, which tend to leak more oils, and the fact that gas stations in the USA and Italy undergo more thorough cleaning. Mijangos-Montiel et al. (2010) also compared COD loads from urban catchments in Malaysia, Canada, the USA, and Mexico. Results are shown in Table 2.2.

Oxygen-demanding substances have threatened or impaired 83,580 of the 935,393 assessed stream miles (~9%) in the USA (US EPA 2011). The pollutant load entering these streams is not necessarily from urban stormwater runoff. Loading may originate from agricultural runoff, combined sewer overflows that result in untreated sewage entering the water body, and other natural and/or anthropogenic sources. For example, BOD concentrations in urban stormwater runoff are often less than sewage treatment plant effluent values, which are typically around 20 mg/L. Detrimental effects of urban stormwater runoff BOD have been documented, however. Lee and Jones-Lee (2003) report that urban runoff from a several inch rainfall event caused the dissolved oxygen levels of the San Joaquin River Deep Water Ship Canal to drop from 7 to 9 mg/L to about 3.5 mg/L, and that fish kills coincident to urban runoff from the same storm were associated with low DO levels in nearby rivers.

2.1.8 Hydrocarbons

Hydrocarbons are organic compounds consisting entirely of hydrogen and carbon molecules. Hydrocarbons can reduce the ability of some organisms to reproduce, can negatively impact the growth and development of various aquatic species, and can be lethal at high concentrations. For example, fish kills have been attributed to high levels of polycyclic aromatic hydrocarbons (PAHs) (Watts et al. 2010). When consumed, hydrocarbons can bioaccumulate in aquatic organisms, and, when collected in bottom sediment, degradation of hydrocarbons can consume oxygen (Stenstrom et al. 1982), which can negatively impact the entire aquatic ecosystem.

In stormwater runoff, hydrocarbons originate from vehicle coolants, gasoline, oils, lubricants, coal tar-based asphalt sealants (a source of PAHs), atmospheric deposition, and other sources. Thus, gas station runoff and vehicles in general are a major source of the hydrocarbon load in runoff (Mijangos-Montiel et al. 2010). Once in stormwater, hydrocarbons are often associated with particulates (Stenstrom et al. 1982).

2.2 Composition of Urban Stormwater

The chemical composition of stormwater varies with time during a storm event. Pollutant concentration is therefore often represented as event mean concentration (EMC), where the EMC is calculated by (2.1):

$$\mathrm{EMC} = \frac{\sum_{i=1}^{n} C_i Q_i}{\sum_{i=1}^{n} Q_i} \tag{2.1}$$

where

Q_i = flow during interval i
C_i = concentration during interval i

Median concentrations of relevant stormwater constituents are provided in Tables 2.3 and 2.4. Two major analyses of urban stormwater throughout the USA (US EPA 1983; Maestre and Pitt 2005) show that EMCs vary significantly among storms and that relationships between annual median EMC and land use are weak. The values in Tables 2.3 and 2.4 should therefore be used only as approximations. Field measurements are required to determine the actual concentration of a given constituent for a particular watershed and rain event.

Table 2.3 Composition of urban stormwater concentrations of major constituents

Metropolitan area	TSS	VSS	TP	DP	COD	BOD	TKN	NO_3–N	NH_4
Minneapolis-St. Paul, MN	184	66	0.58	0.2	169	N/A	2.62	0.53	N/A
Marquette, WI	159	N/A	0.29	0.04	66	15.4	1.5	0.37	0.2
Madison, WI	262	N/A	0.66	0.27	N/A	N/A	N/A	N/A	N/A
USA cities (median for all sites)	100	N/A	0.33	0.12	65	9	1.5	0.68	N/A
USA MS4 discharge sites (median for all land use	58	N/A	0.27	0.13	53	8.6	0.6	N/A	N/A
California highway runoff (median for all sites)	59.1	N/A	0.18	0.06	N/A	N/A	1.4	0.6	N/A

Minneapolis-St. Paul, MN, mean EMC (Brezonik and Stadelmann 2002)
Marquette, geometric means (Steuer et al. 1997)
Madison, geometric means (Waschbusch et al. 1999)
USA cities, medians (US EPA 1983)
USA MS4 discharge sites (Maestre and Pitt 2005)
California highways, medians (Kayhanian et al. 2007)
N/A = Information not reported in the source
Note: All values are in mg/L. *TSS* = total suspended solids, *VSS* = volatile suspended solids, *TP* = total phosphorus, *DP* = dissolved phosphorus, *COD* = chemical oxygen demand, *BOD* = biochemical oxygen demand, *TKN* = total Kjeldahl nitrogen, NO_3–N = nitrate nitrogen, NH_4 = ammonium

Table 2.4 Composition of urban stormwater metals (mg/L) and coliforms (#/100 mL)

Metropolitan area	Total lead	Total zinc	Total copper	Total cadmium	Coliforms
Minneapolis-St. Paul, MN	0.060	N/A	N/A	N/A	N/A
Marquette, WI	0.049	0.111	0.022	0.0006	10,200
Madison, WI	0.032	0.203	0.016	0.0004	175,106
USA cities (median for all sites)	0.144	0.160	0.034	N/A	21,000
USA MS4 discharge sites (median for all land use	0.016	0.117	0.016	0.001	12,000
California highway runoff (median for all sites)	0.0127	0.1112	0.0211	0.00044	N/A

Minneapolis-St. Paul, mean EMC (Brezonik and Stadelmann 2002)
Marquette, geometric means (Steuer et al. 1997)
Madison, geometric means (Waschbusch et al. 1999)
USA cities, medians (US EPA 1983)
USA MS4 discharge sites (Maestre and Pitt 2005)
California highways, medians (Kayhanian et al. 2007)
N/A = information not reported in the source

Stormwater Treatment Processes 3

Abstract

Stormwater treatment practices may reduce runoff volumes, contaminant concentrations, and/or the total contaminant mass load carried by runoff into receiving water bodies. Processes used by treatment practices include physical processes such as sedimentation, filtration, and infiltration, along with thermal, biological, and chemical processes. A single treatment practice may use multiple processes.

This chapter discusses these processes in detail and presents a brief discussion of how to assess the performance or current condition of each process within a treatment practice.

This chapter discusses the physical, biological, and chemical processes that can be used to improve the quality of urban stormwater. Stormwater treatment processes are not the same as stormwater treatment practices. A stormwater treatment practice is something that improves stormwater runoff quality, reduces runoff volume, reduces runoff peak flow, or any combination thereof. Examples of stormwater treatment practices, which will be discussed in the next chapter, include sand filters, infiltration basins and trenches, rain gardens, dry ponds, wet ponds, constructed wetlands, filter strips, swales, wet vaults, and underground sedimentation practices, among others.

A stormwater treatment process is the mechanism by which a stormwater treatment practice improves stormwater runoff quality, reduces runoff volume, reduces runoff peak flow, or any combination thereof. For example, a dry pond holds stormwater and, relative to uncontrolled conditions, releases it slowly to downstream receiving waters. The primary treatment process of a dry pond is sedimentation because most of the pollutants in stormwater that are retained by a dry pond settle to the bottom of the pond. It should be noted, however, that some treatment practices use more than one primary process. In this book these practices are referred to as hybrid practices. Because the treatment process is important when considering assessment and maintenance, stormwater treatment practices in this book are organized by their primary treatment process.

A.J. Erickson et al., *Optimizing Stormwater Treatment Practices: A Handbook of Assessment and Maintenance*, DOI 10.1007/978-1-4614-4624-8_3,

3.1 Physical Processes

3.1.1 Sedimentation

Process: Sedimentation is the process by which solids settle out of a water column. Sedimentation of particles is the primary pollutant retention process in many stormwater treatment practices.

Assessment Considerations: The sedimentation rate (i.e., particle settling velocity) of small particles under quiescent conditions is given by Stokes' (1851) Law, as shown in (3.1):

$$V_s = \frac{g(s-1)d^2}{18\nu} \tag{3.1}$$

where

V_s = terminal settling velocity of the solid particle
s = specific gravity of sediment (2.65 for silica sand)
g = gravitational acceleration (9.81 m/s^2)
d = diameter of particle
ν = kinematic viscosity of the fluid (varies with temperature)

A relationship for settling velocity that incorporates larger particles, such as sands, has been developed by Ferguson and Church (2004), as shown in (3.2). This equation becomes Stokes' Law for small particle diameters and results in a constant drag coefficient for large particle diameters:

$$V_s = \frac{g(s-1)d^2}{18\nu + (0.75\,Cg(s-1)d^3)^{1/2}} \tag{3.2}$$

where

C = constant (0.4 for spheres, 1 for typical sand grains)

A comparison of Stokes' Law (3.1) and the equation by Ferguson and Church (2004) (3.2) is provided in Fig. 3.1.

As shown in (3.1) and (3.2), fluid temperature (via fluid density and viscosity), particle size, and particle density all influence settling velocity. The dependence of settling velocity on particle size is shown in Fig. 3.1. Particle density can have a significant impact on settling velocity. For example, algae growing in wet ponds or wetlands typically have much smaller sedimentation rates than inorganic particles because their density is much smaller. Similarly, particles tend to settle faster as fluid temperature increases. Salinity also affects settling velocity, but the effect is minimal for stormwater salt concentrations, even when snowmelt contains deicing agents. Particle density affects settling velocity more than temperature for the range of particles and fluid temperatures common to stormwater runoff, as shown in Figs. 3.2 and 3.3.

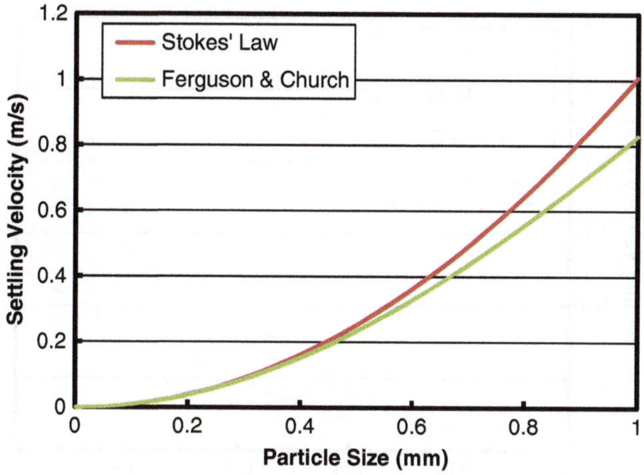

Fig. 3.1 Comparison of settling velocity determination by Stokes' Law and Ferguson and Church (2004) (particle density = 2.65 g/cm^3; temperature = 25°C; $C = 1$)

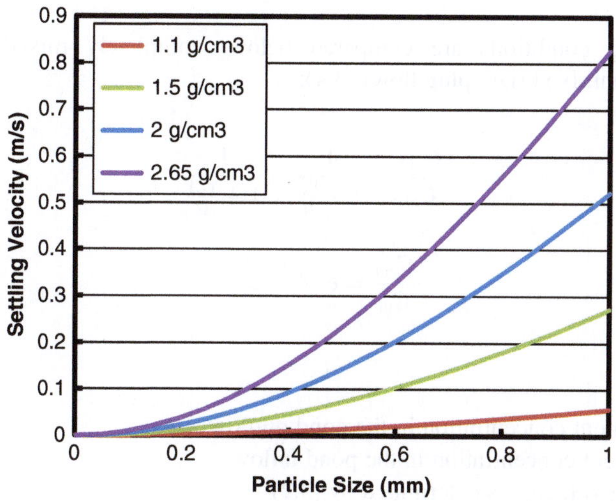

Fig. 3.2 Effect of particle density on settling velocity using (3.2) (temperature = 25°C; $C = 1$)

Turbulent eddies in the stormwater treatment practice reduce particle settling and tend to mix the water column so the resulting sediment concentration profile is a balance between settling and mixing. At the bottom of the stormwater treatment practice, however, turbulence dissipates, so that the particles near the bottom settle out according to (3.2). To illustrate the effect of mixing on settling velocity, two

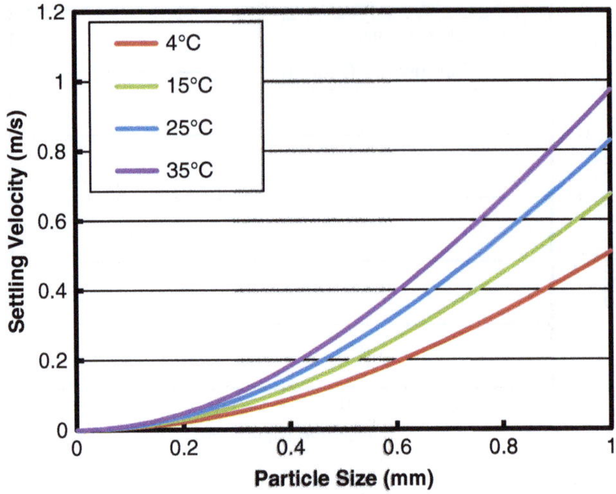

Fig. 3.3 Effect of fluid temperature on settling velocity using (3.2) (particle density = 2.65 g/cm^3; $C = 1$)

ideal mixing conditions are compared below: completely mixed (3.3) and completely unmixed (i.e., plug flow) (3.4):

$$\frac{C_{\text{out}}}{C_{\text{in}}} = \frac{1}{1 + \frac{V_s t}{H}} = \frac{1}{1 + \frac{V_s A}{Q}} \tag{3.3}$$

$$\frac{C_{\text{out}}}{C_{\text{in}}} = \mathrm{e}^{-\frac{V_s t}{H}} = \mathrm{e}^{-\frac{V_s A}{Q}} \tag{3.4}$$

where

C_{out} = sediment concentration in the pond outflow
C_{in} = sediment concentration in the pond inflow
V_s = settling velocity as determined by (3.2)
t = hydraulic retention time = volume/discharge
H = mean settling pond depth = volume/area
A = settling pond area
Q = discharge into settling pond = discharge out of the pond

Under completely mixed conditions, pollutants that flow into the stormwater treatment practice are assumed to be instantly mixed throughout the water contained within the practice. Under plug flow conditions, pollutants that flow into the stormwater treatment practice are assumed to move through the practice under laminar conditions without any mixing. In reality, stormwater treatment

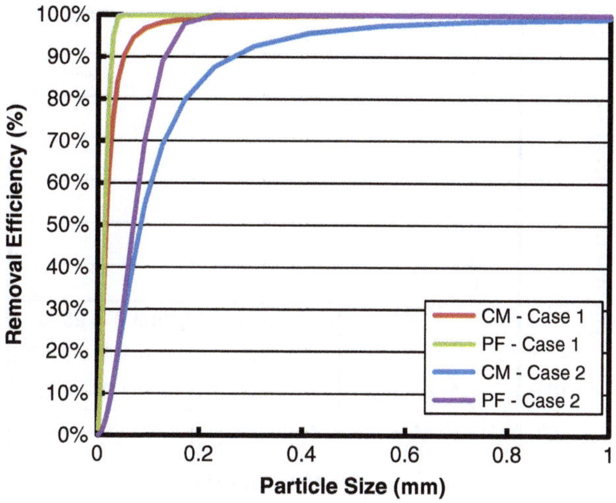

Fig. 3.4 Effect of mixing on stormwater treatment practice removal efficiency by sedimentation. Case 1: peak rainfall intensity = 0.61 in/h, pond: impervious watershed area ratio = 5%. Case 2: peak rainfall intensity = 1.58 in/h, pond: impervious watershed area ratio = 0.5%. Settling velocity by (3.2); *CM* = completely mixed using (3.3); *PF* = plug flow using (3.4); particle density = 2.65 g/cm^3; temperature = 25°C; $C = 1$

practices operate somewhere between completely mixed and plug flow conditions. To illustrate minimum and maximum treatment efficiency, completely mixed (3.3) and plug flow (3.4) conditions are shown in Fig. 3.4.

The effect of temperature and particle density on removal efficiency in stormwater treatment practices can also be examined using one of these mixing models, as shown in Fig. 3.5. As shown in Figs. 3.2 and 3.3, particle density has more effect on sedimentation than fluid temperature.

3.1.2 Filtration

Process: Filtration is the retention of suspended particles while water is passing through granular media. The main mechanism of filtration is straining, in which suspended solids are trapped by media particles. Filtered water is discharged from the filter media where it may be collected by an underground collection system, flow over the surface, and/or travel to another treatment practice or receiving water body.

Assessment Considerations: Filtration removes suspended solids and sediment-bound pollutants from solution, but allows them to accumulate in the filter. Filtered material eventually clogs filters, reducing the flow rate through the filter. Clogging can be measured by the change in head loss across the filter, a reduction in filtration rate, or an increase in the time required to filter a certain volume of runoff.

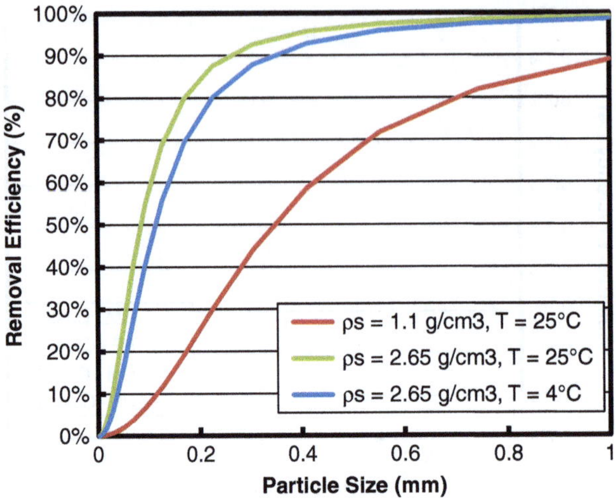

Fig. 3.5 Effect of particle density and fluid temperature on stormwater treatment practice removal efficiency by sedimentation. Settling velocity by (3.2); completely mixed using (3.3); peak rainfall intensity = 1.58 in/h; pond: impervious watershed area ratio = 0.5%; $C = 1$

For the same head (i.e., depth of water), the filtration rate is greater for filter media with large pore spaces (i.e., large grain size such as gravel) than for filter media with small pore spaces (i.e., small grain size such as sand or silt). Filter media with large pores, however, allows larger solids to pass through the filter, which results in a lower fraction of retained solids.

3.1.3 Infiltration

Process: Infiltration occurs when stormwater flows into the ground rather than into a collection system or across the ground surface and into a water body. Once underground, the infiltrated water may continue to the groundwater table or move laterally as subsurface flow, but in contrast to filtration, it is not collected by any system or designed to flow out of the treatment media through which it has traveled. Thus, infiltration reduces peak discharge and is the primary means through which stormwater volume reduction is achieved. It is often assumed that the soil will filter infiltrated water, but this may not be true in some circumstances with subsurface infiltration or in areas with geological formations such as karst or fractured bedrock.

Assessment Considerations: Like filtration, infiltration is limited by clogging of the media (soil), which is typically measured by the time required to infiltrate a known water volume or depth or by measuring the saturated hydraulic conductivity of the soil. Because it is challenging to obtain samples location of infiltrated water, contaminant removal efficiencies for infiltration systems are often difficult to obtain. If such information is desired, infiltration practices should be constructed to allow for convenient sampling of infiltrated water at a known depth and/or location beneath the ground surface.

3.1.4 Thermal Processes

Process: Stormwater with an elevated temperature will cool over time if the surface it is in contact with (i.e., ground, air, etc.) is at a lower temperature than the stormwater itself, and the stormwater is not in direct sunlight. As this temperature differential increases, the rate of cooling increases. Also, a greater area of contact between the warm stormwater and the relatively cool surface will result in more water being cooled and a larger overall impact with regard to temperature mitigation.

Assessment Considerations: Wet detention ponds are often built as mitigation measures to lower the peak flow during storms and to remove pollutants from storm runoff. Water stored in wet detention ponds, however, is heated by the sun during hot summer days, and as new stormwater enters the pond, the heated surface water is displaced. The temperature of water stored in wet detention basins is often warmer than that of streams, so that wet pond effluent becomes a thermal pollution point source for nearby streams. In one study, wet pond effluent was found to be 1.8°C (3°F) greater in temperature than untreated runoff from a parking lot (Herb et al. 2009). Increases in temperature caused by a wet pond, however, are offset by reductions in peak flow rate. Compared to unmitigated runoff from pavement, wet pond mitigation practices tend to reduce the magnitude, but increase the duration, of stream temperature impacts due to stormwater (Herb et al. 2009).

Infiltration practices are very effective for mitigation of thermal pollution, because warm surface runoff is cooled as it passes through soil and may mix with shallow groundwater. Although infiltrating water may locally increase shallow groundwater temperatures, there is little evidence that an infiltration pond with a reasonable buffer distance (40–50 m) can measurably impact the temperature of a stream. For storms that exceed the infiltration capacity, infiltration practices may still be of substantial benefit in reducing thermal loading because some of the initial, highest temperature runoff may still infiltrate into the soil.

Wetlands with substantial shading from emergent vegetation can also have some thermal mitigation capacity, because shaded standing water will be substantially cooler than open water (Herb et al. 2007b). It is difficult, however, to quantify possible thermal loading reductions for wetlands, because the thermal load reduction depends on both the hydraulic characteristics of the wetland and the extent of surface shading that exists. As with contaminants, hydraulic short-circuiting will reduce the thermal mitigation effectiveness of a wetland (Herb et al. 2007b).

3.2 Biological Processes

3.2.1 Degradation of Organic Matter

Process: Several biological processes are involved in pollutant retention. One is microbial respiration, in which organic matter in water is oxidized to CO_2.

Assessment Considerations: In practice, the amount of readily degradable organic matter is quantified as biochemical oxygen demand (BOD). Oxidation of BOD produces CO_2 and water. The BOD decay constant, k, in (3.5), is dependent upon temperature. Although the magnitude of k is dependent upon bacterial population, the temperature effect on decomposition is generally described by (3.6). Thus, the decay constant can be adjusted for different temperatures, as long as the bacterial population is similar:

$$C = C_o \exp^{-kt} \tag{3.5}$$

where

C = BOD concentration at any time
C_0 = BOD concentration at time zero
k = BOD decay coefficient, day^{-1}
t = time in days

$$k_2 = k_1 \theta^{(T_2 - T_1)} \tag{3.6}$$

where

T_1 = Temperature 1
T_2 = Temperature 2
k_1 = BOD decay constant at temperature 1 (typically determined through field measurements)
k_2 = BOD decay constant at temperature 2
θ = empirical constant (a common value is 1.05, Metcalf and Eddy 1991)

Calculations using (3.6) show that it takes three times longer to achieve a 50% BOD reduction at 5°C (41°F) than at 25°C (77°F) (Fig. 3.6). This implies that BOD decay would be slowest during the snowmelt period, when temperatures are just above freezing.

3.2.2 Denitrification

Process: Denitrification is a bacterial reaction that occurs under anaerobic (no dissolved oxygen) conditions, which are typical in sediments. Denitrification

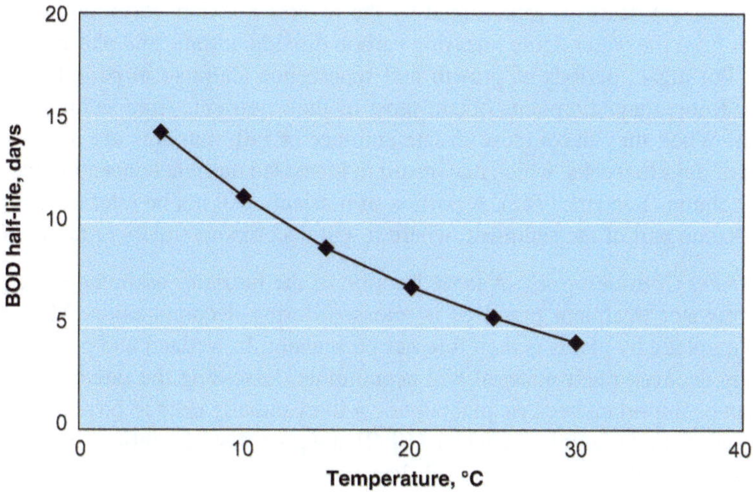

Fig. 3.6 Half-life of BOD (time needed to reduce BOD by 50%) as a function of temperature with $k_{20} = 0.1$ day^{-1} (temperature in °F = ((°C)(1.8) + 32))

converts nitrate (NO_3^-) in stormwater to nitrogen gas (N_2), but denitrification cannot occur without a source of organic matter.

Assessment Considerations: Nitrate (NO_3^-) is generally less than one-third of the total nitrogen in urban stormwater (Table 2.3). Unless additional nitrate is produced by degradation of organic material or nitrification (oxidation of ammonia), denitrification can remove only about one-third of stormwater nitrogen. The end products are harmless gases.

Conditions that allow for denitrification occur in wetlands, where rooted plants supply the carbon, and in pond sediments, where carbon is supplied by dead algae. If the assessment program reveals that nitrate removal efficiencies are less than desired, assessment of the organic carbon supply may be warranted. As with other biological processes, denitrification is also controlled by temperature. For denitrification in treatment wetlands, Kadlec and Knight (1996) suggest a θ value of 1.09 be used in (3.6).

3.2.3 Plant Growth and Nutrient Uptake

Process: Many stormwater treatment practices include plants: algae in ponds; emergent aquatic plants in wetlands and ponds; and grasses and other plants in rain gardens, filter strips, and swales. Plants assimilate (take up) nutrients during growth. Photosynthesis (the forward reaction) converts carbon dioxide (CO_2), nitrate (NO_3^-), phosphate (HPO_4^{2-}), and water to algae, producing oxygen (O_2).

Respiration (or death) is represented by the reverse reaction. Respiration removes oxygen from the water while releasing carbon dioxide, nitrate, and phosphate to the water. For algae, periods of growth and senescence alternate in periods of a few weeks. Rooted aquatic plants obtain most of their nutrients from sediment during growth. When they decompose in late summer or fall, nutrients are released in a matter of days to weeks, which may result in increased nutrient concentrations in the water column (Landers 1982). A portion of nutrients will not be released but rather will become part of the sediment, where it will decompose slowly or not at all.

Assessment Considerations: A large fraction of the nutrients assimilated by plants in stormwater treatment practices is released during decomposition. Most of the nutrient uptake by plants is therefore not permanent. In wetland and pond systems, partially decayed plant material will accumulate. Assessing the rate of accumulation can be important because plant debris will eventually need to be removed from the practice. For example, Schueler (1992) suggests that vegetation removal from wetlands is required at intervals of 2–10 years.

3.3 Chemical Processes

Chemical precipitation and adsorption can be used to remove dissolved constituents from stormwater. Precipitation removes dissolved materials by forming insoluble solid complexes that can be removed by sedimentation or filtration. Adsorption is the process of dissolved ions becoming attached to the surface of a solid particle, which also can be removed by sedimentation or filtration. It is also possible that constituents associated with particles may dissolve (from precipitated particles) or desorb back into soluble forms. These processes are important for metals and phosphate, although each of these pollutants requires a slightly different assessment approach. The discussion that follows applies to the condition of equilibrium. Although these processes are not always at equilibrium, describing nonequilibrium processes is beyond the scope of this book.

3.3.1 Metals

Process: Metals ions (e.g., Cd^{2+}) or metal hydroxides (e.g., $CdOH^{+}$) bind to negatively charged sorbent surfaces. Dissolved cadmium, copper, and zinc are removed from stormwater runoff by sorption to organic material, such as compost (Morgan, 2011). Metal sorption to organic material is dependent on the functional groups such as carboxyl, phenolic-OH, carbonyl, sulfonic, amine, and imide functional groups (Essington 2004), as well as functional group pK_a value, the molecules to which they are attached, and pH (Elliott et al. 1986; Grimes et al. 1999; Davis et al. 2001; Harmita et al. 2009).

The pH will also affect the dissolved metal speciation, and subsequently soprtion. For example, cadmium, copper, and zinc all exist as metals ions (e.g., Cd^{2+}) for pH values less than 10.1, 8.96, and 7.7, respectively (Elliott, 1986; Essington, 2004) and as metal hydroxides (e.g., $CdOH^{+}$) for pH values greater than these values. Elliott et al. (1986) found that more sorption occurs for hydrolyzed metals (e.g., $CdOH^{+}$) than for metal ions.

Assessment Considerations: Assessment of metal retention in stormwater treatment practices is typically done by measuring input and output fluxes of metals, calculating the total influent and effluent mass loads, and performing a mass balance on the practice. The difference between the total mass entering the practice and the mass exiting the practice is assumed to be the mass retained. In infiltration practices, dissolved metals may infiltrate with stormwater and adsorb to the soil media. This typically occurs within the first 50 cm of soil depth, but actual depths can vary depending on soil and metal parameters (Weiss et al. 2008). Ultimately, any soil, even one that initially adsorbs a large fraction of metals, will reach its capacity and stop adsorbing metal ions from the infiltrated water. When this happens, groundwater contamination can occur, and maintenance, such as soil replacement, is necessary.

In cases where greater metal retention is desired, an examination of both the partitioning between dissolved and particulate bound metal fractions and the size and settling velocity of the solid particles is necessary to determine the factor that is limiting removal. For example, metals may be difficult to remove from stormwater with a small suspended solids concentration if the dissolved metal fraction is large. Also, if the particles have a relatively small settling velocity or are small, removal of metals may be difficult because sedimentation and/or filtration may be ineffective at removing the small particles on which the metals have adsorbed.

The metal content of sediments may dictate ultimate disposal or handling methods of the sediments (Polta 2001; Polta et al. 2006). For example, metal concentrations in sediments of stormwater ponds may be high enough to require hazardous waste disposal, or they may be greater than Soil Reference Values for human exposure (Polta et al. 2006) that dictate handling protocol.

Studies have shown that increased concentrations of salt (commonly sodium chloride) in stormwater can cause previous sorbed metals to become desorbed and discharged with the stormwater (Paus et al. unpublished).

3.3.2 Phosphate

Process: Dissolved phosphate in infiltrated stormwater can undergo both precipitation and adsorption to soil particles. Phosphate is sometimes considered immobile in soils, but many studies have shown that phosphate adsorption is limited. Adsorption capacity breakpoints are commonly 20–50 mg PO_4 -P/kg of "Bray P" soil mass (Pote et al. 1999; McDowell et al. 2001; Fang et al. 2002). Addition of more phosphate beyond this point results in breakthrough, which means that soluble phosphate will not be adsorbed but instead will pass through the soil with the water in which it is dissolved. Evidence of phosphate breakthrough has been observed with infiltrated stormwater (Weiss et al. 2008), septic systems (Robertson et al. 1998), urban soils (Zang et al. 2001), and under wastewater-irrigated fields (Zvomuya et al. 2005). Also, stream phosphate concentrations have been correlated with average watershed Bray P (Klatt et al. 2003), indicating that soils do not adsorb all phosphorus.

Although standard sand filters do not remove a significant amount of dissolved phosphorus, dissolved phosphorus can be removed from water via surface adsorption using an enhanced sand filter. In this filter, an enhancing agent of steel wool or elemental iron is mixed in with the regular sand filter media. As the iron rusts and forms iron oxides, phosphate binds to the iron oxides by surface adsorption (Erickson et al. 2007, 2012) and is retained within the filter.

Dissolved pollutants may also be retained by adsorption onto filter media and previously deposited solids or by chemical precipitation reactions within the filter media. Removal of dissolved pollutants such as phosphorus, however, is typically minimal for standard sand filters. Harper and Herr (1993) reported that pilot-scale and full-scale sand filters retained 40–50% particulate phosphorus, but only 5% dissolved phosphorus. Similarly, Herrera Environmental Consultants (1995) reported that typical sand filter media had little capacity (0–28% total, 0–38% dissolved) for phosphorus retention. On the other hand, dissolved phosphorus removal can be significantly enhanced if the sand is amended with iron, calcium, aluminum, or magnesium (Arias et al. 2001). Steel wool improved phosphorus retention capacity of ASTM standard C33 sand by 25–99% in pilot-scale stormwater filters (Erickson et al. 2007). Other amendments, such as peat and compost, can have the opposite effect by releasing nutrients as stormwater passes through the filter and subsequently increasing the load of nutrients downstream (Erickson et al. 2007).

Assessment considerations: When phosphate retention by soil adsorption is important, the buildup of soil phosphorus should be assessed periodically. This can be done with measurements of extractable P using the Bray or Olsen methods. Also, when elemental iron is used in an enhanced sand filter, the media should be allowed to continually pass through cycles of wetting and drying. This allows the iron to rust, which oxidizes more elemental iron and forms more adsorption sites. Ultimately, the capacity of the iron will be exhausted and the media will have to be replaced. All governing regulations should be followed when disposing of the exhausted media.

Stormwater Treatment Practices

4

Abstract

Common stormwater treatment practices include wet ponds, dry ponds, infiltration basins and trenches, constructed wetlands, permeable pavements, and others. In preparation for the remaining chapters, which focus on the assessment and maintenance of these stormwater treatment practices, this chapter introduces and briefly discusses each of the practices covered in this book, which are categorized by their primary operating process (i.e., sedimentation, filtration, biological, etc.) as defined in Chap. 3.

This chapter presents stormwater treatment practices that, by utilizing the processes discussed in Chap. 3, improve stormwater runoff quality or reduce volume or peak flow. It must be noted that many stormwater treatment practices utilize more than one process to manage stormwater; however, one primary process usually plays the most significant role. The practices discussed in this chapter are divided into four categories based on the primary process used: sedimentation, filtration, infiltration, and biological processes.

Although some stormwater treatment practices, such as public education and source reduction, can be implemented through public policy, outreach efforts, and other means, this book focuses on practices that are designed to reduce contaminant concentrations in runoff post-contamination. In the development of a stormwater management plan, however, all practices should be considered and many times the most cost-effective practice may be preventative in nature.

In addition to details and information about each type of stormwater treatment practice, the following sections provide guidance on typical land area requirements and pollutant removal performance. For ease of reference, that information is provided in Table 4.1.

A.J. Erickson et al., *Optimizing Stormwater Treatment Practices: A Handbook of Assessment and Maintenance*, DOI 10.1007/978-1-4614-4624-8_4,

Table 4.1 Typical land area requirements and pollutant removal efficiency for several stormwater treatment practices

		Pollutant removal efficiency			
	Land area required (percent of impervious watershed, unless otherwise noted) (%)	Total suspended solids (TSS) (mean ± 67% confidence interval) (%)[a]	Total phosphorus (TP) (mean ± 67% confidence interval) (%)[a]	Total suspended solids (typical range) (%)[b]	Total phosphorus (typical range) (%)[b]
Dry ponds	0.5–2.0%[c]	53 ± 28%	25 ± 15%	30–65%	15–45%
Wet ponds	2–3%[b]	65 ± 32%	52 ± 23%	50–80%	15–45%
Surface sand filters	<3%[b]	82 ± 14%	46 ± 21%	50–80%	50–80%
Infiltration basins	2–3%[d]	See Table 4.2			
Infiltration trenches	2–3%[b,d]				
Bioretention practices	5.0%[b,d]	N/A	72 ±11%	50–80%	50–80%
Constructed wetlands	3–5%[b]	68 ± 25%	42 ± 26%	50–80%	15–45%
Filter strips	100%[b,d]	75 ± 20%	41 ± 33%	50–80%	50–80%
Swales	10–20%[b,d]			30–65%	15–45%

[a]Weiss et al. (2007)
[b]US EPA (1999a), Tables 5–7 and 6–9
[c]Total watershed area (Urban Drainage Flood Control District 1992)
[d]Claytor and Schueler (1996)
N/A = Information not reported in the source

4.1 Sedimentation Practices

Sedimentation practices capture particles via settled particles, and settled particles eventually need to be removed from the stormwater treatment practice. For this reason, pretreatment is generally recommended because it can simplify maintenance. A pretreatment facility is designed to remove sediment that settles quickly in a relatively small volume, which is easily accessed for maintenance (e.g., vacuum trucks). Typically, a significant portion of the sediment can be removed from the water by pretreatment, which will reduce the maintenance frequency and lengthen the usable life of the stormwater treatment practice.

4.1.1 Dry Ponds

Dry ponds are unlined depressions in the ground surface fitted with inlets and outlets to manage the collection and release of stormwater. Dry ponds are sometimes called dry detention ponds or detention basins. Dry ponds (Fig. 4.1) temporarily store stormwater runoff and release the water through a designed outlet structure at a slower rate than if

Fig. 4.1 Example of a dry pond

the dry pond were not present. Dry ponds are designed to drain completely and should not maintain a pool of water after draining a runoff event.

Infiltration and evapotranspiration may occur in a dry pond, but these processes are usually not the primary modes of water transport out of the pond. In fact, infiltration is often negligible and ignored and most contaminant removal is achieved by settling of the solid particles within the pond. Thus, dry ponds can be effective at reducing the suspended solids concentration and any contaminants (e.g., phosphorus, metals) in the particulate phase.

The prescribed surface area for a dry pond is typically 0.5–2.0% of the total watershed area (Urban Drainage Flood Control District 1992). Historically, the primary function of dry ponds was to reduce the peak runoff flow rate and reduce the risk of flooding downstream due to urbanization of the upstream watershed. More recently, the pollutant retention mechanisms that occur within dry ponds have been investigated. Weiss et al. (2007) reported that on average (±67% confidence interval) dry ponds in the USA retain 53% (±28%) of total suspended solids and 25% (±15%) of total phosphorus. The US EPA (1999a) reported typical ranges of 30–65% for total suspended solids and 15–45% for total phosphorus in dry ponds.

4.1.2 Wet Ponds

Wet ponds (retention ponds) are depressions in the ground with elevated outlets designed to allow water to pond and be stored between runoff events (Fig. 4.2). The

Fig. 4.2 Example of a wet pond

ponded water remains in the pond after the outlet is no longer discharging runoff because the outlet structure opening is at a higher elevation than the pond bottom. The pool of water remains within the wet pond, or is retained, until the next runoff event displaces it or until all the water below the outlet opening infiltrates and/or is removed via evapotranspiration. Although infiltration is possible, some wet ponds do not infiltrate stormwater because they have an impermeable liner or because the groundwater table is too high for the water to infiltrate.

Wet ponds are typically designed with a surface area of 2–3% of the impervious watershed area (US EPA 1999a). One purpose of storing water in a wet pond is to allow more time for solids to settle to the bottom of the pond. Thus, wet ponds typically achieve greater total suspended solids removal rates than dry ponds. On average, (±67% confidence interval), wet ponds in the USA retain 65% (±32%) of total suspended solids and 52% (±23%) of total phosphorus (Weiss et al. 2007). The US EPA (1999a) reported typical ranges of 50–80% for total suspended solids and 15–45% for total phosphorus in wet ponds.

4.1.3 Underground Sedimentation Devices

Underground sedimentation devices treat stormwater without the need for land surface area; therefore, they are often used in urban areas where land area is limited and/or obtaining land is cost-prohibitive. Prefabricated underground sedimentation devices, sometimes called hydrodynamic separators or proprietary devices (Fig. 4.3), are available from many manufacturers and are best used as pretreatment in conjunction with other devices. They can be as simple as standard sumps under

Fig. 4.3 Full-scale fiberglass prototype of a two-chamber hydrodynamic separator undergoing laboratory performance tests (Environment 21 V2B1 unit)

manholes (Fig. 4.4), if they are sized for sediment collection (Barr Engineering 2011). While these products do remove solids through sedimentation, most store minimal stormwater and therefore do not significantly reduce peak flows.

Wet vaults are underground vessels that differ from hydrodynamic separators in that they temporarily store and treat stormwater runoff. Thus, wet vaults do reduce peak flow rates and can remove suspended solids through sedimentation. A common wet vault design consists of large diameter concrete or corrugated metal pipes placed underneath a parking lot. Parking lot and rooftop runoff is routed into the underground pipes for temporary storage and subsequent release to the storm sewer. Some wet vaults are designed with open bottoms (i.e., arch pipes) or perforations in the walls of the vault, which, given the proper underground conditions, can allow stormwater to infiltrate into the surrounding soil.

4.2 Filtration Practices

Filtration practices capture particulate pollutants by physical sieving. Captured particles collect in the pore spaces in the filter media, which can eventually clog the filter and reduce the flow rate through the filter. For this reason, pretreatment is generally recommended, because sediment must eventually be removed by surface

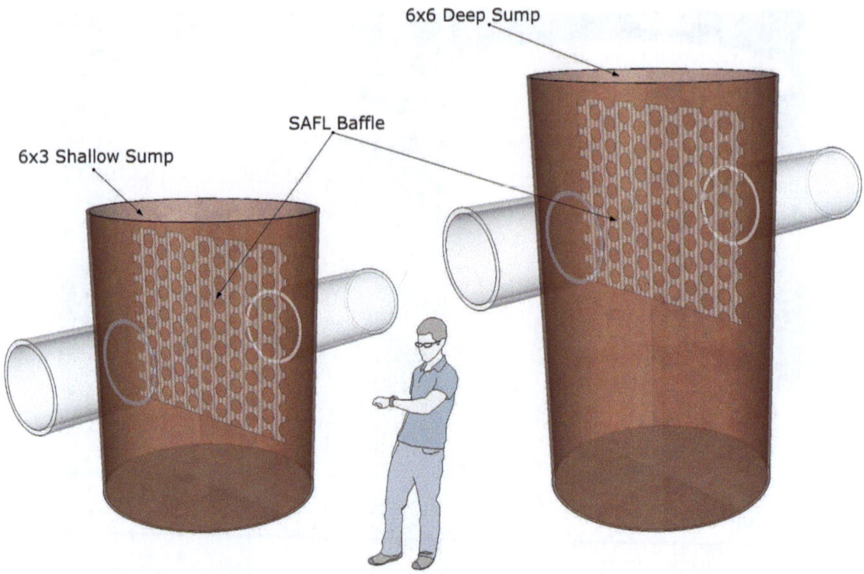

Fig. 4.4 Two sumps below manholes outfitted and sized for sediment collection. The SAFL Baffle is a retrofit that will improve retention of solids during high flows (Howard et al. 2011)

scraping or media replacement. A pretreatment facility is designed to remove sediment that settles quickly in a relatively small volume which is easily accessed for maintenance (e.g., vacuum trucks). Typically, a significant portion of the sediment can be removed from runoff by pretreatment, which will reduce the maintenance frequency and lengthen the usable life of the stormwater treatment practice.

4.2.1 Surface Sand Filters

Surface sand filters (often called "Austin sand filters") have a filter mechanism typically made up of a layer of filter media (18–24 in., 46–61 cm). The gravel bed contains a perforated pipe collection system that collects filtered stormwater and delivers it either downstream in the conveyance system or directly to receiving waters. These systems are installed in depressions (Claytor and Schueler 1996) as shown in Fig. 4.5.

The filter media consists of locally or commercially available sands selected and sieved specifically to meet desired filtration specifications. Standard concrete sand (ASTM 2002) that has a particle size distribution as shown in Fig. 4.6 is a recommended filtration medium that is readily available in the USA (Claytor and Schueler 1996). A photo of an Austin sand filter is provided in Fig. 4.7.

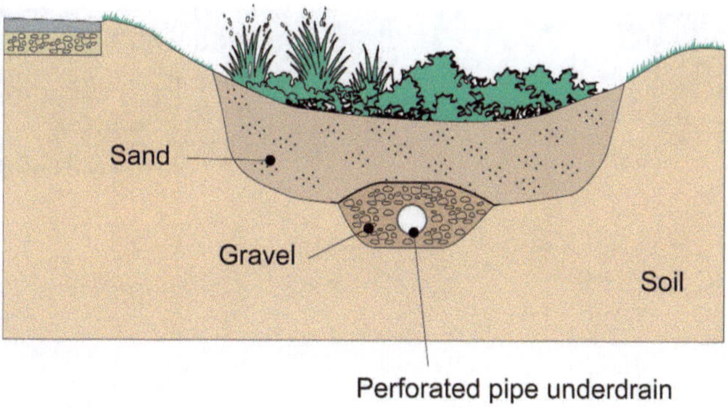

Fig. 4.5 Sand filter design cross section

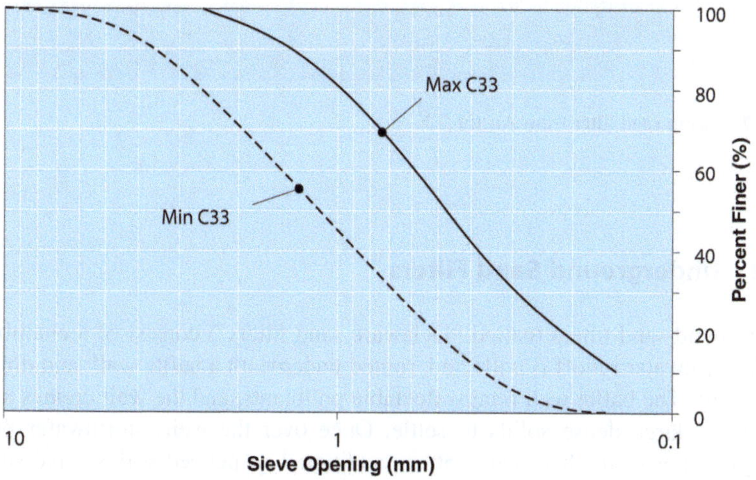

Fig. 4.6 American Standard for Testing Materials for C33 sand (ASTM C33-02a 2002 (1 mm = 0.039 in.)

The surface area typically prescribed for a surface sand filter is 3.0% or less of the impervious watershed area (US EPA 1999a). Weiss et al. (2007) reported that on average (±67% confidence interval), sand filters in the USA retained 82% (±14%) of total suspended solids and 46% (±21%) of total phosphorus. The US EPA (1999a) reported typical ranges of 50–80% for total suspended solids and 50–80% for total phosphorus in surface sand filters.

Fig. 4.7 Austin sand filter from Austin, TX

4.2.2 Underground Sand Filters

Underground sand filters (called "Delaware sand filters") consist of a chamber in which stormwater runoff is collected, routed underneath a baffle wall, and directed over a weir. The baffle wall retains floatable pollutants, and the weir creates a pool that allows large dense solids to settle. Once over the weir, stormwater passes through a filter media bed that captures additional suspended solids. Like surface filters, filter media for underground sand or soil filters consist of native soils or locally or commercially available sands selected and sieved specifically for filtration purposes. Underground sand or soil filters are typically constructed on-site or are prefabricated and purchased from commercial vendors.

An advantage of underground sand filters is that they usually do not require surface land area because the entire practice is underground. They may not, however, have the hydraulic capacity of surface filters if subsurface space is limited. Even if the hydraulic capacity is less, a properly designed underground filter is expected to perform similarly to a surface filter with regard to total suspended solids and total phosphorus retention. Thus, the values reported by Weiss et al. (2007) and US EPA (1999a) for surface sand filters that were provided above also apply to underground filters.

Fig. 4.8 Soil filter at Carver County maintenance facility, Minnesota, with permeable overflow weir in the foreground

4.2.3 Soil Filters

One option that is sometimes practiced is to use native soil instead of sand. Using native soil as the filter media can reduce the overall cost of a filtration practice, but the grain size distribution of native soils is often not appropriate for stormwater filtration (i.e., it will not pass the design storm within 48 h). Soil filters are typically designed similar to infiltration basins, with plants, except that the soil is drained by drain tile due to a relatively impermeable layer at depth below the permeable soil. One example is shown in Fig. 4.8.

4.2.4 Hybrid Filters

Some stormwater treatment practices are designed to filter runoff in part of the practice, but not throughout. For example, a soil filter may be designed with trenches that are backfilled with sand or other nonnative media and contain underdrains to increase filtration rates. This will increase the volume of stormwater that is filtered and reduce the amount that overflows or bypasses the practice with little or no treatment. For an example, see "Case Study: Monitoring a Dry Detention Pond with Underdrains" in Chap. 8. Another example is a wet or dry detention basin in which underdrains are installed in the upper banks to allow filtration through the

Fig. 4.9 Photo of a Minnesota Filter in Maplewood, MN. The large rocks serve aesthetic purposes

banks when stormwater fills the basin during a runoff event. These "hybrid" filtration systems combine filtration with other stormwater treatment processes.

4.2.5 Enhanced Sand Filters

Filters are not generally designed to remove dissolved compounds, which can constitute roughly half of certain pollutants. Dissolved phosphorus removal can be significantly enhanced if the sand is amended with iron, calcium, aluminum, or magnesium (Arias et al. 2001). Steel wool, an iron-based product, improved phosphorus retention capacity of ASTM standard C33 sand by 25–99% in pilot-scale stormwater filters (Erickson et al. 2007) and iron filings have been shown to have similar removal capabilities in full-scale settings (Erickson et al. 2012). Some materials, such as peat and compost, can have the opposite effect by releasing nutrients, thereby becoming a source of nitrogen and phosphorus (Erickson et al. 2007).

Some enhanced filters (called "Minnesota sand filters") are currently being researched and designed to remove dissolved phosphorus (Erickson et al. 2007, 2012). Other enhancements to remove dissolved metals are also being investigated. An example of one enhanced filter, which combines sand filtration with iron filings to remove dissolved phosphorus is shown in Fig. 4.9. The appearance is similar to the sand filter shown in Fig. 4.5, with a gravel underdrain system separated by filter fabric from the enhanced sand layer and topped off with pea gravel to reduce blowing of the finer sand.

4.3 Infiltration Practices

4.3.1 Infiltration Basins

An infiltration basin is a natural or constructed impoundment that captures, temporarily stores, and infiltrates a designed volume of stormwater within a targeted time period. Infiltration basins often contain a flat, densely vegetated floor situated over naturally permeable soils. Nutrients and pollutants are removed from the infiltrated stormwater through chemical, biological, and physical processes, with volume reduction through infiltration usually considered the primary process. Infiltration basins are well suited for drainage areas of 5–50 acres (2.03–20.25 ha) with land slopes that are less than 20%, with typical depths in the basin ranging from 2 to 12 ft (0.61–3.66 m). Claytor and Schueler (1996) report infiltration practices typically occupy 2–3% of the impervious watershed area. A schematic of a typical infiltration basin is shown in Fig. 4.10.

Studies have documented the effectiveness of pollutant removal in properly functioning infiltration practices (Schueler 1987; Schueler et al. 1992, US EPA 2000a, Winer 2000). While these studies do not differentiate between infiltration basins and infiltration trenches, pollutant removal efficiencies determined by these studies (Table 4.2) can be assumed to apply to both.

Fig. 4.10 Typical infiltration basin

Table 4.2 Pollutant removal from stormwater infiltration basins and trenches (Schueler 1987; Schueler et al. 1992; Winer 2000)

Pollutant	Winer (2000)	Schueler (1987)	Schueler et al. (1992)
Sediment	95%	99%	90%
Total P	65%	65–75%	60%
Total N	50%	60–70%	60%
Trace metals	95%	95–99%	90%
Bacteria	N/A	98%	90%
BOD	N/A	90%	70–80%

N/A = Information not reported in the source

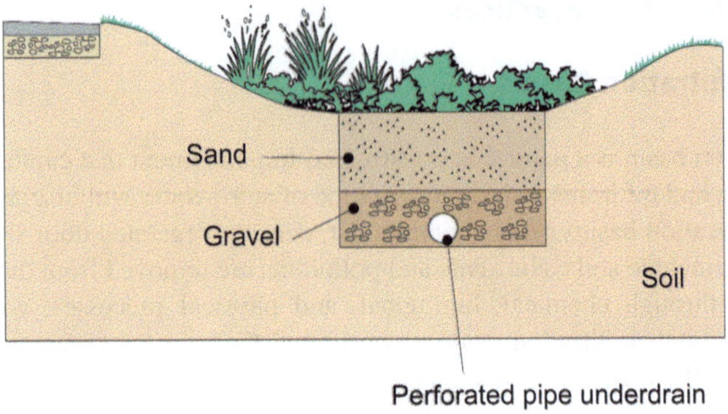

Fig. 4.11 Typical infiltration trench design

4.3.2 Infiltration Trenches

An infiltration trench is a shallow excavated trench, typically 3–12 ft deep (0.91–3.66 m), typically backfilled with a coarse stone aggregate, allowing for the temporary storage of runoff in the void space of the aggregate. Discharge of the stored runoff occurs through infiltration into surrounding permeable soil. Infiltration trenches are well suited for drainage areas of 5 acres (2.03 ha) or less and are often used in parking lots. Infiltration trenches typically occupy 2–3% of the impervious watershed area (US EPA 1999a; Claytor and Schueler 1996). A schematic of a typical infiltration trench is shown in Fig. 4.11.

4.3.3 Permeable Pavements

Asphalt and concrete can be designed and constructed to be permeable and infiltrate stormwater runoff. According to Ferguson (2005), there are nine categories of permeable (or porous) pavement. These include permeable aggregate, permeable turf, plastic geocells, open-jointed paving blocks, open-celled paving grids, permeable concrete, permeable asphalt, soft permeable surfacing, and decks.

For cases where the permeable pavement is either asphalt or concrete, the pavement system is designed so that stormwater infiltrates through the permeable upper pavement layer and then into a reservoir of stone or rock below. Water from the reservoir then either percolates into the underlying soil, where it may recharge groundwater, or is collected by a perforated pipe underdrain system and carried to a

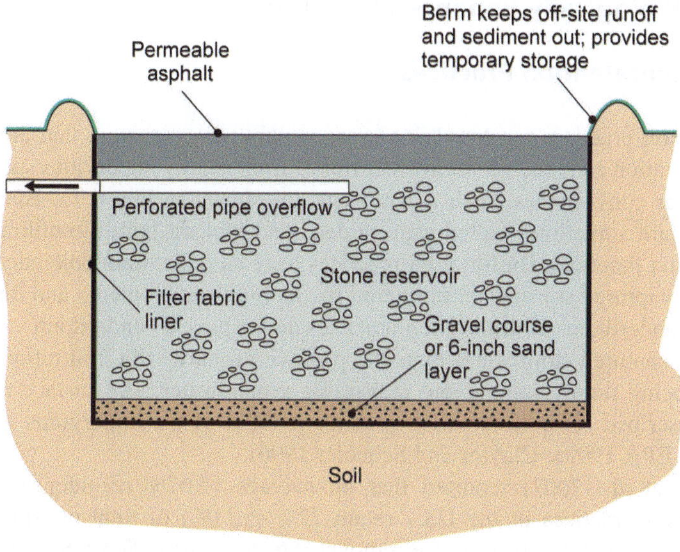

Fig. 4.12 Typical permeable pavement installation (1 in. = 2.54 cm)

surface discharge location. An illustration of a vertical section through an asphalt permeable pavement is presented in Fig. 4.12.

The performance of permeable pavers in removing contaminants from the fraction of infiltrated stormwater is not well documented. Scholz and Grabowiecki (2007) reviewed literature on permeable pavement systems and found that if permeable pavements are designed and constructed correctly, they can remove suspended solids and nitrogen. If, however, an underground collection system withdraws infiltrated water from the soil, nitrogen removal is limited because plant uptake and denitrification do not occur. The authors also found that permeable pavements can be effective at retaining metals in the soil underneath the pavement and that the soil can act as a bioreactor and remove a significant fraction of infiltrated hydrocarbons.

Collins et al. (2010) investigated four different types of permeable pavers for their ability to remove nitrates and nitrites and compared their ability to that of asphalt. The pavers investigated were pervious concrete (PC), two kinds of permeable interlocking concrete pavement (PICP), and concrete grid pavers (CGP) that were filled with sand. The runoff from all of the pavers except CGP had larger nitrate and nitrite concentrations than asphalt, probably due to nitrification occurring in the soil profile, and the CGP had the largest total nitrogen concentration. It was theorized that the CGP acted similarly to a sand filter because it had 4 in. of sand base. All of the pavements investigated buffered runoff, probably due to calcium carbonate and magnesium carbonate in the pavement and aggregate.

4.4 Biologically Enhanced Practices

4.4.1 Bioretention Practices

Bioretention practices are low-lying areas, natural or excavated, that are planted with vegetation and receive stormwater runoff from nearby impervious surfaces via stormwater conveyances, such as curb cuts, as shown in Fig. 4.13. Bioretention practices are sometimes called rain gardens and include both bioinfiltration and biofiltration practices. Biofiltration practices have an underdrain collection system such that captured stormwater is filtered by the biofiltration media and discharged from the underdrain. Bioinfiltration practices do not have an underdrain system and therefore captured stormwater exits the practice primarily via infiltration into the soil, reducing runoff volume and recharging groundwater. The surface area typically prescribed for a bioretention practice is 5.0% of the impervious watershed area (US EPA 1999a; Claytor and Schueler 1996).

Weiss et al. (2007) reported that on average (±67% confidence interval), bioretention practices in the USA retain 72% (±11%) of total phosphorus, but data were unavailable to report a removal rate for solids. The US EPA (1999a) reported solids and phosphorus removal rates for infiltration practices that may or may not have vegetation, but did not report specific values for bioretention

Fig. 4.13 Example of a curb cut out that allows road runoff to enter a rain garden

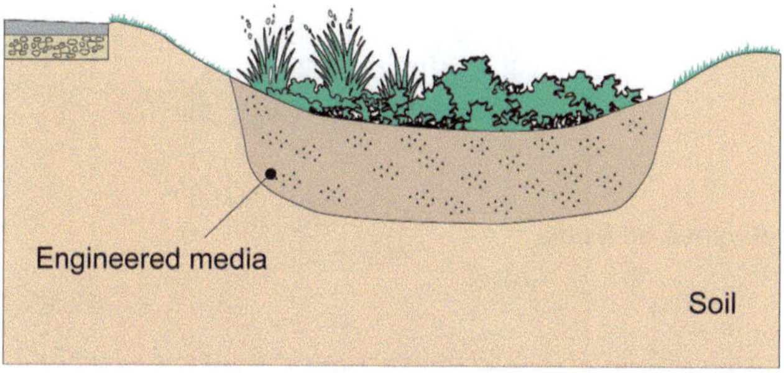

Fig. 4.14 Typical bioinfiltration (rain garden) cross section

practices. The US EPA's reported range of removal for infiltration practices is 50–80% for total suspended solids and 50–80% for total phosphorus.

4.4.2 Bioinfiltration Practices

Bioinfiltration practices are bioretention practices without underdrains (Fig. 4.14) and are sometimes called rain gardens. While excess stormwater can exit the practice through a designed overflow outlet, the runoff design volume exits the practice primarily via infiltration into the soil, reducing runoff volume and recharging groundwater. Because bioinfiltration practices infiltrate stormwater into the underlying soil, it is often difficult to collect water samples that have passed through the practice.

4.4.3 Biofiltration Practices

If the infiltration capacity of the surrounding soil is insufficient, or if infiltration is not desired, a vegetated depression may be designed to include underdrains to collect and remove stormwater that has passed through a layer of soil. Such practices are classified as biofiltration practices and are constructed by excavating the soil, placing a drain tile or perforated pipe collection system at the bottom, backfilling with a soil that has the desired saturated hydraulic conductivity, and then planting with vegetation (Fig. 4.15). In biofiltration practices, the soil-filtered water captured by the drain tile or perforated pipe collection system is then delivered downstream in the conveyance system or to receiving waters. Biofiltration practices can denitrify the water that is retained between storm events when a drain is designed with an upturned inlet (Hunt et al. 2006)

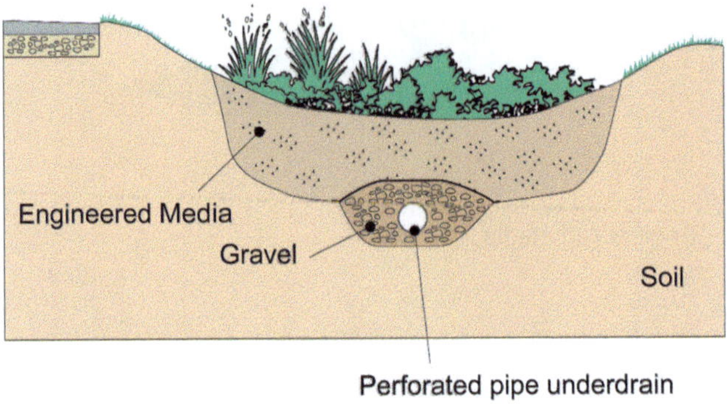

Fig. 4.15 Typical biofiltration system cross section

4.4.4 Constructed Wetlands

Wetlands are lowland areas where the groundwater level is higher than the ground surface elevation such that persistent shallow pools are present. Constructed wetlands are designed to mimic natural wetlands. Shallow pools, vegetation, and microorganisms remove pollutants from stormwater runoff through sedimentation, filtration, and biodegradation, respectively. Plants can also take up pollutants such as nutrients (e.g., N and P) and micronutrients (e.g., Cu and Zn) and store them in the wetland by converting them to plant biomass. Eventually, however, collected nutrients and metals will need to be removed from the wetland, or the wetland may become a source of these pollutants. Constructed wetlands are typically designed with a surface area of 3–5% of the impervious watershed area (US EPA 1999a).

Constructed wetlands reduce runoff peak flow by temporarily storing stormwater runoff. Runoff volume is reduced by evapotranspiration, but due to the high water table, infiltration is usually not significant. On average (±67% confidence interval), constructed wetlands in the USA retain 68% (±25%) of total suspended solids and 42% (±26%) of total phosphorus (Weiss et al. 2007). These average values fall within the ranges reported by the US EPA (1999a) of 50–80% for total suspended solids and 15–45% for total phosphorus.

4.4.5 Filter Strips and Swales

Filter strips, also called buffer strips or buffers, are vegetated areas specifically designed and positioned for overland sheet flow. Overland sheet flow occurs when stormwater flows on the land surface in a thin layer with a relatively slow velocity. The vegetation filters solids and reduces runoff velocities, which allows for more infiltration and sedimentation to occur. Sheet flow is required for filter strips to effectively treat stormwater runoff.

Swales are vegetated canals or trenches used to convey stormwater runoff, which allow solids to settle while also filtering suspended solids with vegetation. During conveyance, infiltration into the swale sides and bottom may occur. Swales may also be called drainage ditches, grassed channels, dry swales, vegetated swales, wet swales, biofilters, or bioswales. Permeable structures (e.g., check dams) are sometimes installed in swales to reduce flow velocities, which increases settling and infiltration. Grassed swales are typically designed with a surface area of 10–20% of the impervious watershed area (US EPA 1999a).

On average (±67% confidence interval), filter strips and grassed swales in the USA retain 75% (±20%) of total suspended solids and 41% (±33%) of total phosphorus (Weiss et al. 2007). The total suspended solids average value is within the range of 50–80% reported by the US EPA (1999a), but the average value for total phosphorus is less than the US EPA reported range of 50–80%.

Visual Inspection of Stormwater Treatment Practices

5

Abstract

Assessment of a stormwater treatment practice can be accomplished by visual inspection, testing, or monitoring. The least complex method of assessment is visual inspection, which involves a site visit, inspecting the practice for any evidence of malfunction, and documenting site conditions and results. This chapter presents and discusses key visual indicators of malfunction, such as evidence of erosion, undesired water in the practice, soil and vegetation conditions, and various other indicators.

5.1 What Is Visual Inspection?

The first and least complex level of assessment is visual inspection. Visual inspection involves inspecting a stormwater treatment practice for evidence of malfunction and can be accomplished with a brief site visit. Visual inspection can be used to quickly and cost-effectively determine if, and potentially why, a stormwater treatment practice is not operating properly, but it only provides a qualitative assessment. Even if there are no outward signs of malfunction, visual inspection cannot guarantee that the practice is operating properly.

If a stormwater treatment practice is determined to be nonfunctional based on visual inspection, a higher level of assessment may be necessary to determine or verify the cause of failure and, based on those results, schedule appropriate maintenance. In other instances, such as with erosion or structural damage to a treatment practice, visual inspection may be enough to schedule the appropriate maintenance activity.

The qualitative information gathered by visual inspection is often a valid indicator as to whether the stormwater treatment practice is malfunctioning and potential causes. As mentioned above, however, visual inspection alone cannot provide quantitative information about stormwater treatment practice performance

A.J. Erickson et al., *Optimizing Stormwater Treatment Practices: A Handbook of Assessment and Maintenance*, DOI 10.1007/978-1-4614-4624-8_5,

Fig. 5.1 Examples of visual inspection for a rain garden that is not functioning properly (*left*) and a rain garden that may be operating properly (*right*) (Photos by B. Asleson)

such as peak flow reduction, runoff volume reduction (e.g., infiltration), and pollutant removal efficiency.

As an example of the visual inspection process, photographs of two rain gardens are shown in Fig. 5.1. The rain garden on the left contains standing water even though there has been no recent rainfall and the rain garden on the right has no standing water and contains healthy, non-wetland plants. It is visually obvious that the rain garden on the left in Fig. 5.1 is malfunctioning because rain gardens should infiltrate all of their stored water within a few days (usually 2 days). Thus, visual inspection alone has indicated that maintenance, additional assessment, or some other corrective action is required for the rain garden on the left in Fig. 5.1. Visual inspection cannot, however, provide quantitative evidence that the rain garden on the right is operating as designed or expected, especially if it has not rained recently. Even if it had rained recently and the rain garden was dry, it could be that infiltration was occurring too quickly through preferential flow paths. Thus, the rain garden on the right may also be in need of maintenance. As evidenced in this example, the timing of visual inspection relative to recent rainfall or snowmelt events can be very important.

Considering the minimal effort and cost required for visual inspection, it is recommended that visual inspection be used as the initial assessment tool for all stormwater treatment practices. Quantitative information on performance will require additional assessment via capacity testing (level 2a), synthetic runoff testing (level 2b), or monitoring (level 3).

To ensure that stormwater treatment practices continue to function properly over time, visual inspections should be scheduled once per year or more frequently and should occur during the beginning of the rainy season after the snow has melted (if applicable). Photographs should be taken as part of any visual inspection to document conditions of the stormwater treatment practice for future reference. As with any field work, safety is of utmost importance and should always be addressed when conducting visual inspections.

Because each stormwater treatment practice, watershed, and assessment goal varies, some stormwater treatment practices may require visual inspections more frequently than once per year. Therefore, the reader should consider the specific recommendations based on the treatment process utilized by the practice, which are discussed in the following sections. For example, if an assessment program is being developed to assess a dry pond, the reader should follow the discussion and recommendations under Sedimentation Practices, because dry ponds' primary treatment process is sedimentation. In addition, visual inspection checklists specific to stormwater treatment practices are provided in Appendix A.

As previously stated, visual inspection is a relatively quick, qualitative assessment of the state of a stormwater treatment practice that should be performed at least once per year or more, if conditions warrant. This chapter provides a discussion on what aspects of a stormwater treatment practice should be investigated when performing a visual inspection and why these aspects are important. Items that are common to the visual inspections of most or all stormwater treatment practices are presented and discussed first. More specific items are presented and discussed in subsequent sections based on the major treatment process of the practice.

5.2 Common Visual Inspection Items

5.2.1 History

The history of previous visual inspections and other assessment actions at the site should have been documented and should be reviewed as part of the current assessment. It is important to determine whether the practice has been previously assessed so that current assessment efforts are cost-effective (i.e., neither duplicated nor wasted). If previous assessment has occurred, the current assessment should verify, if possible, that any actions suggested by the previous assessment were completed. Also, if testing or monitoring have previously occurred, the results of these previous assessment efforts may indicate that certain aspects of the practice need to be investigated with particular care and attention. For example, if a previous monitoring effort revealed that a filtration practice may be short-circuiting a portion of the flow, but the extent of possible short-circuiting was not enough to warrant maintenance or further assessment, particular attention could be paid to inspect for signs that the short-circuiting has worsened.

The recent history of rainfall in the watershed is also important to all visual inspections. For example, many treatment processes are designed to drain the design storm volume (i.e., water quality volume or maximum storage volume) within a certain amount of time (typically 48 h in the USA). Thus, it is important to know how long it has been since the most recent runoff event. For example, if a practice is designed to be dry (i.e., a dry pond) and the time since the last runoff event exceeds the design drain time, there should be no standing water in the practice. Thus, the time since the last runoff event will alter how answers to subsequent questions in the visual assessment are interpreted.

5.2.2 Access

The condition and extent of access to the site must be evaluated and recorded. Access to the areas upstream and downstream of the site as well as the site itself is needed in order to properly assess and maintain the practice. This is true regardless of the level of assessment. If access is not complete, the quality of the assessment may be jeopardized.

5.2.3 Inlet and Outlet Structures

If present, the condition of any structure where runoff is designed to enter or exit the practice (including overflow structures) must be visually inspected. Inlet and outlet structures should be free of debris, sediment, vegetation, and other obstructions so that stormwater runoff can enter and exit the practice as designed. If an inlet structure is even partially clogged, suspended solids may be deposited in the upstream conveyance system, or upstream areas may flood because the conveyance of the system is reduced by the obstructions. If an outlet structure is partially or completely clogged, the treatment rate may be reduced and stormwater runoff may not pass through the practice within the design time. This can result in flooding and untreated stormwater runoff bypassing the practice. Any obstructions should be removed immediately to ensure proper operation of the practice. Figure 5.2a, b shows a clogged outlet and inlet structure. Note that the trash, debris, and dead vegetation trapped in the screen have prevented water in the channel from entering the outlet structure.

Figure 5.3a, b shows a staged outlet structure overcome with excessive vegetation. The excess vegetation has blocked the low-flow entrance, which will cause elevated water levels in the pond during rainfall events.

The structural integrity and alignment of all inlet and outlet structures should also be inspected. Large cracks, severe dents, corrosion, missing bolts, and other malformations may indicate lack of sound structural integrity and impending failure. Inlet and outlet structures can become misaligned for several reasons, including frost heave of the soil, vehicular impact, and geotechnical failure. Misaligned inlet or outlet structures may allow stormwater runoff to enter or exit

Fig. 5.2 (**a**) A clogged outlet structure that has caused standing water to develop within the practice (©Stormwater Maintenance, LLC, www.SWMaintenance.com, reprinted with permission). (**b**) Inlet structure clogged with debris

Fig. 5.3 (**a**) A staged outlet structure overcome with excessive vegetation (©Stormwater Maintenance, LLC, www.SWMaintenance.com, Reprinted with permission). (**b**) Close-up of the low-flow outlet showing that the outlet has been partially blocked by vegetation that has also trapped litter and debris (©Stormwater Maintenance, LLC, www.SWMaintenance.com, reprinted with permission)

a practice by means other than those intended by design, or they may prevent stormwater runoff from entering or exiting the practice at all. This condition can result in erosion, channelization, or flooding of surrounding areas, which can further exacerbate misalignment, or create other problems. Figure 5.4 shows a misaligned outlet structure that would not be able to carry the design flow. Also, if water were to enter the hole in the concrete face, erosion of the soil around the misaligned pipe would be exacerbated. The same is true for the old brick and mortar face with cracks shown in Fig. 5.5.

Figure 5.6 shows the flared-end section of a concrete outlet pipe whose structural integrity has been undermined by excessive erosion. Misaligned inlet and outlet structures should be repaired or replaced as soon as possible to reduce detrimental effects.

Fig. 5.4 A misaligned outlet structure pipe (©Stormwater Maintenance, LLC, www.SWMaintenance.com, reprinted with permission)

Fig. 5.5 A cracked brick and mortar outlet (©Stormwater Maintenance, LLC, www.SWMaintenance.com, reprinted with permission)

5.2.4 Multicomponent Systems

Some stormwater treatment practices are designed with two or more components. For example, treatment wetlands are sometimes designed with two or more wetland cells in series. Also, when space is limited, it is not uncommon for dry ponds or other practices to be placed in series in an effort to conserve space and/or work

Fig. 5.6 A misaligned flared-end concrete pipe section that is also subject to excessive erosion (©Stormwater Maintenance, LLC, www.SWMaintenance.com, reprinted with permission)

within the limitations of the site. In these cases, it is important to recognize multicomponent systems and perform visual inspection on each of the components in the system to ensure the entire practice is functioning properly.

5.2.5 Water in the Practice

Some practices are designed to hold a permanent pool of water. Others, however, are designed to completely drain within a specified period of time. Standing water in a practice that is designed to be dry is the result of one of three possibilities: (1) rainfall has occurred so recently that the design drainage time has not elapsed and stormwater runoff has not had enough time to pass through the practice, (2) the treatment rate of the practice is unusually slow, such that stormwater runoff does not pass through the practice within the design time, but the practice does eventually drain completely, or (3) the outlet structure(s) are clogged such that stormwater runoff cannot exit the practice. If the last runoff event has occurred within the design drain time then the practice may be functioning properly, although proper operation cannot be verified with visual inspection. If, however, the design drainage time has elapsed since the most recent runoff event, it is likely that the state of the practice is either scenario (2) or (3), as described above. Figure 5.1, which was previously discussed, presents two examples, a rain garden with standing water that is not functioning properly, and a rain garden without standing water that may be functioning properly.

The condition of water in any practice is important to investigate and document. This is true both for practices with a permanent pool of water and for those that are designed to be dry within a certain time after a runoff event. For example, surface sheen on the water within a practice is caused by hydrocarbon substances such as automotive oil and gasoline and may indicate that an illicit discharge has occurred. If the possibility of an illicit discharge exists, an illicit discharge manual such as the manual authored by Brown et al. (2004) should be consulted to help verify, locate, and eliminate the illicit discharges.

If no illegal discharge of hydrocarbons has occurred within the watershed, then a surface sheen may indicate that stormwater runoff is stored in the practice such that the small amounts of hydrocarbons typically found in stormwater runoff are accumulating. If this is happening, then the practice may be failing and maintenance to restore the practice may be necessary. If the practice is designed to collect and accumulate hydrocarbons, maintenance to remove the sheen may be warranted.

If the water in the practice is murky or cloudy, it is likely that the runoff contains a large suspended solids concentration with a substantial portion of fine particle sizes, such as clays and silts. If the suspended solids load in runoff consisted of larger particles, the water would not be cloudy for an extended period of time because larger particles settle out of standing water relatively rapidly. Stormwater runoff that is cloudy can indicate that the watershed contains a significant source of fine particle suspended solids, which may quickly clog some practices such as filtration and infiltration practices.

Finally, if there is standing water in a practice or if runoff is entering a practice even though there has not been a recent rainfall or snowmelt event, water could be entering the stormwater conveyance system from a leak, spill, or surface application such as lawn watering. The source of the runoff should be determined.

5.2.6 Illicit Discharges

All visual inspections should determine if conditions suggest an illicit discharge may have occurred or may be occurring within the watershed. This process, however, is beyond the scope of this text. An illicit discharge manual (e.g., Brown et al. 2004) should be consulted for identifying and locating illicit stormwater discharges.

5.2.7 Erosion and Sediment Deposition

The interior of the practice and the conveyance system upstream and downstream of the practice should be examined for evidence of erosion. Erosion or channelization indicates that flow velocities are too large or that stormwater runoff is entering or exiting the practice by means other than those intended by design. If this is occurring, the cause of the erosion should be determined. Erosion and

Fig. 5.7 A channel that exhibits signs of erosion through bare soil and exposed tree roots (©Stormwater Maintenance, LLC, www.SWMaintenance.com, reprinted with permission)

channelization can increase the suspended solids concentration anywhere in the conveyance system, including within the practice itself. Locations of erosion or channelization should be restored to the original design condition and stabilized and/or the flow velocities should be reduced to prevent erosion. Figure 5.7 shows a channel bottom exhibiting signs of erosion (note the bare soil and exposed tree roots). Figure 5.8 shows a different channel that has experienced erosion. Note the bare soil on the side of the channel and the small tree that appears to have recently fallen over.

Excess and rapid sediment deposition within the practice (as determined by comparing the current state of the practice to the documented results of previous assessments) may indicate a significant source of sediment in the watershed that requires remediation to prevent downstream pollution. Sediment deposition reduces the volume of stormwater that can be stored in a practice and can cause sediments to become resuspended during subsequent storm events. Excess stored sediment may need to be removed.

Excessive sedimentation downstream of a practice may indicate that erosion is occurring within or upstream of the practice. If the practice is a sedimentation facility, the practice not be operating properly or is too small to capture the sediment. Figure 5.9 shows the downstream end of an outlet structure that has been filled in with sediment. This warrants further inspection and/or assessment.

Fig. 5.8 A channel that exhibits signs of erosion through bare soil and a fallen tree (©Stormwater Maintenance, LLC, www.SWMaintenance.com, reprinted with permission)

Fig. 5.9 An outlet to a stormwater treatment practice that is clogged with excessive sediment deposition (©Stormwater Maintenance, LLC, www.SWMaintenance.com, reprinted with permission)

Fig. 5.10 Bare soil that indicates excessive periods of inundation with water (©Stormwater Maintenance, LLC, www.SWMaintenance.com, reprinted with permission)

5.2.8 Soil and Vegetation

The condition of the soil surface and vegetation, if any, within the practice should be examined as part of any visual inspection. Bare soil or lack of healthy vegetation significantly different from the original design may indicate that the practice is not operating properly. For example, if the practice was designed to include vegetation and that vegetation has died or is unhealthy, it could indicate that standing water has remained in the practice for excessively long periods of time. Vegetation that is transitioning from the original vegetation to wetland species may also be indicative of long periods of standing water. Figure 5.10 shows an inlet to a stormwater conveyance system that is surrounded by bare soil. This indicates that the area around the intake is inundated with water for excessive periods of time such that the vegetation has died. Note that the elevation of the top surface of the intake is above the ground surface. Such a condition would cause water to pond around the intake.

Vegetation, especially with deep roots, can increase and maintain infiltration rates into soils that do not have impermeable surfaces (e.g., concrete). If the surface of a practice capable of infiltration becomes clogged or sealed, the roots of vegetation can provide pathways for stormwater runoff to penetrate the surface and subsequently infiltrate into the underlying soils, increasing runoff volume reduction.

Fig. 5.11 Side slope erosion at an unintended flow entrance. Note also the tree growing on a constructed slope, which may cause soil failure (©Stormwater Maintenance, LLC, www.SWMaintenance.com, reprinted with permission)

5.2.9 Litter and Debris

Litter and debris in a practice are indications that pretreatment practices are failing or not present. Litter and debris may limit the effectiveness of a practice by reducing the stormwater storage volume and therefore the retention time or by plugging the outlet or inlet structures. When present, litter and debris should be removed from a practice.

5.2.10 Banks or Sides of Practices

The banks or sides of any practice should be thoroughly inspected for changes that may alter the stability of the soils or evidence that the sides are failing. Various items to consider are discussed below.

The presence of erosion or channelization on the banks of a practice indicates that stormwater runoff is entering at a large velocity by means other than those intended by design. Erosion and channelization on the banks can fill the practice with sediments from the bank and subsequently reduce its effectiveness by reducing the volume available for stormwater storage and treatment. Figure 5.11 shows bank erosion from water entering the practice at an unintended location; this problem needs to be dealt with immediately. The next rainstorm could create excessive

Fig. 5.12 (**a**) A bank of a stormwater pond that has subsided (©Stormwater Maintenance, LLC, www.SWMaintenance.com, Reprinted with permission). (**b**) A bank of a stormwater pond that has no evidence of soil instability or erosion (©Stormwater Maintenance, LLC, www.SWMaintenance.com, reprinted with permission)

erosion that may threaten the practice and stability of the road, in addition to depositing substantial amounts of sediment in the practice.

Soil slides or bulges indicate that the soil is, or potentially will be, unstable and further sliding or bulging may lead to complete bank failure. If this occurs, the practice could lose storage volume and/or infiltration capacity, and the collapsed soil could be washed downstream. Figure 5.12a shows the bank of a practice that has subsided, whereas Fig. 5.12b shows a practice with no signs of bank instability. The subsided bank should trigger additional investigation, as it may indicate slope instability and/or erosion under the water surface.

Animal burrows may also lead to soil failure, which, as described above, could reduce the performance of the practice or be washed downstream. Figure 5.13 shows a groundhog and its burrow in the bank of a stormwater treatment practice.

Seeps and wet spots indicate subsurface flow into a practice and could lead to soil slides or erosion and channelization on the banks of the practice.

Poorly vegetated areas can lead to increased erosion and the collapse of the bank.

Trees on constructed slopes can cause soil instability and the loss of leaves in the autumn can lead to clogging of inlet and outlet structures. In addition, the root system, if extensive enough, can damage inlet, outlet, and underdrain structures, if present. Figure 5.11 also shows a tree on a constructed slope. The bare soil indicates erosion possibility due to large water velocities and/or weak soil. Note the bare patch of soil that exists around the trunk of the tree.

5.3 Visual Inspection for Sedimentation Practices

In addition to items common to most or all stormwater treatment practices, sedimentation practices have additional issues that should be considered.

Fig. 5.13 A groundhog and burrow within a stormwater treatment practice (©Stormwater Maintenance, LLC, www.SWMaintenance.com, reprinted with permission)

5.3.1 Water in the Practice

In a dry pond, stormwater that is green from algae or biological activity indicates that stormwater has been stored in the pond for an extended period of time such that microorganisms have developed. Typically, if a dry pond is draining correctly, microorganisms and algae growth will not have enough time to develop.

Invasive, tolerant fish species like carp (*Cyprinus carpio*) or shiner minnows (*Notropiscornutus*) in a wet pond are indications of poor water quality in the pond (e.g., low dissolved oxygen, turbid, limited habitat) such that tolerant and invasive species are present. More information should be gathered to determine the cause of the poor water quality, and remediation should be performed.

5.3.2 Erosion and Sediment Deposition

Erosion and sediment deposition within a dry pond can reduce the storage volume of the pond and lead to a reduction in performance of both peak flow attenuation and suspended solids removal. Additionally, previously captured sediments can become entrained by stormwater that is short-circuiting through the pond and be carried out of the practice. Figure 5.14 provides a photo of a pond that is ready for dredging. Sediment deposition will hinder water from entering the remainder of the pond, causing higher water level than the pond design specifications.

Visual inspection of sedimentation practices should include inspection and documentation of the amount and distribution of retained solids, if possible. For

Fig. 5.14 Sediment deposition in a retention pond (©John Chapman, reprinted with permission)

example, a large deposit of solids near the inlet of a dry pond may alter the inflow conditions and resuspend previously settled solids.

5.3.3 Soil and Vegetation

Dry ponds are typically designed to have surface vegetation. If vegetation has died or is unhealthy, or there is bare soil in some or all locations, it indicates that standing water has remained in the pond for excessively long time periods such that the vegetation has died.

The roots of vegetation provide pathways into the soil for infiltration, which can help reduce runoff volumes. Thus, vegetation in dry ponds should only be controlled to reduce the plant density or if it is undesirable for aesthetic or nuisance reasons.

If vegetation in a wet pond has died, is unhealthy, or has transitioned to another species, the pond is not functioning as designed. Vegetation in a wet pond should be selected to withstand the permanent pool of water and temporary periods of inundation above the permanent pool. If the vegetation has not survived or is not healthy, the pond may not be draining properly or water conditions may have changed from the original design.

Fig. 5.15 Sedimentation downstream of a stormwater treatment practice indicates a potential malfunction (©Stormwater Maintenance, LLC, www.SWMaintenance.com, reprinted with permission)

5.3.4 Downstream Conditions

Conditions downstream of a dry or wet pond can provide evidence useful for assessment. Properly designed and functioning ponds should remove most sand-sized particles (0.125–2 mm) from stormwater runoff. Sediment deposition downstream of a pond could indicate that the pond is failing to remove sediment as designed. This could result from short-circuiting within the pond due to excessive sediment storage, inlet or outlet malfunctioning, or some other reason. Short-circuiting can reduce residence times, which can reduce sediment removal and cause resuspension and scour of previously trapped sediment. Figure 5.15 shows sedimentation downstream of a pond that would warrant further assessment or inspection of the pond. As previously discussed, erosion downstream of the pond outlet could be contributing to the sediment deposition.

5.3.5 Structural Integrity

Underground sedimentation devices are encased practices placed totally underground, typically along roadways and parking lots. Thus, they are subject to traffic and soil loads and must be structurally sound. If the unit has any cracks, leaks, joint failures, or any other avenues for water to pass through, stormwater may prematurely exit the system or groundwater may infiltrate into the unit. In either case, the unit needs further assessment or maintenance. The unit may also not be performing as designed if it is corroded, has dents or other abnormalities on interior components, or if the orientation of any components deviates from its original alignment.

5.4 Visual Inspection for Filtration Practices

Visual inspection is useful for identifying a filtration practice that is not performing as designed. Specific items that should be part of a visual inspection are discussed in the following sections.

5.4.1 Water in the Practice

Filtration practices are not designed to have a permanent pool of water. If the filter media is significantly clogged, however, the permeability of the practice will be reduced and water will remain in the practice for excessive periods of time. To determine if a filtration practice is clogged, visual inspection should occur after the design drain time, typically 48 h, has elapsed following a large runoff event. If, at this time, standing water is present, the practice is not draining properly. Runoff in excess of the filter design capacity should overflow the filter through the emergency spillway or bypass the filter. Thus, the stormwater runoff captured by the filtration practice should still drain within the design time regardless of the total runoff volume.

If water in a filter is green from algae or biological activity, it is an indication that the filter is clogged and stormwater has remained in the practice long enough that microorganisms have developed. Typically, if a filter is draining correctly, microorganisms and algae will not have enough time to develop.

5.4.2 Erosion and Sediment Deposition

A filter should be inspected for the presence of a visible layer of fine material (i.e., mud) on the surface of the filter media (see Fig. 5.16a). A layer of fine material on the surface is an indication that:

1. Stormwater was present for an extended period of time such that fine material was allowed to settle to the filter surface or that the stored stormwater runoff evaporated and/or infiltrated through the sides and into the surrounding soils.
2. The filter media may be clogged. Filtration practices collect particles in the pore spaces of the media. If silts, clays, or both are present on the surface of the filter, the pore spaces within the filter media may be full.

In case 1, once on the surface, the layer of fines can significantly reduce the filtration capacity by providing extra resistance to flow. An example of this surface layer is shown in Fig. 5.16a. In case 2, the entire depth of media is clogged.

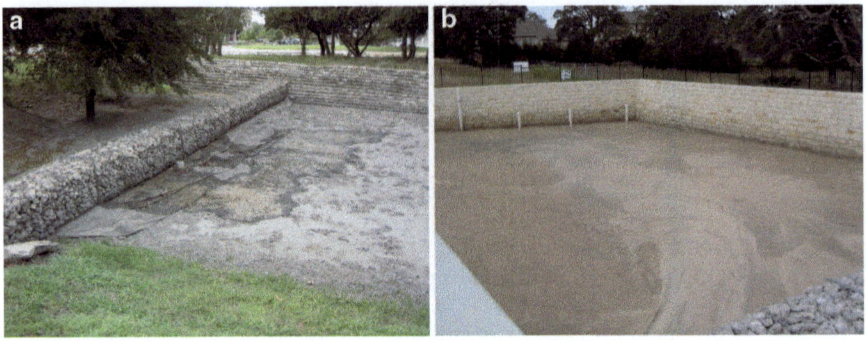

Fig. 5.16 (**a**) A sand filter covered with a layer of fine silt, limiting the filtration of stormwater. (**b**) A clean sand filter with no evidence of sediment accumulation on the surface

An example of a clean sand filter with no evidence of sediment accumulation on the surface or in the media is shown in Fig. 5.16b.

5.4.3 Downstream Conditions

Conditions downstream of a filtration practice can provide evidence of the function of the practice itself. Properly designed and functioning filtration practices remove a large percentage of suspended solids from stormwater runoff. If sediments are present in the effluent such that downstream deposition is occurring and erosion in the conveyance system downstream of the filter is not occurring, the geotextile fabric or the subsurface collection system is likely failing.

If the filtration unit is operating properly, downstream erosion is typically not a problem because the filter media passes runoff relatively slowly and downstream velocities are small. If downstream erosion is occurring for a relatively small filter, it is an indication that stormwater is passing through the filter too quickly or a large flow rate of water has bypassed the filter altogether. As a result, further inspection or assessment may be necessary. For filters with a large surface area and large total flow rates, however, runoff velocities may be large enough to cause erosion. In this case, the downstream conveyance system may need to be reinforced or otherwise stabilized.

5.4.4 Soil and Vegetation

Some filtration practices are designed to include vegetation on the surface of the filter. If vegetation was included in the original design, a lack of vegetation may indicate that stormwater has inundated the filtration practice longer than the original surface vegetation could withstand and longer than desired.

5.5 Visual Inspection for Infiltration Practices

Visual inspection of an infiltration practice involves investigation of the practice for indications of inadequate infiltration capacity, excessive infiltration through preferential flow paths, erosion, and other aspects. The scope of the visual inspection depends on the specific objectives and the type of stormwater treatment practice involved.

5.5.1 Water in the Practice

To determine if an infiltration practice is clogged, visual inspection should occur after the design drain time, typically 48 h, has elapsed following a large runoff event. If standing water is present at this time, the practice is not draining properly. Runoff in excess of the design volume should bypass the practice or overflow the filter through the overflow structure. Thus, the stormwater runoff captured by the infiltration practice should still drain within the design time regardless of the total runoff volume. A lack of vegetation in the practice typically is a sign that the practice is not draining properly, and standing water has killed the plants, as shown in Fig. 5.17.

Fig. 5.17 Infiltration basin without plants indicates that the basin is not sufficiently infiltrating and standing water has prevented plants from growing

Green water in an infiltration practice, whether from algae or biological activity, indicates that stormwater has been stored for an extended period of time such that microorganisms have developed. Typically, if a practice is infiltrating properly, microorganisms and algae growth will not have enough time to develop.

More involved observations include examining the soil profile for signs of persistent wet conditions in the surface soil or shallow subsurface soil. Such wet conditions indicate poor drainage conditions, which mean that infiltration capacities are lower than designed. Signs of persistent wet conditions in the soil are discoloration of the soil to a grayish tone and soil mottling. Mottling is an indication of anaerobic conditions resulting from persistent saturated or very wet conditions.

For permeable asphalt or concrete pavements, indicators of poor infiltration performance are persistent standing water on the pavements following rainfall or evidence of sediment deposition on the surface.

5.5.2 Erosion and Sediment Deposition

A visible layer of silts, clays, or both on the surface of an infiltration practice is an indication that the infiltration practice may be clogged. Infiltration practices collect particles on the surface and in the pore spaces of the soil. The presence of silts, clays, or both on the surface of the practice indicates that the pore spaces within the soil may be filled so that the infiltration rate has been reduced. Thus, the infiltration practice is not likely infiltrating stormwater runoff within the design time.

One of the main indications of poor infiltration capacity for an infiltration practice is a crust or layer of fine sediment that lies on the surface, which indicates that water was pooled for a substantial time period. If a crust is present, even if it shows signs of desiccation cracking, it could become a barrier to infiltration if it gets wet again.

While it might not be obvious during visual inspection, the subsurface pores of an infiltration practice can be clogged, even if the surface is clear of sediments. Closer examination by examining the soil just beneath the surface with a trowel or shovel might reveal clogging.

5.5.3 Soil and Vegetation

Vegetation in the bottom of an infiltration practice can increase infiltration effectiveness. Plants typically lose 30–50% of their root structures annually, which eventually produces macropores that can increase the infiltration rate of the practice so that more stormwater runoff is infiltrated. Additionally, vegetation can reduce overland flow velocities and can therefore reduce erosion and resuspension of captured solids. Compared to infiltration basins, infiltration trenches typically have a larger grain size so that vegetation cannot grow without clogging the pores.

Vegetation can also be an indication of the drain time of an infiltration basin practice. Terrestrial vegetation often cannot withstand long periods of inundation, and some terrestrial species cannot even withstand short periods of inundation. If an infiltration practice has an abundance of terrestrial vegetation, it is likely that the practice infiltrates stormwater runoff quickly (< 48 h) so that the terrestrial vegetation is not submerged for long periods of time. If, however, the infiltration practice has signs of aquatic vegetation, the practice may not be infiltrating stormwater runoff properly.

The presence of terrestrial vegetation, however, gives no indication regarding the presence of preferential flow paths. It is possible that an infiltration practice that drains quickly may be malfunctioning by channeling runoff through preferential flow paths. This scenario cannot be verified by visual inspection alone and must be verified with a higher level of assessment.

Although no vegetation should grow inside an infiltration trench, an indication of poor infiltration in a trench is poor vegetative growth (or wetland vegetation) in the area surrounding the trench. If the infiltration rate into the trench is small, water may pond around the trench for a relatively long period of time, increasing the chances that desired resident vegetation will suffer and/or wetland vegetation will take hold.

If a permeable pavement is designed to be vegetated, then the visual inspection issues discussed for other vegetated stormwater treatment practices typically apply to the permeable pavement.

5.6 Visual Inspection for Biological Practices

Visual inspection of biologically enhanced practices focuses on the vegetation (species, condition, abundance, etc.) and the condition of the soil. The species found in the practice and their condition and abundance can provide visual clues as to functionality of the practice. For example, abundant terrestrial vegetation in a rain garden indicates adequate soil moisture and quick drainage of stored runoff. Conversely, standing water and wetland vegetation (cattails, water lilies, etc.), or no vegetation at all in a bioretention practice shows that stormwater runoff does not infiltrate within the design drainage time.

5.6.1 Inlet and Outlet Structures

Bioretention and bioinfiltration practices typically have overflow structures instead of outlet structures. Outflow for a bioretention practice is intended to go into the soil such that deep percolation or evapotranspiration occurs. The overflow structure should only be active if a runoff event larger than the design runoff event occurs. All overflow structures should be free of debris, sediment, vegetation, and other obstructions so that stormwater runoff can easily exit the bioretention practice in the event of a large storm event. If the overflow structure is partially or completely clogged, surrounding areas may be flooded. Any obstructions should be removed immediately to ensure proper operation of the bioretention practice.

5.6.2 Water in the Practice

Constructed wetlands are designed to have a permanent pool of water. The absence of standing water in constructed wetlands is the result of one of three possibilities: (1) rainfall has not occurred in a length of time such that all stored stormwater runoff has been lost to evapotranspiration (i.e., drought conditions), infiltrated, or both, (2) the outlet structure is damaged or malfunctioning such that stormwater runoff is allowed to drain out of the constructed wetlands, or (3) the inlet structure is clogged or misaligned such that stormwater runoff is not entering the constructed wetlands. If it has rained in the last 48 h, then the constructed wetlands should have received or will soon receive stormwater runoff and therefore drought conditions should not be of concern. If approximately 48 h has elapsed since the last runoff event and standing water is not present in the constructed wetlands, it is likely that either scenario (2) or (3) are the cause.

Filter strips and swales are designed for stormwater conveyance and not stormwater storage. Standing water in a filter strip or swale is an indication of failure by (1) downstream flooding or (2) blockage that is preventing stormwater runoff from being conveyed downstream. Areas downstream of the filter strip or swale should be inspected for signs of flooding or obstruction, and the filter strip or swale should be inspected for obstructions. Figure 5.18 shows a drainage swale with standing water. Note the algal blooms adjacent to the banks algal blooms indicate [that] standing water has been present for a relatively long time period.

5.6.3 Soil and Vegetation

Vegetation, especially vegetation with deep roots, can increase and maintain infiltration rates in bioretention practices with permeable surfaces. If the surface of the bioretention practices becomes clogged or sealed, vegetation can provide pathways for stormwater runoff to penetrate the surface and subsequently infiltrate into the underlying soils, increasing runoff volume reduction. Thus, vegetation in bioretention practices should only be controlled if it is undesirable for aesthetic or nuisance reasons.

Vegetation in the bottom of a bioretention practice is designed to dry out the soil between runoff events because this can help maintain infiltration capacity. Every year plants can lose 30–50% of their root structures. This can create macropores that increase the infiltration rate into the soil. Additionally, vegetation can reduce overland flow velocities, which can reduce erosion and resuspension of captured solids.

Fig. 5.18 Algal bloom in a drainage swale with standing water (©Stormwater Maintenance, LLC, www.SWMaintenance.com, reprinted with permission)

Vegetation can also be an indication of the drain time of a bioretention practice. Terrestrial vegetation often cannot withstand long periods of inundation, and some cannot even withstand short periods of inundation. If a bioretention practice has an abundance of terrestrial vegetation, it is likely that the practice infiltrates stormwater runoff quickly (< 48 h) and is therefore operating properly. If, however, the bioretention practice has signs of aquatic vegetation or has little vegetation, it is likely the practice is not adequately infiltrating stormwater runoff and is therefore failing.

Species of vegetation in planting plans for bioretention practices, wetlands, filter strips, and swales are selected based on desirable characteristics that a particular species of plant may exhibit. During the construction and operational life of a biologically enhanced practice, the vegetation may deviate from the original design and possibly affect the performance of the practice. If planting designs are available, the vegetation in the practice at the time of inspection should be compared to the vegetation designated in the design plans. Particular items to investigate include species that are not healthy or have disappeared as well as the introduction of weeds, wetland vegetation, and/or other invasive vegetation.

The health of the vegetation may indicate that conditions are too wet/dry, that the site is too sunny/shady, or that the soil lacks nutrients, has become compacted, or contains toxic pollutants, etc. The survival of the vegetation is critical to maintaining proper function of a bioretention practice, wetland, filter strip, or swale. The apparent visual health of the vegetation in the practice should be assessed during the growing season. Some indications of unfavorable conditions are wilted leaves/stem, discoloration of leaves, lack of flowering buds developing, stunted growth, and a decrease in the number of plantings present.

Under optimal site conditions, the vegetation should have an appropriate size and density for its species. Underdeveloped vegetation can be an indication of poor health, while overdevelopment can hinder the development of other plant species in the bioretention practice, constructed wetland, filter strip, or swale.

Vegetation in constructed wetlands should be consistent with native or design-specified wetland vegetation. The absence of vegetation anywhere in or around constructed wetlands may be an indication of poor water quality or excessive infiltration that has previously dried the wetland.

For guidance on vegetation identification and other related items, please refer to Plants for Stormwater Design: Species Selection for the Upper Midwest (Shaw and Schmidt 2003).

Capacity Testing of Stormwater Treatment Practices

6

Abstract

A stormwater treatment practice can be assessed by testing, which involves making a series of measurements under conditions that are not a result of a natural runoff event. Capacity testing involves either the measurement of sediment surface elevations within a stormwater treatment practice or making measurements to determine the saturated hydraulic conductivity of soil within the practice. This chapter discusses how capacity testing can be applied to various stormwater treatment practices and also includes examples demonstrating how the obtained data can be used to schedule maintenance. The chapter concludes with a case study involving the assessment of infiltration rates in a bioinfiltration practice (i.e., rain garden).

Testing (level 2) of stormwater treatment practices consists of making a series of measurements under conditions *other than* a natural runoff event. Testing is more complex than visual inspection but typically is not as complex or time consuming as monitoring, which involves assessing the performance of a practice during a natural runoff event. Testing should be considered as an assessment option if filtration/infiltration rates, sediment storage, or pollutant removal are important aspect of the stormwater treatment practice. Testing, however, should only be considered after visual inspection has been performed and revealed no obvious malfunctions of the practice. There are two levels of testing that can be used to assess a stormwater treatment practice: capacity testing and synthetic runoff testing. Capacity testing is discussed in this chapter and synthetic runoff testing is discussed in Chap. 7.

Capacity testing uses a series of point measurements to determine either the filtration/infiltration capacity at the surface of various locations within a treatment practice or the remaining sediment storage capacity of an entire practice. The surface infiltration capacity at specific locations within a practice is quantified by measuring the saturated hydraulic conductivity (K_s) at each location. The individual values can be used to identify areas where the practice is clogged or is experiencing unusually large infiltration rates. This enables maintenance efforts to be focused where they will be most effective. Because capacity testing for saturated hydraulic

A.J. Erickson et al., *Optimizing Stormwater Treatment Practices: A Handbook of Assessment and Maintenance*, DOI 10.1007/978-1-4614-4624-8_6,

conductivity only accounts for soil properties at the surface, any soil properties below a depth of approximately 20 cm (8 in.) will not be represented in the results.

Sediment storage capacity is determined by measuring the existing sediment surface elevation at a series of locations to estimate the total volume of retained sediment stored within the practice. Comparing this value with the total design sediment storage capacity (or as-built plans) of the practice allows for an estimate of the remaining sediment storage capacity of the entire practice to be made.

6.1 Measuring Sedimentation

Sediment accumulation tests can be applied to any stormwater treatment practice that collects sediment and allows sediment surface elevations to be measured, such as dry ponds, wet ponds, wetlands, wet vaults, and underground sedimentation devices. A major advantage of sediment accumulation testing as compared to synthetic runoff testing is that it can be performed for all sizes of stormwater treatment practices. Synthetic runoff testing is dependent upon an adequate water supply of synthetic runoff, which restricts its use to smaller stormwater treatment practices. Compared to monitoring, testing for sediment accumulation requires less time and is less expensive. Another advantage is the ability to use patterns of sediment accumulation as a diagnostic test for maintenance procedures because the source of the accumulation can be more easily identified. For example, if sediment accumulation is primarily near the inlet to the practice, the source is likely in the upstream watershed, and maintenance can be targeted near the inlets. If, however, the accumulated sediment is primarily in the practice away from the inlet and possibly near the banks, the source of the sediment may be erosion of the banks or nearby contributing area. In this case, maintenance should not only be targeted to remove the accumulated sediment but should also address the cause of the erosion by stabilizing the banks or eroded areas.

Annual testing can be used to estimate the sediment accumulation rate over time (e.g., kg/year). Sediment accumulation testing cannot be used, however, to assess pollutant removal efficiency, because it only measures the amount of sediment captured and does not measure the amount of sediment entering or exiting a stormwater treatment practice. Therefore, if the assessment goals include pollutant removal efficiency, synthetic runoff testing or monitoring must be considered.

6.1.1 Measuring Sediment Surface Elevations

Sedimentation practices remove, on average, over 50% of the influent suspended solid mass load (Weiss et al. 2007). These practices have a designated storage volume in which to store the captured sediment, but as the practice fills with sediment over time, the available storage volume is reduced and removal efficiency drops. Eventually, sediment will need to be removed from the practice in order to maintain the desired level of sediment removal.

Sediment accumulation tests estimate the sediment accumulated in a stormwater treatment practice. Using surveying equipment or global positioning system (GPS) units, sediment surface elevation is measured at known locations throughout the stormwater treatment practice and the data are entered into three-dimensional computer-aided drafting software. For small areas like manholes or underground sedimentation devices, only a few measurements may be required. The data can then be compared to similar data of initial surface elevation measurements (or as-built plans or design drawings) to determine the amount of sediment that has accumulated. The amount of accumulated sediment can then be compared to the design sediment storage volume to determine the available capacity for additional sediment storage. The rate of sediment accumulation in mass per time can also be calculated for a given time period using (6.1):

$$Rate\ of\ sediment\ retention = \rho_s \frac{(V_2 - V_1)}{t_2 - t_1} \tag{6.1}$$

where

ρ_s = density of sediment
V_2 = volume of accumulated sediment measured at time t_2
V_1 = volume of accumulated sediment measured at time t_1
t_2 = time of measurement of V_2
t_1 = time of measurement of V_1

6.1.2 Scheduling Maintenance

These tests should be performed soon after construction is complete to develop a benchmark for future assessment. The sediment accumulation rate can then be used to estimate when the sediment storage volume will be near capacity (roughly 30–50% of the original water storage capacity) and when sediment removal will be necessary.

Capacity testing can also be used to determine the extent of targeted maintenance. For example, sediment tends to accumulate first near the inlet to the practice and subsequently fills the remaining storage volume. A proactive maintenance schedule could use the results from capacity testing to determine the volume of sediment to be removed and provide specific instructions for maintenance workers about the location of the sediment to be removed.

6.1.3 Dry Ponds

The amount of retained sediment can be compared to the design capacity to determine the available sediment retention capacity and, if the sediment accumulation rate is known, to estimate when the pond will require maintenance (i.e., sediment cleanout). One to three days are typically required to perform sediment retention assessment for a single dry pond.

6.1.4 Wet Ponds

Sediment retention tests can be performed on a wet pond to estimate the depth and, subsequently, the volume of sediment retained. Bottom elevations in a wet pond are measured either with a level and level rod (from a boat) or with a sonar depth measurement device. The water surface can be used as a local elevation standard if a staff gauge has been installed in the pond to measure water surface elevation. Sonar depth measurements can be made in the winter when the wet pond is covered with sufficient ice to traverse or in the summer from a boat. Using waders to enter or cross a wet pond is not recommended because bottom sediments can be soft and therefore a safety hazard.

Corresponding longitude and latitude are recorded either with GPS or with a total station. Using the basin topography and the original topography (from as-built plans or design drawings), the amount of sediment retained in the pond can be estimated. The volume of retained sediment can be compared to the design capacity to determine the available sediment retention capacity, and, if the sediment accumulation rate is known, to estimate when the pond will require maintenance (i.e., sediment cleanout). As with dry ponds, these tests should be performed soon after construction is complete to develop a benchmark for future assessment.

6.1.5 Underground Sedimentation Devices

If the sediment collection area can be accessed, sediment retention testing can be performed by utilizing staff gauges or visual benchmarks and as-built plans to determine the volume of sediment collected in an underground sedimentation device. These measurements can be used with estimates or measurements of sediment inflow rates to develop a maintenance or cleanout schedule. When the collected solids volume meets or exceeds the solids storage capacity of a wet vault or underground sedimentation device, solids will no longer be removed at desired levels. Furthermore, resuspension of retained solids can result in negative pollutant removal efficiencies.

6.1.6 Constructed Wetlands

Sediment retention tests can be performed on a wetland to estimate the depth, and subsequently the volume, of sediment retained in a constructed wetland. Sediment surface elevations in a wetland are measured either with a level and level rod (from a boat) or with a sonar depth measurement device. The water surface can be used as a local elevation standard if a staff gauge has been installed in the pond to measure water surface elevation. Sonar depth measurements can be made in the winter when the wetland is covered with a layer of ice that is strong enough to walk on safely or in the summer from a boat. Using waders to enter or cross a constructed wet land is not recommended because bottom sediments can be soft and therefore a safety hazard.

Corresponding longitude and latitude are recorded either with GPS or a total station. The amount of sediment retained in the constructed wetland can be estimated from the measured basin topography and the original basin topography (from as-built plans or design drawings). If sediment capacity testing is performed periodically, the rate of sediment accumulation can be estimated from (6.1). The amount of retained sediment can also be compared to the design capacity to determine the available sediment retention capacity, and, if the sediment accumulation rate is known, to estimate when the wetland will require maintenance (i.e., sediment cleanout). These capacity tests should also be performed following construction to develop as-built plans.

6.2 Measuring Infiltration/Filtration

Infiltration capacity testing estimates the saturated hydraulic conductivity (K_s) at the soil surface at specific locations within the practice. A single point measurement with an infiltrometer can take between 30 s and several hours, depending on the soil characteristics of the practice. Measurements of K_s should be performed shortly after construction to establish a baseline for future tests and to investigate or identify construction impacts such as soil compaction on infiltration capacity. Examination of individual K_s values can reveal locations where the K_s values are too small or too large. This allows for localized maintenance to be performed on the practice, which can be more cost-effective than performing maintenance on the entire practice. Small infiltration rates may be attributed to clogging of the surface layer with captured sediments or a relatively impermeable subsurface layer. In this case, a soil core can be examined for the presence of relatively impermeable layers to determine the cause of small K_s values. Large infiltration rates indicate areas that may pass water through preferential flow paths that may avoid treatment, such as macropores, cracks, and fissures. These areas can be further inspected to determine the most effective course of action.

Using capacity test results to accurately estimate the time required for the practice to infiltrate/filtrate a given runoff volume currently requires three-dimensional, time dependent modeling. A simpler, and likely more accurate, approach would be to apply synthetic runoff testing (Chap. 7) or monitoring (Chap. 8) to make these estimates.

Infiltration capacity tests can be performed on the following stormwater treatment practices: dry ponds, bioretention practices (rain gardens), sand filters, infiltration trenches, infiltration basins, filter strips, swales, and permeable pavement. Saturated hydraulic conductivity (K_s) may vary based on climatic season, soil conditions, etc.; therefore, infiltration capacity tests could be performed at several different times throughout the year to get a representative estimate of K_s. In order to compare results over time and season, tests should be performed at the same location each time measurements are made. An example schedule for K_s tests could include testing in the spring after the ground thaws, in midsummer, and in late fall before the ground freezes. For more information on data analysis for capacity testing, see Chap. 12.

An advantage of infiltration capacity tests as compared to synthetic runoff testing is that they can be performed for all sizes of stormwater treatment practices. Synthetic runoff testing is dependent upon an adequate supply of synthetic runoff, which limits its use to smaller stormwater treatment practices. The advantage of capacity tests compared to monitoring (level 3) is that less time and expense is required to perform the assessment. Another advantage is the ability to evaluate maintenance procedures. The cause of reduced infiltration capacity can be easily identified using capacity testing and specific locations within the stormwater treatment practice with small (or excessively large) infiltration capacity can be identified. These locations can be repaired, as opposed to repairing the entire practice. Infiltration capacity testing conducted annually can also be used to estimate the change in saturated hydraulic conductivity (K_s) with respect to time, if the tests are performed under similar conditions and at the same locations.

6.2.1 Measuring Saturated Hydraulic Conductivity

Several devices are available to determine soil saturated hydraulic conductivity (K_s), as discussed in Chap. 9. Most infiltration measurement devices also require soil moisture to be measured, procedures for which are discussed in Chap. 11. If the tools necessary to determine the bulk density and the gravimetric moisture content are not available, then the initial volumetric moisture content of the soil can be estimated using Table 6.1 or Fig. 6.1. Note that Fig. 6.1 provides estimates for soil moisture based on soil texture and relative moisture content varying between low moisture (i.e., wilting point), average moisture (i.e., available water), and high moisture (i.e., field capacity) conditions. Although it can be more accurate to select initial soil moisture using Fig. 6.1 based on field-observed moisture conditions and soil texture, it has been found that the change in moisture content has a less than 20% effect on the calculated value of K_s, which can be considered as minor relative to the orders of magnitude in spatial differences (Regalado et al. 2005).

After the identification of the soil type and the initial moisture content from Table 6.1 or other means, the final moisture content is assumed to be the effective porosity for that soil, which can be estimated from Table 6.2.

A positive capillary pressure indicates that the soil is hydrophobic (repels water) and a negative capillary pressure value indicates the soil is hydrophilic (attracts water). Most soil is hydrophilic.

Saturated hydraulic conductivity (K_s) values can vary spatially by orders of magnitude depending on many factors, such as soil texture, plant root structure, porosity, and soil moisture, amoung others. (Warrick and Nielsen 1980; Asleson et al. 2009). It is thus essential to take many measurements of K_s on an infiltration practice in order to accurately represent the variation in surface infiltration capacity over the entire practice. As discussed in Chap. 9, it is recommended that a falling head method be used to measure K_s because many measurements can be collected quickly and in some cases simultaneously. The Modified Philip-Dunne (MPD) infiltrometer is a falling head device that can be used to perform capacity testing of stormwater treatment practices to measure K_s.

Table 6.1 Guide for estimating soil water moisture content based on soil feel and appearance for several soil textures (Wright and Bergsrud 1991)

Loamy sand	Θ (%)	Sandy loam	Θ (%)	Loam	Θ (%)	Clay loam	Θ (%)
Leaves wet outline on hand when squeezed	15%	Very dark color, leaves wet outline on hand when squeezed, makes a short ribbon	20%	Very dark color, leaves wet outline on hand when squeezed, will ribbon out greater than one inch	28%	Very dark color, leaves slight moisture on hands when squeezed, forms ribbons to about 2″	29%
Appears moist, forms a weak ball	12.5%	Quite dark color, forms a hard ball	17.5%	Dark color, forms plastic ball, feels smooth and slippery with glossy appearance when rubbed	25%	Dark color, feels smooth and slippery with glossy appearance when rubbed, forms ribbons easily	27%
Appears slightly moist, cohesive	10%	Fairly dark color, forms a good ball	15%	Quite dark color, forms a hard ball	22%	Quite dark color, forms thick ribbon, may feel smooth and slippery with glossy appearance when rubbed	25%
Appears dry, will not form a ball under pressure	7.5%	Slightly dark color, forms a weak ball	12.5%	Fairly dark color, forms a good ball	19%	Fairly dark color, forms a good ball	23%
Dry, loose, single grained, flows through fingers	5%	Lightly colored by moisture, will not form a ball	10%	Slightly dark color, forms a weak ball	16%	Forms a ball, small clods will flatten out	21%
		Very slight color due to moisture, loose, flows through fingers	7.5%	Lightly colored by moisture, small clods crumble fairly easily	13%	Slightly dark color, clods crumble	19%
				Slightly colored by moisture, powdery, dry, sometimes slightly crusted but easily broken down in powdery condition	10%	Some darkness due to unavailable moisture, hard, baked, cracked sometimes has loose crumbs on surface	17%

6.2.2 Dry Ponds

As with other practices, saturated hydraulic conductivity (K_s) testing of dry ponds is used to estimate the rate at which stored water infiltrates into the soil at the surface and how this rate varies with location over the surface of the pond. Results can be

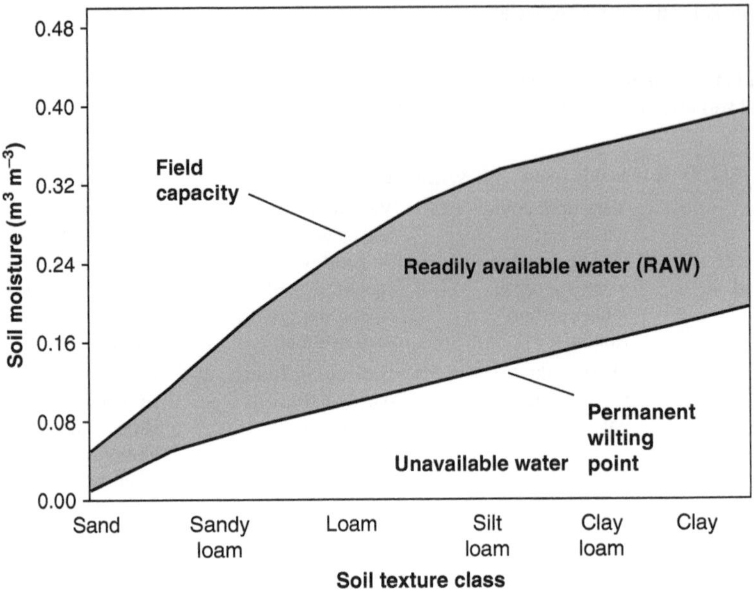

Fig. 6.1 Relationship of soil moisture, soil texture, field capacity, and wilting point (From Horttechnology 20:1 p. 133–142 (2010), reprinted with permission, © American Society for Horticultural Science)

Table 6.2 Typical soil measurements

Soil type	Porosity	Effective porosity	Capillary pressure (cm)	Saturated hydraulic conductivity (cm/s)
Sand	0.437 (0.374–0.5)	0.417 (0.354–0.48)	−4.95 (−0.97 to −25.36)	2.5–7.5 × 10^{-3}
Loamy sand	0.437 (0.363–0.506)	0.401 (0.329 to 0.473)	−6.13 (−1.35 to −27.94)	1–4 × 10^{3}
Sandy loam	0.453 (0.351–0.555)	0.412 (0.283–0.541)	−11.01 (−2.67 to −45.47)	0.3–3 × 10^{-3}
Loam	0.463 (0.375–0.551)	0.434 (0.334–0.534)	−8.89 (−1.33 to −59.38)	1–14 × 10^{-4}
Silt loam	0.501 (0.42–0.582)	0.486 (0.394–0.578)	−16.68 (−2.92 to −95.39)	0.9–18 × 10^{-4}
Sandy clay loam	0.398 (0.332–0.464)	0.33 (0.235–0.425)	−21.85 (−4.42 to −108)	0.8–6 × 10^{-4}
Clay loam	0.464 (0.409–0.519)	0.309 (0.279–0.501)	−20.88 (−4.79 to −91.1)	0.2–2 × 10^{-4}
Silty clay loam	0.471 (0.418–0.524)	0.432 (0.347–0.517)	−27.3 (−5.67 to −131.5)	1–2 × 10^{-4}
Sandy clay	0.43 (0.37–0.49)	0.321 (0.207–0.435)	−23.9 (−4.08 to −140.2)	0.2–3 × 10^{-5}
Silty clay	0.479 (0.425–0.533)	0.423 (0.334–0.512)	−29.22 (−6.13 to −139.4)	5–10 × 10^{-5}
Clay	0.475 (0.427–0.523)	0.385 (0.269–0.501)	−31.63 (−6.39 to −156.5)	5–8 × 10^{-5}

Roughly two-thirds of the measurements are within the values given in parenthesis (Rawls et al. 1983; Rawls et al. 1998; Saxton and Rawls 2005)

used to focus on localized areas that are not performing as designed. Areas with exceedingly large or small K_s values may be in need of maintenance or should be further inspected. Because dry ponds typically infiltrate into the existing soil, K_s values that trigger further inspection and/or maintenance will vary with local soil properties. Thus, K_s test results should be compared to design values and/or results from previous tests to determine areas where further inspection or maintenance may be necessary.

6.2.3 Filtration Practices

Capacity testing for the assessment of saturated hydraulic conductivity (K_s) of filtration practices is similar to capacity testing for infiltration practices and involves a series of K_s point measurements that can be used to determine areas that are clogged and where localized maintenance is needed. For design purposes, a minimum value of K_s of 1.07 m/day is recommended (Claytor and Schueler 1996). Thus, areas with a K_s value less than this value should be considered for maintenance.

Saturated hydraulic conductivity (K_s) tests can also be used to detect the presence of macropores within a filtration practice. Macropores allow stormwater runoff to flow quickly through the filtration media, resulting in minimal solids removal. If the results from the measurements indicate that any values of K_s within the practice are larger than the K_s of gravel (85 m/day, 280 ft/day), then it is likely that minimal treatment by filtration is occurring as a result of macropores. K_s values less than 85 m/day do not preclude the presence of macropores, but indicate that the number of macropores, if present, may not necessarily be significant.

Saturated hydraulic conductivity (K_s) tests are applicable to surface, underground, and hybrid filtration practices. K_s tests in hybrid filtration practices, however, should be interpreted differently from tests as applied to surface or underground filtration practices. Hybrid filters (discussed in Chap. 4) use sand or soil filtration with subsurface collection in part of the practice, but not throughout. In locations where stormwater runoff filters through media and is collected by the underdrain system, K_s tests indicate the rate at which stormwater runoff is filtered. In areas where stormwater is stored and/or allowed to infiltrate, K_s tests indicate the rate at which stormwater runoff infiltrates into the native soil, which is the rate that stormwater runoff volume is reduced by infiltration.

6.2.4 Infiltration Basins

Testing the infiltration capacity of the surface of infiltration basins involves a series of saturated hydraulic conductivity (K_s) point measurements. Due to large spatial variation in K_s values even at locations in close proximity to each other, an accurate representation of the infiltration capacity of the entire basin currently requires numerous measurements and three-dimensional, time-dependent modeling. As with dry ponds, infiltration is usually into natural soils and test results should be compared to design parameters and/or results from previous tests. Even in infiltration basins with engineered soil, K_s often has a large variability (Asleson et al. 2009). There are many techniques available for measuring K_s, as described in Chap. 9.

6.2.5 Infiltration Trenches

Infiltration trenches are generally filled with coarse gravel or crushed rock. Therefore, it is not appropriate to use infiltrometers or permeameters for infiltration trenches. Synthetic runoff testing should be used to evaluate the infiltration rate of infiltration trenches.

6.2.6 Permeable Pavements

It is not currently feasible to use soil infiltration measurement devices for permeable pavements because of the structure of the pavement material. Individual devices have been developed for falling head tests of permeable pavement (Cooley 1999; Gilson Co. 2003; ASTM 2009).

6.2.7 Bioretention Practices (Rain Gardens)

Bioretention practices are often designed to draw down their storage volume (design storm volume) in less than 48 h. If visual inspection indicates that a bioretention practice does not infiltrate the design storm volume in less than 48 h, the soil media may be clogged. Clogged media may cause flooding of surrounding areas or force untreated stormwater to bypass the bioretention facility. Conversely, if the design storm volume drains in less than 6 h, large macropores or preferential flow paths may be present. Macropores can short-circuit the filtration process, passing untreated (or minimally treated) stormwater directly to the effluent structures or to groundwater.

If the bioretention practice is not infiltrating stormwater at the desired rate, the soil profile should be inspected for soil properties, including texture, color, moisture, and bulk density (see Chap. 11). Alternatively, research has shown that saturated hydraulic conductivity (K_s) can be estimated from soil texture classification (Rawls et al. 1998; Saxton and Rawls, 2005). Table 6.2 lists K_s values based on USDA soil texture from various authors. Note that each textural class has a range of possible values.

Saturated hydraulic conductivity (K_s) testing throughout the bioretention practice can be used to assess the spatial range of infiltration capacity and to identify areas of small or large K_s. An infiltrometer and/or permeameter should be chosen and used throughout the bioretention practice.

Measured K_s for the bioretention practice should be compared to design specifications or previous test results to determine if the practice is performing effectively. An example of capacity testing applied to bioretention facilities based on the work by Asleson et al. (2009) is given at the end of this section in a case study. K_s tests should be performed periodically to determine if the bioretention practice performance is stable or changing significantly.

6.2.8 Filter Strips and Swales

Filter strips and swales (without check dams) rarely maintain standing water because they are designed for stormwater conveyance, not storage. Nevertheless, infiltrometer and/or permeameter tests can be performed on filter strips and swales to determine saturated hydraulic conductivity (K_s) values within the practice. Some swales have berms or check dams to reduce flow velocities and store stormwater runoff temporarily, which also increases sedimentation and infiltration. K_s tests should be focused on the locations where infiltration occurs or is likely to occur based on the design, such as upstream of a berm or check dam.

6.3 Case Study: Capacity Testing of Infiltration Rates at a Bioinfiltration Practice

Five bioretention practices (basins A–E), three of which (basins B, C, and D) are in series, were evaluated for performance with infiltration capacity testing. Basins C and B serve as overflow basins and are connected to basin D by two drop structures consisting of bricks (see Fig. 6.2). The assessment was conducted on the basin D bioretention practice. A thorough assessment of basin D was conducted in the 2 years after the basins were installed.

The basin D bioretention practice is approximately 67 square meters (716 square feet) in size, with a ponding depth of 0.15 m (0.5 ft). It is designed to provide storage for the maximum amount of water the space would allow. Stormwater runoff is directed to the bioretention practice using two inlets, a curb cut along the northwest corner of the bioretention practice and an inlet pipe located at the center of the north border of the bioretention practice, which is connected to the stormwater sewer system. The storm sewer inlet pipe has a 12.5 cm by 30 cm (5 in. by 12 in.) subgrade of wall stone to prevent erosion. The native soil was excavated and filled with a sand trench to a depth of 1–1.3 m (3–4 ft) and a width of 1 m (3 ft) in the center of the basin. Clean sand with only 5% passing through a 200 μm sieve was used for the sand trench. The basin D bioretention practice was designed to infiltrate the maximum storage volume within 24 h at an estimated infiltration rate of 1.2 cm/h (0.5 in./h). The basin was then filled with planting topsoil to a depth of 20 cm (8 in.) and planted with selected vegetation. The plant design plan is shown in Fig. 6.2.

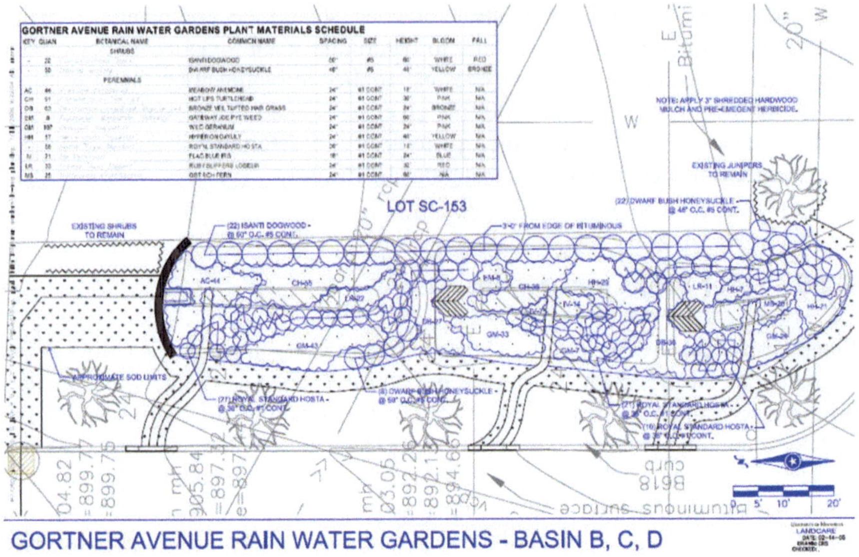

Fig. 6.2 Plant design plan

6.3.1 Assessment Goals

The purpose of the assessment was to determine if the bioretention practice had the ability to infiltrate stormwater runoff at the appropriate rate. The soil in the practice was primarily sandy loam, which, from Table 6.2, is expected to have saturated hydraulic conductivity (K_s) values ranging from 0.0003 to 0.003 cm/s. Thus, without design guidelines or results from prior capacity tests, the performance of areas with K_s values within this range would be considered satisfactory. Areas with values significantly outside this range would be subject to further inspection or assessment, as they may be clogged ($K_s < 0.0003$ cm/s) or may contain macropores with preferential flow paths ($K_s > 0.003$ cm/s).

6.3.2 Assessment Techniques

After a thorough visual inspection assessment revealed no malfunctioning, the saturated hydraulic conductivity (K_s) of the soil was measured using the MPD infiltrometer to determine the bioretention practice's capacity for infiltrating water. Locations where point measurements of K_s were to be made were distributed evenly throughout the entire bioretention practice and marked using orange utility flags. These locations varied in their proximity to the vegetation but were never placed directly over the base of the plant. Additional locations were marked at the low

Fig. 6.3 Flags marking locations of permeability tests

point of the site to better represent areas inundated by frequently occurring small runoff events. Figure 6.3 is a photograph of the bioretention practice with orange utility flags marking test locations. A total of 40 locations were marked in this site to evaluate the spatial variability of K_s within the practice. The coordinates of each location and the perimeter of the bioretention practice were recorded using a GPS device.

The MPD infiltrometer is a falling head infiltrometer. The device was uniformly pounded into the soil to a depth of 5 cm. The initial soil moisture was measured at five locations around the base of the MPD infiltrometer. Mulch from the bioretention practice was placed inside the device to prevent erosion; water was then poured into the device to the desired height, which was 17 in. for this site. Two sets of the change in water level with time measurements were made for additional data. The first set was the visual method, which requires an initial height of water at time zero, a time at the half way point (approximately 8 in.), and the time when the MPD becomes empty. The second method made continuous measurements using an ultrasonic sensor (Fig. 6.4).

Fig. 6.4 Modified Philip-Dunne permeameter with ultrasonic sensor

6.3.3 Assessment Results

The 40 locations used for point measurements were positioned using GPS and input into digital mapping software (ArcView). Figure 6.5 is an ArcMap that shows the measurement locations and their corresponding saturated hydraulic conductivity (K_s) values obtained with the MPD infiltrometer. The figure illustrates how K_s varies spatially throughout the stormwater treatment practice.

The MPD infiltrometer was used for this assessment because it is relatively easy to use, results could be obtained quickly, and multiple MPD infiltrometers could be operated at the same time. As a result, saturated hydraulic conductivity (K_s) values at many locations could be obtained in a day or less. To increase time efficiency, multiple MPD devices can be used with little additional staff hours invested. This level of assessment (i.e., level 2) was determined to be the most beneficial technique for understanding the spatial variability of the site and developing a maintenance schedule for the practice.

Observation of Fig. 6.5 reveals that locations marked by the two largest dots (i.e., green and blue) have K_s values in range expected for sandy loam soils. All other locations have lower than expected values with the two smallest dots (red and orange) being significantly lower (i.e., greater than an order of magnitude lower).

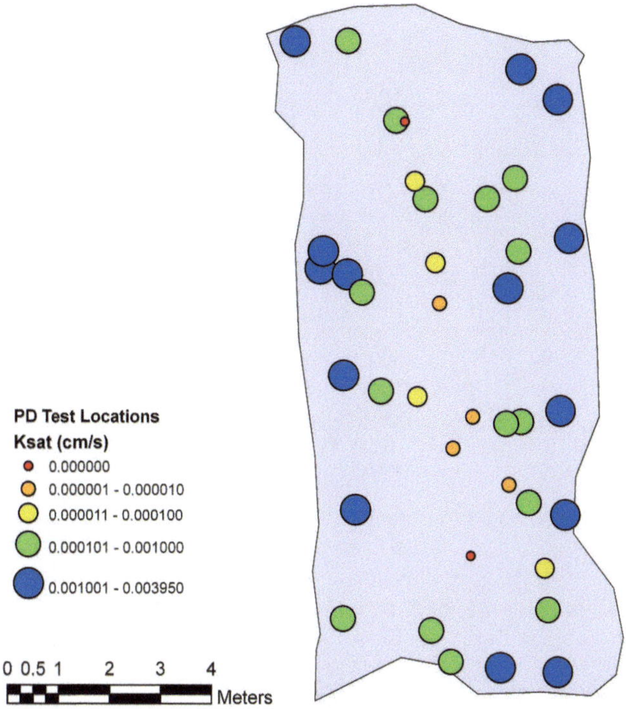

Fig. 6.5 ArcMap of K_s using the Modified Philip-Dunne (MPD) permeameter measurements

If other assessment efforts reveal that the practice is not infiltrating runoff within the design time (typically 48 h), maintenance efforts can be focused on areas with small K_s values.

6.3.4 Scheduling Maintenance

Infiltration capacity testing can provide information necessary to schedule maintenance. For example, measurements of saturated hydraulic conductivity (K_s) in a bioretention practice can indicate where poor infiltration is occurring. As seen in the previous case study, the infiltration rate in the bottom of stormwater treatment practices begins to decrease as sediment clogs the pore spaces. Infiltration capacity testing can quantify this change spatially within the practice and over time as the facility ages. This information can be used to schedule appropriate maintenance, such as removing accumulated sediment, replacing groundcover management like mulch, and replacing engineered soils within the practice. Also, information gathered from infiltration capacity testing can be used to optimize maintenance efforts by identifying specific locations where maintenance is needed. As such, maintenance efforts can focus on areas that need attention and no resources will be wasted on areas that are performing adequately.

Synthetic Runoff Testing of Stormwater Treatment Practices

7

Abstract

Synthetic runoff testing involves filling a stormwater treatment practice with water from a fire hydrant, water truck, or other available water source. Thus, synthetic runoff testing is limited to smaller practices. This chapter presents details related to using a water source to fill a practice, including determining if the water source is adequate and estimating the time it will take to fill the practice to a desired level. Analysis methods related to obtaining infiltration capacities and contaminant removal performance are given, as are examples, and a case study of synthetic runoff testing applied to an underground sedimentation practice.

If visual inspection reveals no malfunctioning of a stormwater treatment practice and the assessment plan warrants further assessment, synthetic runoff testing may be considered. Synthetic runoff testing is limited to smaller practices, however, because a water supply that can fill a significant portion of the practice is required. If an adequate water supply exists, synthetic runoff testing can be used to determine the overall effective saturated hydraulic conductivity (K_s) and, if additional requirements are satisfied, the pollutant capture efficiency of a stormwater treatment practice.

Synthetic runoff testing differs significantly from capacity testing in that the overall performance of the practice is measured instead of a series of point measurements. Thus, synthetic runoff testing does not give localized results and cannot be used to pinpoint localized areas where the practice may need maintenance. Synthetic runoff testing does allow for the overall effective saturated hydraulic conductivity (K_s) value for the entire practice to be estimated and, with this value, the required time to infiltrate or filter a given runoff volume can be estimated.

With synthetic runoff testing, a prescribed amount of synthetic stormwater, possibly with a known designated dose of pollutant, is applied to a stormwater treatment practice under controlled conditions. The overall effectiveness of a practice with regard to volume reduction can be estimated by measuring the change

A.J. Erickson et al., *Optimizing Stormwater Treatment Practices: A Handbook of Assessment and Maintenance*, DOI 10.1007/978-1-4614-4624-8_7,

in water volume within the practice, the corresponding elapsed time, and possibly other variables such as effluent flow rate. Also, if the synthetic stormwater is dosed with a known mass of pollutant, measuring the total mass of pollutant retained by the practice or the mass of pollutant that passes through the practice allows for the effectiveness of the practice with regard to pollutant capture to be assessed. Results from synthetic runoff testing can also be used to calibrate watershed models for simulation of performance during natural rainfall events.

Synthetic runoff testing can be used to evaluate the infiltration rate or the removal of pollutants by a stormwater treatment practice. Synthetic runoff testing uses a clean water source (e.g., a fire hydrant or water truck) that may contain targeted pollutants at predetermined concentrations or pollutant loads, which is applied to the stormwater treatment practice under well-controlled conditions. Because adding targeted pollutants to synthetic stormwater may require authorization from local governments (municipal, watershed districts, or state), authorization requirements should be investigated before performing synthetic runoff tests with pollutants. If the required discharge or volume of water is outside the reasonable discharge of the water source, then synthetic runoff testing is not likely to be feasible.

7.1 Measuring Sediment Retention

Synthetic runoff testing can also be used to measure the sediment retention by stormwater treatment practices. Wilson et al. (2009) has shown this technique to be repeatable and accurate for underground sediment retention structures, but it has not been used on most other treatment practices. Manholes, grit chambers, and many proprietary devices can be classified as underground sediment retention structures. These structures are often suitable for synthetic runoff testing because of their relatively small design discharge. At a specific water discharge, a given quantity and size of sediment can be fed into the sediment retention structure. After the structure has been drained, the sediment retained is then extracted from the structure, dried, and weighed. The difference in mass between the sediment fed and the sediment retained is presumed to have passed through the facility, and sediment retention efficiency can be computed for each sediment size and water discharge. Synthetic runoff testing with sediment can be conducted relatively accurately and is an effective means of determining how well a device will remove various sizes of sediment and verifying that a device is functioning as designed.

Sediment retention testing can be applied to other stormwater treatment practices, provided there is an adequate water supply. Some stormwater treatment practices (e.g., dry ponds) are constructed from soil, and in such cases, separating sediment added to synthetic runoff from the soil that makes up the bottom of the stormwater treatment practice can be difficult. An alternative solution for such stormwater treatment practices may be to use automatic samplers to capture synthetic stormwater samples at the outflow for comparison to sediment that was added to the influent synthetic runoff. Another alternative solution may be to paint sediments (or use sediment of a different color) added during synthetic runoff testing so that they can be separated from sediments already in the stormwater treatment practice or that are part of the original treatment practice construction. With these alternatives available, sediment retention testing can be applied to most stormwater treatment practices, including sand and soil filters, underground filters, hybrid filters, dry ponds, wet ponds, underground sedimentation devices, wet vaults, rain gardens with a measurable outflow, constructed wetlands, filter strips, and swales.

7.2 Measuring Infiltration/Filtration Rate

One application of synthetic runoff testing is to assess total drain time of stormwater treatment practices. The entire basin is filled with water and the change in water level in the basin is measured over time, which is a direct measure of the drain time. Another application of synthetic runoff testing is to measure pollutant removal efficiency. Pollutant removal efficiency can be evaluated by adding a well-characterized pollutant (e.g., suspended solids and phosphorus, among others) to the influent water at a desired concentration and measuring the amount of pollutant retained by the stormwater treatment practice, the concentration exiting the stormwater treatment practice, or both. Whether measuring drain time or pollutant removal efficiency, the goal of synthetic runoff testing is not to mimic natural storm events, but to accurately measure the rate of infiltration or pollutant removal under controlled conditions.

For filtration or infiltration rate assessment, the following three conditions must be met for synthetic runoff testing to be feasible:

1. There must be a water supply that can provide the required discharge and total volume of runoff needed (see next section).
2. Outflow paths other than infiltration must be either measurable or can be temporarily plugged.
3. The water surface elevation in the stormwater treatment practice must be continuously measurable during the test.

When a stormwater treatment practice can be filled rapidly with synthetic stormwater, there is no need to measure the rate at which water is added because the infiltration rate is relatively small in comparison to the inflow rate. When the rate at which water is infiltrating is not negligible compared to the rate at which the stormwater treatment practice is filled, both the rate at which water is added to the stormwater treatment practice and the rate at which water is infiltrating into the stormwater treatment practice must be measured or estimated.

Synthetic runoff testing to assess drain time can be performed on the following stormwater treatment practices: bioretention practices (rain gardens), dry ponds, infiltration basins, sand and soil filters, underground sand filters, and underground wet vaults. Large stormwater treatment practices, however, may require a water volume or flow rate in excess of the available water supply.

Saturated hydraulic conductivity (K_s) in stormwater treatment practices may vary based on climatic season, soil conditions, etc., and therefore synthetic runoff testing for K_s should be performed at several different times throughout the year if the performance of the practice as a function of season is desired or to obtain an estimate of the overall yearly effective K_s. An example schedule includes testing in the spring after the ground thaws, in midsummer, and in late fall before the ground freezes. If knowledge about the performance of the practice over a number of years is desired, individual test results should only be compared to previous tests under similar conditions (e.g., season of year and water temperature, among others).

The primary differences between measurement results from capacity testing and synthetic runoff testing for K_s relate to the size, vegetation, and subsurface characteristics of the stormwater treatment practice. Synthetic runoff testing, as outlined above, is limited to stormwater treatment practices that are small enough to be filled with a user-supplied water source. Synthetic runoff testing, however, accounts for the increased infiltration that occurs near and around the stems of vegetation that cannot be measured using capacity testing. Additionally, synthetic runoff testing will show when filtration is limited by the subsurface collection system and not by the surface or near-surface layers or, in infiltration practices, where a relatively impermeable layer beneath the surface restricts infiltration.

As with visual inspection (level 1) and capacity testing (level 2a), the procedure for synthetic runoff testing varies for each stormwater treatment practice and assessment goal. Therefore, the reader should consider the recommendations discussed in this chapter. As with any field work, safety is an important concern and must be addressed when conducting synthetic runoff testing.

7.3 Finding an Adequate Water Source

The primary constraint for synthetic runoff testing is the available water volume and discharge that can be provided by nearby fire hydrants or available water trucks. Fire hydrants can typically produce between 2 cfs (0.056 m^3/s) and 4 cfs (0.112 m^3/s) and water trucks can produce up to approximately 1 cfs (0.028 m^3/s). Prior approval is usually required in order to use a fire hydrant for synthetic runoff testing, and it is not uncommon for fire hydrants to be limited to 30 min of use, due to concerns of reducing pressure in the distribution system and corresponding decreases in firefighting capabilities. Longer times may be approved, however. Most commercial water trucks hold approximately 500 ft^3 (14,160 L) of water, but a large water truck can hold up to 1,000 ft^3 (28.3 m^3), allowing the maximum discharge for approximately 20 min.

In order for synthetic runoff testing to be possible, a water supply that can fill the practice to the desired depth in a relatively short time is required. In order to determine the time required to fill a practice, the water supply flow rate into the practice and the infiltration rate out of the practice as a function of time must be known or estimated. With this information, a water mass balance can be performed on the practice if all other flow rates into the practice are negligible. In the mass balance, the change in water volume with time within the practice is equal to the water supply flow rate into the practice minus the total flow rate of water out of the practice via infiltration. Other flow paths such as evapotranspiration will be negligible over the time frame needed to fill a practice and can be ignored. Thus, the mass balance becomes:

$$\frac{dV}{dt} = Q_{in} - fA_i \tag{7.1}$$

where

V = volume of water in the practice
t = time
Q_{in} = flow rate into the practice (e.g., from fire hydrants or water trucks)
f = infiltration rate out of the practice
A_i = the area available for infiltration

The infiltration rate into the practice varies with time as the soil below the practice becomes saturated and the head (water elevation) within the practice increases. The infiltration rate can be modeled mathematically with an equation such as the Green–Ampt equation (Dingman 2002).

Given a water source, the time and total water volume required to fill a practice to a desired depth can be determined by solving (7.1) incrementally with a spreadsheet in time steps of Δt. Replacing the derivatives in (7.1) with Δ's and rearranging gives:

$$\Delta V = Q_{in}\Delta t - fA_i\Delta t \tag{7.2}$$

When using the Green–Ampt equation to estimate the infiltration rate, f, often times the head (or depth) of water on top of the soil is assumed to be insignificant and is ignored. In the case of synthetic runoff testing, however, the practice must be filled to a significant depth and the head cannot be ignored. Without the assumption that the depth of water above the soil surface is zero, the Green–Ampt equation becomes:

$$f = K_s\left(\frac{(\theta_i - \theta_f)(\psi_f + z_w)}{F} + 1\right) \tag{7.3}$$

where

f = infiltration rate into the soil
K_s = overall effective saturated hydraulic conductivity of the practice
θ_i = initial volumetric moisture content of the soil
θ_f = final volumetric moisture content of the soil (after infiltration)
ψ_f = soil suction at the wetting front (a positive value)
z_w = head of water on the soil (i.e., water depth in the practice)
F = cumulative depth of surface water infiltrated

Note that F is not the depth the wetting front has infiltrated into the ground; rather, it is the equivalent depth of water (when above the soil surface) that has infiltrated. If the depth the wetting front has infiltrated into the ground is L and the fraction of the soil is available to be filled with water is $(\theta_f - \theta_i)$, then $F = L(\theta_f - \theta_i)$.

Substituting (7.3) into (7.2) gives:

$$\Delta V = Q_{in}\Delta t - K_s\left(\frac{(\theta_i - \theta_f)(\psi_f - z_w)}{F} + 1\right)A_i\Delta t \tag{7.4}$$

The change in water volume within a practice, ΔV, over an incremental time step, Δt, can be estimated using (7.4) if soil properties and the stage–storage relationship (i.e., the depth of water in the practice as a function of total volume of water in the practice) are known. The overall effective saturated hydraulic conductivity of the soil (K_s) can be estimated from previous assessment efforts. The value of K_s, however, should not be assumed to be equal to a typical value based on the soil type because the solution technique is sensitive to K_s and an inaccurate value can significantly impact results. The area of infiltration, A_i, changes with time as the practice fills with water, but it can be assumed constant and equal to the water surface area when the practice is at mid-depth without significantly affecting the results. Another assumption that can simplify the solution technique without losing much accuracy is that infiltration occurs vertically downward at every location.

Soil moisture can be measured as discussed in Chap. 11. If the tools required to determine moisture content are not available, then the initial volumetric moisture content of the soil can be estimated by using Table 6.1. After the identification of the soil type and the initial moisture content, the final moisture content can be assumed to be the effective porosity for that soil, as listed in Table 6.2. It has been found that the change in moisture content has a less than 20% effect on the calculation of saturated hydraulic conductivity (Regalado et al. 2005).

If the practice is rectangular in shape, the volume is calculated by (7.5):

$$V = z_w A + \frac{S}{2}Pz_w^2 + \frac{16}{3}z_w^3 \tag{7.5}$$

where

V = volume of water in the practice
z_w = water depth in the practice
A = bottom area of the practice
S = side slope (H:V)
P = perimeter of the practice

For any given volume of water in the practice, (7.5) can be used to find the water depth that corresponds to the given water volume. Thus, (7.5), or a similar equation for a different practice with different geometry, gives the stage–storage relationship.

Equation (7.4) can be used to find the incremental change in storage within the practice for successive time steps of Δt. The cumulative storage in the practice is found by summing each successive change in incremental storage volume. By keeping track of the total volume as a function of time, the time required to fill a practice with a water supply may be determined. This process is demonstrated in Example 7.1.

Example 7.1: Determining if a water source is adequate for synthetic runoff testing

Alan, the director of the county environmental services department, would like to use synthetic runoff testing on a rectangular dry pond with a bottom area that is 5 m by 7 m and with 3:1 side slopes. Alan knows that a fire hydrant in the area can provide 0.05 m^3/s for 2 h and, based on assessment results from previous years, he expects the saturated hydraulic conductivity (K_s) to be about 10 cm/h. Alan has already estimated the porosity of the underlying soil to be 0.45 with a wilting point of 0.04 and a suction head of 2 cm. Alan would like to know if the pond can be filled to a depth of 1.5 m in 2 h, assuming it is initially empty and the test will occur in late July with no recent rainfall such that he can assume the wilting point will be equal to the initial volumetric moisture content.

First, Alan assumes the initial volumetric moisture content, θ_i, to be 0.04 and the final volumetric moisture content, θ_f, equal to the porosity of 0.45. Alan uses a 5-min time step to calculate the volume of water that enters the practice during each time step:

$$\Delta V_{\text{in}} = 0.05\frac{\text{m}^3}{\text{s}}(5\,\text{min})\left(60\frac{\text{s}}{\text{min}}\right) = 15\,\text{m}^3$$

Now the infiltration rate into the soil during the first time step must be determined from the Green–Ampt equation (7.3), but if F is assumed to be 0, the equation cannot be solved because of division by 0. Thus, Alan assumes a small and reasonable value of $F = 0.10$ cm at $t = 0$. If Alan assumes F to be extremely small, the infiltration rate is excessively large and cannot be assumed to remain constant over the entire time step. Thus, Alan takes note that values of 0.1–0.5 cm are recommended because values in this range typically produce realistic infiltration rates that can be assumed constant during the first time step if the time step is not too large. Furthermore, Alan knows that the results are not sensitive to initial values of F in this range. Using (7.3) and an initial infiltrated depth (F) of 0.1 cm, Alan determines an initial infiltration rate of

$$f = 10\frac{\text{cm}}{\text{h}}\left(\frac{(0.45 - 0.04)(2 + 0)}{0.10} + 1\right) = 92\,\text{cm/h}$$

Alan assumes that this infiltration rate is constant for the entire 5-min time step and then calculates the incremental total volume of water infiltrated during the time step, V_{inf}, to be:

(continued)

Example 7.1: (continued)

$$V_{\text{inf}} = 92\frac{\text{cm}}{\text{h}}\left(\frac{1\text{m}}{100\,\text{cm}}\right)(5\ \text{min})\left(1\frac{\text{h}}{60\ \text{min}}\right)(35\,\text{m}^2) = 2.68\,\text{m}^3$$

Alan notes that this value is feasible because it is less than the total volume of water that entered the practice during the time step (15 m^3). Alan knows that if the calculation yields an infiltrated volume of water that is larger than the amount that entered the practice, the infiltrated volume should be assumed to be the volume of water that entered the practice or the initial depth of infiltrated water, F, should be increased and the calculation repeated.

Based on these results, Alan calculates that there is 15−2.68 = 12.32 m^3 of water in the practice at the end of the first time step (i.e., 5 min). Using (7.5), Alan determines that this volume corresponds to a depth, z_w, of 0.346 m. Alan then uses this depth in (7.3) to find the infiltration rate for the second time step and he adds the change in volume of water in the practice to that of the first time step to find the cumulative volume of water in the practice. Alan repeats this process until the desired depth, total time, or the desired total volume of water in the practice is achieved. Alan performs these calculations in spreadsheet software program and the results are shown in Table E7.1. For this example, Alan calculates the water surface area that is infiltrating, A_i, as a function of water depth.

As shown in Table E7.1, Alan finds that the water depth in the pond at 2 h is only 1.17 m, which is less than the desired depth of 1.5 m. Alan tries continuing the calculation assuming the hydrant can provide the constant flow rate for an additional 30 min, and finds that the pond depth will still only reach about 1.2 m.

Alan also sums the incremental infiltrated volumes for every time step to find that 86.8 m^3 of water would infiltrate during the 2.5 h and, by summing the incremental pond volumes (ΔV) over the entire 2.5 h, that a total of 450 m^3 of water would be delivered to the pond. Thus, Alan estimates that approximately 19% of the water delivered to the pond will infiltrate over 2.5 h.

Alan considers the hypothetical case in which no water is infiltrated during the filling of the basin and determines that the 450 m^3 delivered to the pond in 2.5 h would only fill the pond to a depth of 1.22 m. Alan realizes that this is due to the side slopes which rapidly increase the surface area of the water as the pond is filled. Alan concludes that unless the hydrant flow rate can be increased, it is impossible to fill the pond to 1.5 m in less than 2.5 h.

From the calculations, Alan realizes that filling the pond to a depth of one meter in approximately 30–35 min (Table E7.1) would most likely still yield accurate and valuable synthetic runoff testing assessment results. Alan decides to perform a synthetic runoff test with a shorter duration that fills the basin to 1 m depth.

(continued)

Example 7.1: (continued)

Table E7.1 Spreadsheet solution to Example 7.1 (incremental influent volume (ΔV) = 15 m^3)

Time (t) (min)	Water depth (z_w) (m)	Pond total volume (m^3)	Infiltrated depth (F) (cm)	Infiltration rate (f) (cm/h)	Water surface area (A_i) (m^2)	Incremental infiltrated volume (V_{inf}) (m^3)
0	0.000	0.00	0.100	92.000	35.0	2.683
5	0.346	12.32	7.767	29.301	64.2	1.567
10	0.678	25.75	10.208	38.048	100.4	3.184
15	0.845	37.57	13.379	36.508	121.5	3.698
20	0.917	48.87	16.421	33.387	131.3	3.652
25	0.960	60.22	19.204	30.913	137.2	3.535
30	0.990	71.68	21.780	29.011	141.5	3.422
35	1.013	83.26	24.197	27.510	144.9	3.323
40	1.033	94.94	26.490	26.292	147.7	3.237
45	1.049	106.70	28.681	25.280	150.1	3.163
50	1.063	118.54	30.787	24.422	152.2	3.098
55	1.075	130.44	32.823	23.684	154.1	3.041
60	1.087	142.40	34.796	23.039	155.7	2.990
65	1.097	154.41	36.716	22.470	157.3	2.945
70	1.106	166.46	38.589	21.963	158.7	2.904
75	1.114	178.56	40.419	21.508	160.0	2.867
80	1.122	190.69	42.211	21.096	161.2	2.833
85	1.130	202.86	43.969	20.721	162.3	2.802
90	1.137	215.06	45.696	20.378	163.4	2.774
95	1.143	227.28	47.394	20.062	164.4	2.748
100	1.149	239.54	49.066	19.771	165.3	2.723
105	1.155	251.81	50.714	19.500	166.2	2.701
110	1.161	264.11	52.339	19.249	167.1	2.680
115	1.166	276.43	53.943	19.014	167.9	2.660
120	1.171	288.77	55.527	18.794	168.7	2.642
125	1.176	301.13	57.093	18.587	169.4	2.624
130	1.180	313.51	58.642	18.393	170.1	2.608
135	1.185	325.90	60.175	18.209	170.8	2.592
140	1.189	338.30	61.692	18.036	171.5	2.578
145	1.193	350.73	63.195	17.871	172.2	2.564
150	1.197	363.16	64.685	17.715	172.8	
Total volume		450.0				86.8

Example 7.1 shows that synthetic runoff testing is generally applicable for only smaller stormwater treatment practices, but even if the maximum design depth cannot be achieved with the available water supply, synthetic runoff testing may still be the most cost-effective assessment method.

7.4 Dry Ponds

If all outflow locations of a dry pond can be plugged, synthetic runoff testing can be used to determine the overall effective saturated hydraulic conductivity (K_S) of the pond soil. With all outlets plugged, the pond will act as an infiltration basin during the synthetic runoff test. Therefore, the discussion of synthetic runoff testing of infiltration basins presented later in this chapter applies to dry ponds if all outlets are plugged. If all outlets are not plugged, synthetic runoff testing cannot be successfully completed on a dry pond.

If a dry pond has no outlet except for an emergency overflow spillway, it is classified as a retention pond. These ponds are designed to infiltrate and evapotranspirate the water quality volume (WQV) of runoff. In this case, synthetic runoff testing can be used to determine the K_S and the runoff volume reduction capability of the pond. Retention ponds act essentially as infiltration basins; therefore, just as a dry pond with plugged outlets, they can be assessed using synthetic runoff testing in the same manner as an infiltration basin, as long as an adequate water supply is available.

7.4.1 Determining Pollutant Removal Efficiency

Pollutant removal efficiency of a dry pond can be estimated by adding a well-characterized and known total mass of pollutant (e.g., sediment) to the influent synthetic stormwater and collecting samples of the effluent of the dry pond so that the event mean concentration (EMC) of the effluent can be determined. Multiplying the effluent EMC by the total effluent volume gives the total mass of pollutant that left the pond in the effluent. This value can be used along with the known total mass of pollutant put into the pond to find the percent removal achieved by the pond.

Effluent concentrations are typically measured using automatic samplers. The accuracy of synthetic runoff tests for pollutant removal efficiency may, however, be limited by difficulties in achieving a representative suspended solids concentration through automatic sampling (see Chap. 10 for more information). Also, for this type of synthetic runoff testing, the water source must be able to provide the design discharge for a period of time that allows flow through the sedimentation practice to fully develop and equilibrate. A comparison of the maximum flow rate of the sedimentation practice and the available water source can be used to determine if the water source is adequate.

As previously discussed, synthetic runoff tests (level 2b) can estimate the infiltration rate of a dry pond, its pollutant removal efficiency, or both, and therefore determines very different stormwater treatment practice parameters than capacity testing (level 2a).

7.5 Wet Ponds

Synthetic runoff testing can be used to estimate the retention of pollutants in wet ponds. Wet ponds do not infiltrate a significant amount of stored runoff because they maintain a permanent pool of water even during long dry periods. Therefore, wet ponds only minimally reduce runoff volume by infiltration. Pollutant removal efficiency can be determined by adding a known total mass of pollutant to the influent synthetic stormwater that has a known constant flow rate and then measuring the effluent flow rate and collecting effluent samples as water is discharged from the pond. The accuracy of these tests may be limited by difficulties in achieving a representative suspended solids concentration through sampling (see Chap. 10 regarding sampling suspended solids for more information).

Synthetic runoff tests using a conservative tracer (e.g., chloride, rhodamine) can be used to investigate the hydraulic behavior of a wet pond. Tracer studies involve adding a tracer to the influent and measuring the tracer concentration in the effluent as a function of time during a synthetic runoff event. Results from tracer studies, as illustrated in Example 7.2, can be used to determine if stormwater is short-circuiting through the pond or if dead zones are present.

Short-circuiting in stormwater treatment occurs when stormwater passes through the pond with minimal or no treatment because of incomplete mixing. A poorly located inlet or outlet may result in a portion of the influent bypassing the treatment volume available in the pond. Another cause of short-circuiting may occur during the winter in cold climates. Runoff or snowmelt that enters a frozen wet pond may flow over the ice directly to the outlet structure. Short-circuiting can result in minimal pollutant removal efficiency and is a common cause of wet pond failure.

Dead zones are areas in a stormwater treatment practice where water becomes trapped and does not pass through as intended. This results in the practice having a smaller effective volume. For example, a pond may have areas where stormwater circulates but is not released until the storm event is over or nearly over. Pollutants in the trapped water may or may not be removed by the pond and, if not removed, they may appear in the effluent samples at the end of the runoff event. Pollutants can become trapped in dead zones between storm events and be released during subsequent storm events, which may result in negative removal efficiency (i.e., effluent pollutant load is larger than the influent pollutant load) for individual runoff events.

When conducting tracer studies for wet ponds, the density of the tracer injection must be the same as the density of the receiving water. Fluid temperature and tracers such as chloride and rhodamine can change the density of the tracer injection, resulting in an injection that does not disperse appropriately (e.g., floats or sinks). Nontoxic additives such as methanol can be used to counterbalance density changes caused by tracers.

Example 7.2: Tracer study of a wet pond (Data modified from Shilton et al. 2000)
Alan, the director of the county environmental services department, is reviewing the results of a tracer study that was performed on a stormwater wet pond to examine the hydraulic conditions of the pond. In the study, a perfectly mixed wet pond was modeled ($C = C_0e^{-kt}$) for the same residence time (represented by k) and initial tracer concentration (C_0), as shown in Fig. E7.1.

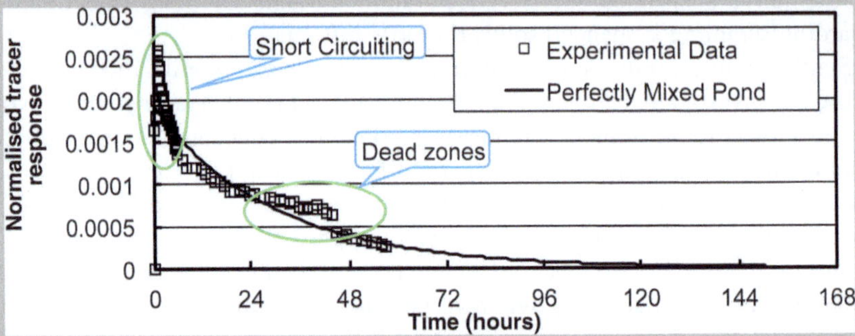

Fig. E7.1 Data obtained from a tracer study

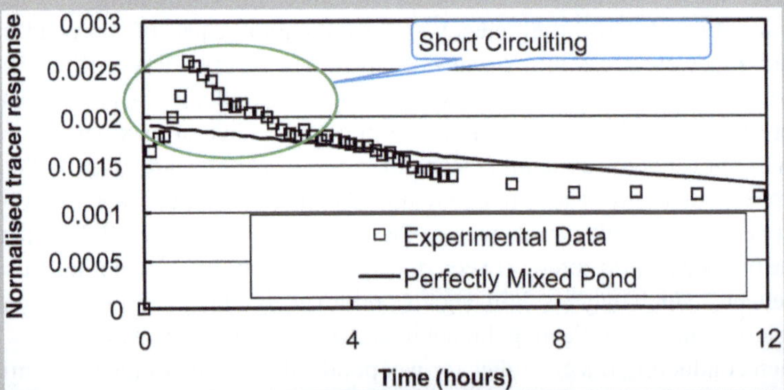

Fig. E7.2 Data from a tracer study illustrating short circuiting

Alan draws two important conclusions from inspecting the data in Fig. E7.1. First, there is evidence of short-circuiting in the system in the early stages of the tracer study, which is enlarged and shown in Fig. E7.2. Alan concludes that short-circuiting is occurring because tracer concentration increases sharply in the beginning of the experiment, up to 38% more than the perfectly mixed pond. This indicates that some of the tracer is

(continued)

Example 7.2: (continued)
exiting the pond more quickly than expected, which is typically the result of influent short-circuiting to the exit. Second, Alan concludes that dead zones are present in the stormwater wet pond because the experimental tracer response is larger than the modeled concentration for the perfectly mixed pond after 24 h. This could indicate that some tracer is temporarily captured in dead zones and released at a later time. The presence of dead zones is confirmed by the sudden drop in tracer response that occurs after approximately 44 h, which indicates that any tracer that had been retained in dead zones was flushed out.

7.6 Underground Sedimentation Devices

Synthetic runoff testing can be used to estimate the retention of solids in a wet vault or underground sedimentation device. Typically, underground sedimentation devices neither infiltrate runoff nor have sufficient storage volume to reduce peak flow; thus, flow volume or peak flow reduction assessment of these devices is not relevant. The synthetic runoff, however, can be dosed with sediment to assess solids capture performance (Wilson et al., 2009). The solids capture performance is determined either by collecting and measuring sediment concentrations in effluent samples or by extracting and measuring the sediment captured by an initially clean device. The latter method is likely to be more accurate, because all of the solids are collected and weighed, whereas the former analyzes only the sediment in discrete effluent samples from water exiting the device. In this situation, it may be difficult to achieve representative suspended solids samples. See Chap. 10 on sampling suspended solids for a discussion of solids sampling.

7.6.1 Case Study: Synthetic Runoff Testing of an Underground Sedimentation Device

A proprietary underground sedimentation practice, as shown in Fig. 7.1, was evaluated using synthetic runoff testing. The device receives stormwater runoff from a 4.2-acre (1.7-ha) residential watershed that is approximately 55% vegetated and 45% impervious.

The device is a dual manhole system consisting of a 5-ft (1.5-m) diameter swirl chamber and a 5-ft (1.5-m) diameter floatables trap. Stormwater influent is introduced tangentially to the swirl chamber by a 15-in. (38-cm) PVC pipe,

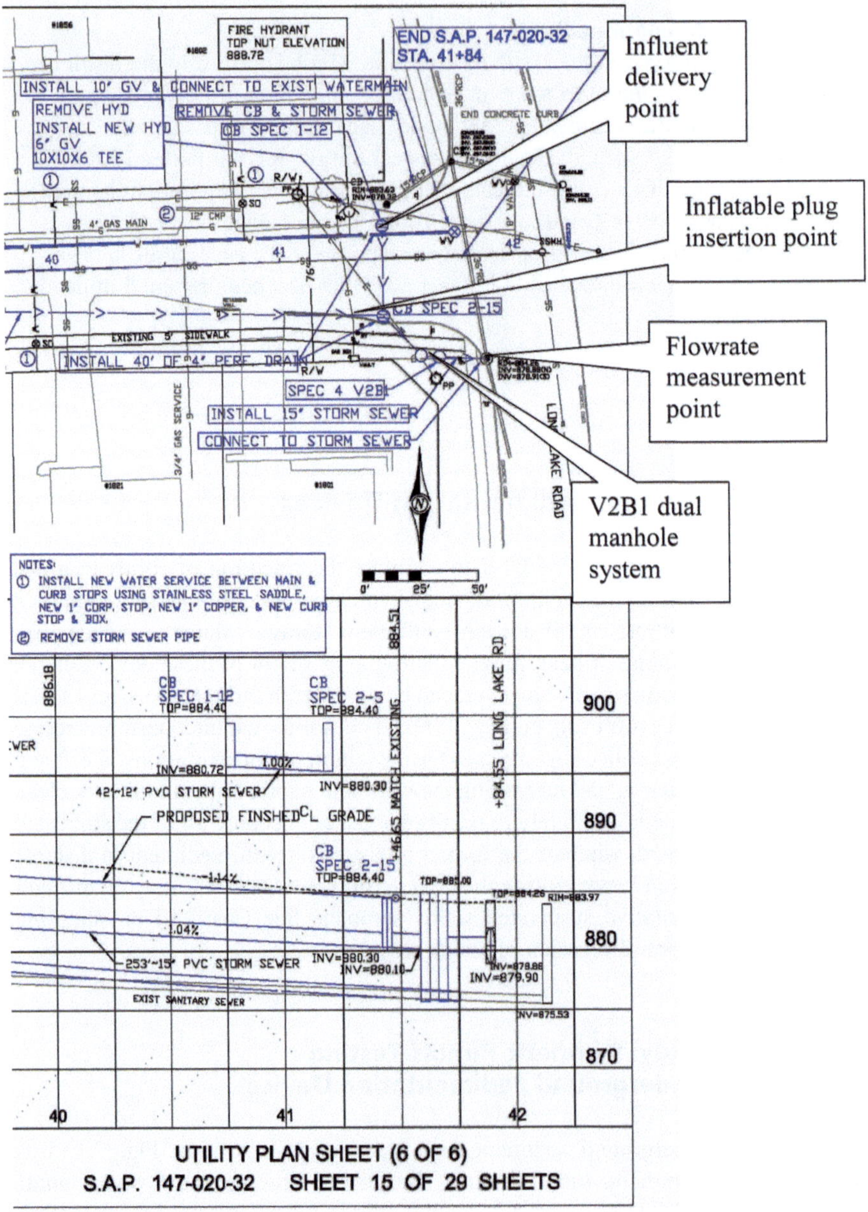

Fig. 7.1 Plan and profile of installation site

inducing a swirling motion inside the manhole. Relatively heavier particulates contained in the stormwater (sands, trash, etc) settle out of suspension in the swirling chamber. Stormwater escapes the swirling chamber by overflowing an 18-in. (46-cm) diameter PVC standpipe in the middle of the manhole, where the water is conveyed to the floatables trap. The floatables trap manhole contains an underflow baffle wall with a 1 ft by 3 ft (0.3 m by 0.91 m) rectangular hole at its base. Buoyant material (hydrocarbons, cigarette butts, some organic matter, etc.) that passes through the swirling chamber via the overflow standpipe is retained in the floatables trap, because water must travel beneath the baffle wall to escape the system through a 15-in. (38-cm) PVC pipe. Downstream of the device, the effluent from the device discharges into a 36-in. (91-cm) reinforced concrete pipe (RCP), which delivers runoff downstream. There is an overall drop of 0.2 ft (0.06 m) across the system, from the inlet invert to the outlet invert. The distance between pipe inverts and manhole inverts is approximately 4.5 ft (1.37 m) in each treatment manhole. One access point is provided to the swirl chamber and one access point on each side of the baffle wall in the floatables trap, as illustrated in Fig. 7.2.

The unit was designed to accommodate a maximum hydraulic flow rate equivalent to a 10-year event with an intensity of 4.6 in./h (11.7 cm/h) without flooding the street. According to calculations provided by the manufacturer, the corresponding discharge is 6.7 cfs (0.19 cms), which serves as the capacity of the storm drain conveyance system around the device. The device is in line with the conveyance system and has no bypass, meaning the device will receive all flows traveling through the system. However, even though all storm flows travel through the device, treatment is not intended to be provided above the water quality event, defined to be 0.8 in. (2 cm) of rainfall. A runoff coefficient of 0.46 was estimated for the 4.2-acre (1.7-ha) watershed. According to calculations provided by the manufacturer, the water quality flow rate is 1.37 cfs (0.039 cms), which corresponds to the maximum treatment rate for performance assessment.

7.6.1.1 Assessment Goals

The goals of this assessment were twofold:

1. Investigate the practicality of controlled field testing as an alternative to field monitoring
2. Evaluate the sediment removal capability of the device when subject to field testing with a wide range of sediment sizes and influent flow rates

Another result of the assessment was a performance curve for the device in which removal efficiency is plotted vs. a dimensionless parameter. This performance curve serves as a tool to reliably predict the removal performance for a wide range of device sizes, influent flow rates, and pollutant size characteristics. The performance curve can also be used as a tool to accurately size a new stormwater treatment structure, given a target removal efficiency, a target particle size for removal, and a design flow rate.

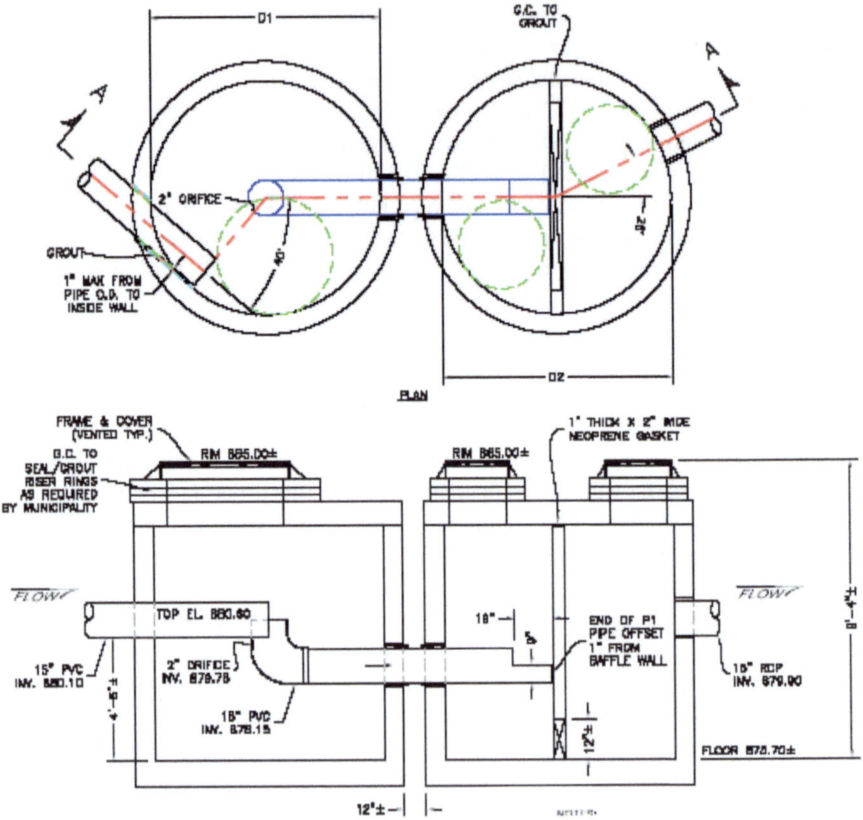

Fig. 7.2 Plan and section of dual manhole stormwater treatment

7.6.1.2 Assessment Techniques

Synthetic runoff testing was used to assess the sediment removal performance of the unit. A fire hydrant supplied a constant flow rate of sediment-free water which was dosed with a known total mass of well-characterized sediment supplied at a specified rate. At the completion of a test, personnel entered the device and removed the sediment retained during the test, allowing for a bulk solids analysis on a known quantity of delivered and retained sand. In addition to providing a more certain performance assessment, the synthetic runoff approach enables comparison of results for a particular device across different watersheds, climates, land uses (i.e., different pollutant loading), influent flow rates, and treatment unit size. This comparison can be accomplished by plotting the removal efficiency as the dependent variable vs. the appropriate dimensionless parameter, as explained in the following paragraphs. Synthetic runoff testing is thus related to the performance of the device and not to the particular watershed. The runoff from the watershed can then be routed through the device using a computer simulation based on the characteristics of the watershed and the results of synthetic runoff testing.

Fig. 7.3 Precalibrated 15-in. (38-cm) circular weir installed downstream of the device. Pressure transducer and transducer anchoring are not shown. This weir location provided free outfall conditions at all flow rates due to the PVC pipe's favorable elevation vs. the existing 36-in. RCP it discharged into

Before this test site was selected, prospective sites were identified, screened, and evaluated for field testing potential based on a variety of characteristics:

1. Location of out-of-vehicle traffic lanes for safety and traffic handling concerns
2. Proximity to a fire hydrant for use as a water source
3. Maximum treatment rate of the device due to finite maximum discharges from hydrants
4. Device allowing for human access to treatment chamber sump for sediment removal and maintenance

The system to be tested also needed to provide a suitable location within the storm drain system for flow rate measurement using a precalibrated weir and pressure transducer. Appropriate permits were obtained from governing agencies.

One of the sites chosen for field testing was the device depicted in Figs. 7.1 and 7.2. Prior to beginning testing activities, the site required several preparation procedures:

1. For real-time flow rate measurement, a precalibrated, 15-in. (38 cm) diameter circular weir and a pressure transducer (Fig. 7.3) were installed approximately 20 ft (6.1 m) downstream of the floatables trap manhole depicted in Fig. 7.2. The pressure transducer measured water depths, which, based on conduit geometry, were used to calculate flow areas and therefore discharge.
2. The device was dewatered and several months' worth of solids accumulation was removed with the assistance of vacuum trucks.
3. A piping system was customized for the delivery of hydrant water as influent test water, using the hydrant's 4-in. (10-cm) connection and a series of fittings, a 4-in. (10-cm) gate valve, and a 4-in. (15-cm) PVC pipe (Fig. 7.4).

Fig. 7.4 Piping system from hydrant to influent injection point and stainless steel sediment feeder

4. Sand was previously sieved into three size fractions for use in each synthetic runoff event, with median sizes: 107 μm (ranging from 89 to 125 μm), 303 μm (ranging from 251 to 355 μm), and 545 μm (ranging from 500 to 589 μm), starting with F110 sand (d_{50}=110 μm), AGSCO 40-70 sand (d_{50}=225 μm), and AGSCO 35-50 sand (d_{50}=425 μm) as supply.
5. An inflatable 15-in. (38-cm) diameter plug was secured to seal off storm drainage upstream of the treatment system but downstream of the influent to prevent nuisance flows in the system from contaminating the controlled influent delivered to the device and to avoid controlled influent from leaving the test system prematurely.

The procedure for field testing the device included the following steps:

1. Establishing a safe work zone, following confined space entry regulations.
2. Installing and inflating with a portable air compressor the 15-in. (38-cm) rubber plug upstream of the device to seal off the upstream reaches of the storm drain system (Fig. 7.5).
3. Connecting piping system from hydrant to influent injection point.
4. Flushing clean hydrant water through the system prior to initial device cleanout.
5. Dewatering the device with sump pumps and removing solids with a wet/dry vacuum cleaner.
6. Establishing an appropriate flow rate through the system using real-time level measurements from a pressure transducer and data logger, and conditioning the flow with a gate valve on the hydrant. The data logger recorded 60-s average

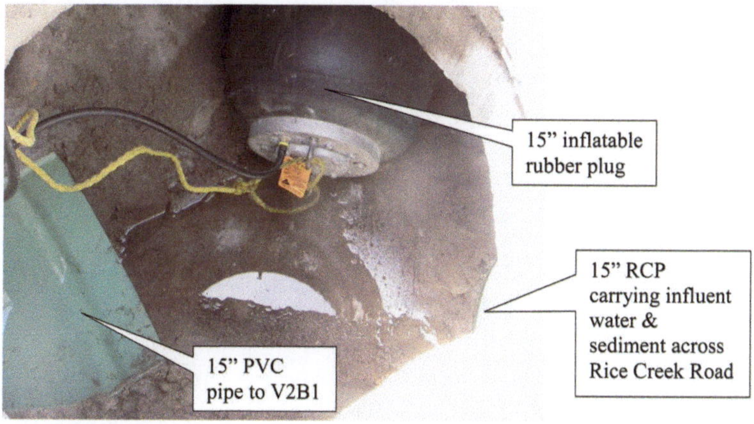

Fig. 7.5 Installation of 15-in. (38-cm) inflatable plug in upstream concrete pipe to seal off nuisance and/or extraneous flows from impacting the assessment

levels and provided an updated readout every second when connected to a laptop computer loaded with the associated software.

7. Introducing 22-33 lbs (10–15 kg) of pre-sieved sand (equal parts of 107, 303, and 545 μm sands) to the influent hydrant water at 200 mg/L using a precalibrated sediment feeder.
8. Recording water temperature, mass of sediment delivered, and test duration.
9. Following a 20-min period to allow sand particle settling, dewatering the device with sump pumps, and removing retained solids from each manhole separately with a wet/dry vacuum cleaner.
10. Oven drying and sieving the collected sediment into size fractions, and weighing each fraction of retained solids for comparison to the known quantity of each size fraction fed to the device during the test.

The data in step 10 above, divided by the known quantity of sand delivered to the device during the test, provided the removal efficiency of the device for each sand size fraction at a particular flow rate. Thus, each test produced three data points, because three discrete sand size ranges were utilized. The testing protocol called for a device to be tested under four flow rate conditions in triplicate, at approximately 25, 50, 75, and 100% of the maximum treatment rate (MTR), for a total of 12 tests. So under ideal test conditions, each device's removal efficiency can be described by 36 data points.

A device's removal efficiency can be plotted as a dependent variable against an appropriate dimensionless independent variable. The dimensionless parameter used as an independent variable was the Peclet number (Pe), which is the ratio of advection to diffusion (Dhamotharan et al. 1981, Wilson et al. 2009). Advection is calculated as particle settling velocity, V_s, times a length scale L_1. Diffusion can be simplified to flow rate Q divided by length scale L_2. Putting advection and diffusion together yields $Pe = (V_s \times L_1 \times L_2)/Q$, where L_1 and L_2 are assumed to be a device's treatment chamber diameter and settling depth.

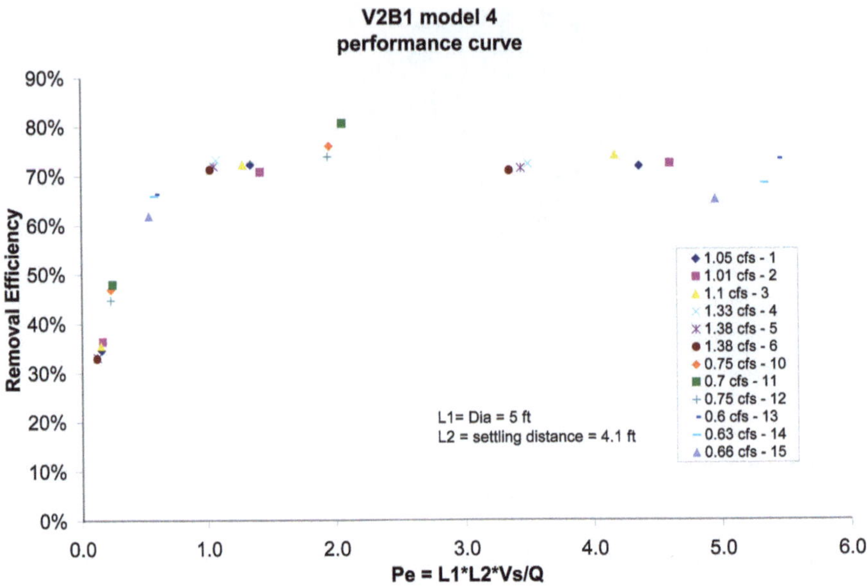

Fig. 7.6 Performance curve of removal efficiency vs. Pe# for the device swirl chamber *only*

As often as possible, the field team attempted to complete more than one test per day in order to maximize the effort in traveling to the site, setting up equipment, and preparing the device for testing, which were relatively constant 'costs' of testing whether 1 or 3 tests were performed. Construction activity adjacent to the stormwater quality test site presented difficulty with coordinating field testing. Additionally, a leaking swirl chamber was repaired to ensure proper hydraulics and system operation.

7.6.1.3 Assessment Results

At large Pe, V_s (i.e., large particles and therefore large settling velocities), coupled with low-flow rate Q_s, a stormwater treatment device can be expected to be successful removing particles from an influent. If the Pe number was allowed to approach infinity (approximating a large detention pond or lake), very near 100% removal could be achieved. The data appear to exhibit this trend, but the required Pe to achieve such removal is unknown. Conversely, at small Pe, V_s (i.e., small particles and therefore small settling velocities), coupled with large flow rate Q_s, a device can be expected to remove fewer particles from influent. This has been upheld in the results obtained, illustrated by the device performance curve depicted in Fig. 7.6.

The first several tests using the different particle sizes and relatively low flows indicated there was a problem carrying out tests with all of the sands designed for use during the experiment. Under low-flow rates, the influent water velocity is small

Fig. 7.7 Illustration of particle settling phenomenon inside the swirl chamber's influent delivery pipe. It is clear that a sandbar has formed, which is believed to contribute to further settling by reducing the vertical setting distance in this pipe

enough that the largest sand particles are not in suspension for the entire distance from the injection point to the device (approximately 45 ft (13.7 m)). Thus, heavier sands dropped out of the water column and settled at the bottom of the pipe, a typical result of which is illustrated in Fig. 7.7. The experiment was modified such that the relatively low-flow rates were increased (which therefore increased influent water velocities in the pipe) and the largest sand sizes removed from the mixture delivered to the device during these low-flow rate tests, producing a total of 30 data points in Fig. 7.6.

7.6.1.4 Conclusions and Recommendations

Understanding how devices perform under varying flow rates, sediment sizes, and treatment chamber sizes is important and helpful for consultants, local governments, and state agencies when selecting, designing, and evaluating stormwater treatment technologies for public infrastructure improvement projects. However, the effectiveness of proprietary underground stormwater treatment devices depends upon the settling velocity of influent solids (i.e., solid size and density) in addition to the size and design of the device. Using Pe to predict a device's performance over a wide range of device model sizes, storm events, and pollutant size characteristics is possible because Pe relates two length scales and particle settling velocity to influent flow rate.

This research showed that controlled field tests are a practical, robust, and accurate means of determining an underground device's performance, based upon the solid size distribution and influent density, in addition to the water discharge and temperature. The results from this research have been successfully verified on three other devices in field tests and other devices in laboratory tests.

More specifically, these efforts have demonstrated that the device is capable of removing coarse solids relatively well (70% +), but is less efficient at removing fine sands (~32 to 48%). If the trend is projected to a lower Pe, one would expect that the device would be even less successful with finer particles such as silt, and remove few, if any, clay particles.

7.7 Filtration Practices

Synthetic runoff testing (level 2b) can be used to measure the filtration rate of filtration practices if the available water supply can provide a sufficient water volume and discharge.

Given accurate contours, computer-aided drafting (CAD) software can be used to calculate the volume of a stormwater treatment practice, or the volume can be approximated by the method demonstrated in Example 7.3, which is not limited to filtration practices. If the practice is initially empty, the volume of water required to fill the practice is the storage volume of the stormwater filter (or the WQV) plus the estimated volume of water that will pass through the filter while the practice is being filled. The flow rate through the filter can be approximated as the flow rate calculated by Darcy's Law using a water depth equal to one-half of the depth of which the filter will be filled to start of the synthetic runoff test. This value of depth represents the average water depth that exists while the practice is being filled, and the flow rate calculated using this depth can be assumed to be a constant flow rate that is leaving the practice as it is filled. Given this outflow, the water supply must provide an influent discharge that can fill the practice in an acceptable amount of time. The process of determining if a water supply is adequate is demonstrated in Example 7.4.

Example 7.3: Estimating the volume of an irregularly shaped practice
Alan, the director of the county environmental services department, is considering synthetic runoff testing for an irregularly shaped sand filter basin. Alan has a 1-ft (0.3 m) interval contour map of the filter basin and knows that the bottom elevation is 612 ft (186.5 m) and that the filter has a maximum WQV depth of 5 ft (1.5 m), with an additional 1-ft (0.3 m) of freeboard. Alan determines the area circumscribed by each contour line as given in Table E7.2. Alan wants to estimate the design WQV of the filter basin.

(continued)

Example 7.3: (continued)

Table E7.2 Elevation and surface area data

Contour (ft)	Area within (ft^2)
612	68
613	159
614	313
615	531
616	784
617	933
618	1,105

Table E7.3 Elevation and storage volume data

Contours (ft)	Storage volume (ft^3)
612–613	113.5
613–614	236
614–615	422
615–616	637.5
616–617	863.5
617–618	1,044

Alan knows from geometry that he can determine an estimate of storage volume between two adjacent contour lines by multiplying the elevation difference between the contours (i.e., 1 ft (0.3 m) in this example) by the average area circumscribed by the same two contour lines. For example, Alan can estimate the storage volume between 612 and 613 ft (186.5 and 186.8 m) as:

$$\left(\frac{68\,\text{ft}^2 + 159\,\text{ft}^2}{2}\right) 1\,\text{ft} = 113.5\,\text{ft}^3$$

and the storage volume between 613 and 614 ft (186.8 and 187.1 m) as:

$$\left(\frac{159\,\text{ft}^2 + 313\,\text{ft}^2}{2}\right) 1\,\text{ft} = 236\,\text{ft}^3$$

Alan repeats this process for all contour line pairs (Table E7.3). Alan can then determine the total storage volume of the pond, or the WQV, by summing the storage volumes available from 612 and 617 ft (186.5 and 188.1 m). Alan excludes the volume between 617 and 618 ft (188.1 and 188.4 m) because this volume is associated with freeboard, not the WQV. Thus, Alan finds the total WQV as:

$$113.5 + 236 + 422 + 637.5 + 863.5 + 1,044 = 2,272.5\,\text{ft}^3$$

Example 7.4: Determining if a water supply is adequate for synthetic runoff testing of a filtration practice

Alan would like to use synthetic runoff testing on a sand filter that has a maximum depth of 6 ft (1.8 m), a design WQV of 3,200 ft^3 (90.6 m^3), and a filter surface area (A) of 750 ft^2 (69.7 m^2). Based on previous assessment results, Alan knows that this filtration practice has an overall effective saturated hydraulic conductivity (K_s) of 4 in./h (10 cm/h). Alan needs to determine if a nearby hydrant that can supply 2 cfs (0.057 m^3/s) of water for 1 h can be used for a synthetic runoff test of the filter if the filter media (L) is 18 in. (46 cm) thick.

First, Alan must estimate the discharge through the filter as the filter is being filled. Using Darcy's equation (12.1)–(12.4), Alan calculates the filtration flow rate when the filter is filled to half of its maximum depth (note the negative symbol from (12.1) has been removed because we know the direction of flow is in the negative vertical (downward) direction).

$$Q = K_s A\left(\frac{z_w + L}{L}\right)$$

$$= \left(4\frac{\text{in.}}{\text{h}}\right)\left(\frac{1\,\text{ft}}{12\,\text{in.}}\right)(750\,\text{ft}^2)\left(\frac{6/2\,\text{ft} + 18/12\,\text{ft}}{18/12\,\text{ft}}\right)\left(\frac{1\,\text{h}}{3,600\,s}\right) = 0.21\,\text{cfs}$$

Therefore, Alan determines that the net flow that will fill the filter will be influent flow rate minus the rate at which water flows through the filter: 2.0–0.21 cfs = 1.79 cfs = 6,450 ft^3/h (0.057–0.006 m^3/s = 0.051 m^3/s = 182.6 m^3/h).

Alan can then calculate the time required to fill the practice by dividing the WQV by the filling rate: 3,200 ft^3 / 6,450 ft^3/h (90.6 m^3 / 182.6 m^3/h) = 0.5 h = 30 min. Thus, Alan knows that with the given assumptions, the fire hydrant will be able to provide the necessary flow for synthetic runoff testing.

Synthetic runoff tests may detect the presence of macropores within a filtration practice. Macropores allow stormwater runoff to flow quickly through the filtration media, resulting in minimal solids removal. If the results from the synthetic runoff tests indicate that the saturated hydraulic conductivity (K_s) for the filtration practice is larger than the K_s of gravel (85 m/day, 280 ft/day), then it is likely that minimal treatment by filtration is occurring as a result of macropores. K_s values less than 85 m/day do not preclude the presence of macropores, but indicate that the number of macropores, if present, may not necessarily be significant. Specific locations can be tested for the presence of macropores using capacity testing (level 2a).

For filtration practices, synthetic runoff testing (level 2b) may require less effort than capacity testing (level 2a). In other words, as long as the water supply is sufficient, it may be easier and require less time to fill a filtration practice and measure the change in water level with respect to time than to perform multiple point infiltration measurements. This is especially true of underground filtration practices, which are typically small systems that have limited access. The information gained from capacity tests and synthetic runoff tests will differ, however, as described in the following paragraphs.

Synthetic runoff testing may be used to estimate an overall effective saturated hydraulic conductivity (K_s) for an entire practice, whereas the results of capacity testing will result in different K_s values for each test location. For example, synthetic runoff testing of a filtration practice may indicate that the practice is able to drain a synthetic storm event that is equivalent to the WQV in less than 48 h and therefore meets design requirements. Capacity testing, however, may indicate that 25% of the filtration practice is not filtering water at all, while the remaining 75% is filtering all of the incoming stormwater water. Furthermore, capacity testing can indicate where malfunction is occurring in the filtration practice, which allows for localized maintenance to restore the practice before the entire filtration practice fails.

7.8 Infiltration Basins

To conduct a synthetic runoff test of the infiltration capacity of an infiltration basin, the basin must be filled with water and the water surface elevation recorded as a function of time. The data can then be used to estimate an overall effective saturated hydraulic conductivity (K_s) value for the entire basin, which can be used to estimate the time it will take the practice to infiltrate the WQV. If the volume of water required to fill the basin is significantly more than is available, then it may be necessary to use monitoring (Chap. 8) to assess the basin. In this case, one must delay the test until an actual runoff event of sufficient magnitude fills the basin. In this situation, the water surface elevation as a function of time is recorded just as if the basin was filled with a fire hydrant, and the collected data are analyzed using the same methods.

To determine the overall effective K_s of an infiltration basin, the basin should be filled with water to its design WQV. During the test, all other inflows of water, if any, should be eliminated and all discharge locations other than infiltration should be plugged. If the water supply is unable to provide the WQV in a reasonable time, a smaller volume of water may be used. After the practice is filled to the desired depth and the water supply is shut off, the water surface elevation will begin to drop due to infiltration into the soil. The water surface elevation should be recorded as a function of time (e.g., every 10 or 15 min) until the basin is empty or almost empty.

7.9 Infiltration Trenches

For infiltration trenches, synthetic runoff testing is conducted by applying synthetic stormwater such that the trench is filled to its WQV and infiltration into the trench is initiated. This could be accomplished by filling the trench with water or by applying synthetic runoff to a clean area directly upstream of the trench so that the applied water flows into the trench. The trench must be filled with its WQV because, given the geometry of a trench, there will be horizontal and vertical infiltration with different pressure heads, and modeling such a scenario or determining the hydraulic saturated conductivity (K_s) is extremely difficult. Thus, in order to assess the infiltration capacity of an infiltration trench, the trench must be filled with its WQV so that the time required to infiltrate this volume can be measured directly. The time to infiltrate the WQV can be compared to governing regulations, previous assessment results, or design standards to assess the condition of the trench.

Conducting a synthetic runoff test for a trench requires a sufficient water supply. To determine if the available water supply is adequate, see the section 7.3 and Example 7.1.

7.10 Permeable Pavements

Synthetic runoff testing of permeable pavements is feasible if water can be stored on the surface of the permeable pavement to a measurable depth (> 6 in. (15 cm)) for a minute or more so that the water surface elevation as a function of time can be recorded. This will provide enough data to allow for an analysis similar to that of infiltration basins. The data collection and analysis of the data is done in the same manner. Generally, the pavement surface will be planar and sloped on a uniform grade, so curbs or some form of berm around the boundaries of the pavement will be required to store and infiltrate a measurable depth of water. The base layer underlying the permeable pavement is generally very porous, so it is necessary to measure only the saturated hydraulic conductivity (K_s) of the pavement layer itself. Knowing the thickness of this layer and the rate of drop of the infiltrating ponded water, it is possible to compute K_s from (7.3).

7.11 Scheduling Maintenance

Scheduling maintenance from synthetic runoff testing results will depend on the type of test performed. If synthetic runoff testing for sediment retention is performed and the stormwater treatment practice is not performing as expected, then more information may be required to determine the most appropriate maintenance. Check the practice for the possibility of short-circuiting, excessive sediment deposition resulting in very little storage volume, and erosion within the practice. If any of these are present, perform appropriate maintenance to correct the situation.

If synthetic runoff testing for infiltration/filtration rate is performed and the stormwater treatment practice is treating water at less than the expected rate, check the practice for excessive sediment deposition. Also check that any vegetation, if present, is healthy and allows appropriate infiltration or filtration into the soils.

If synthetic runoff testing is used to determine pollutant removal efficiency, more information may be required to schedule appropriate maintenance. Consider the pollutant of concern and the function of the practice in general (e.g., filtration) before determining what maintenance actions are required, if any. For example, if synthetic runoff testing determines that a wet detention pond is not capturing phosphorus as expected, determine if the practice is performing in its primary function (sedimentation) and identify potential sources of phosphorus such as pond sediments and illicit discharges.

Monitoring of Stormwater Treatment Practices

8

Abstract

Monitoring, the most comprehensive level of assessment, is achieved by collecting and analyzing influent and effluent runoff samples and/or measuring influent and effluent flow rates as a function of time over the course of one or more natural runoff events. Monitoring, which is not limited by the size of the stormwater treatment practice, can be used to assess the performance of a practice with regard to reduction in contaminant load or concentration and reduction of runoff volume. This chapter discusses and explains the techniques of monitoring and how to carry out a monitoring program for various stormwater treatment practices. It also provides guidance about which stormwater treatment practices are best suited for monitoring and which are not. It ends with a case study of a monitoring effort on a dry detention pond.

8.1 What Is Monitoring?

If capacity testing (level 2a) and synthetic runoff testing (level 2b) are not feasible assessment approaches for a specific location or do not achieve the goals of the assessment program, monitoring should be considered. Monitoring is the most comprehensive assessment technique and can be used to assess water volume reduction, peak flow reduction, and pollutant removal efficiency for most stormwater treatment practices. Monitoring is accomplished during natural runoff events by measuring all influent and effluent flow rates over the entire runoff event and, if pollutant removal is to be assessed, collecting influent and effluent samples to determine pollutant concentrations. To assess runoff volume reduction, peak flow reduction, or both by monitoring a stormwater treatment practice, the inflow(s) and outflow(s) must be measured or estimated according to the techniques described in Chap. 9. The peak influent and effluent flow rates can be compared to determine the peak flow reduction and the total volume of influent can be compared to the total volume of effluent to determine the runoff volume reduction. Additional information

A.J. Erickson et al., *Optimizing Stormwater Treatment Practices: A Handbook of Assessment and Maintenance*, DOI 10.1007/978-1-4614-4624-8_8,

about monitoring stormwater treatment practices is available in the report "Urban Stormwater BMP Performance Monitoring" (US EPA 2002).

As stated above, pollutant removal efficiency can also be determined by monitoring if, in addition to measuring inflow and outflow discharges, all inflow and outflow locations are sampled according to the techniques described in Chap. 10. Pollutant removal efficiency can then be determined as the difference between the influent and effluent pollutant mass loads or event mean concentrations (EMCs), as defined and described in Chap. 12.

Natural runoff events vary in discharge and duration, so they require continuous flow measurement (or estimation). Pollutant removal assessment also requires sampling of all flows entering and exiting a stormwater treatment practice. For accurate estimates of performance, monitoring takes more time to complete (typically 14 or more continuous months), more equipment, and more labor. It, therefore requires larger expenditures than the first two levels of assessment. Monitoring is the only method that accurately measures the quantity and quality of runoff from a specific watershed, and the response of a stormwater treatment practice to that runoff. Capacity testing (level 2a) and synthetic runoff testing (level 2b) measure the ability of a stormwater treatment practice to perform specific processes (e.g., infiltration, sediment retention). These data can be used in models to estimate how a stormwater treatment practice would perform in a given watershed during natural runoff events.

Monitoring has more potential for uncollected or erroneous data as compared to synthetic runoff tests for the following reasons:

1. Weather is unpredictable and can produce various runoff volumes of various durations with varying pollutant concentrations at various times. In order for a storm event to be monitored correctly and accurately, all the monitoring equipment must be operating correctly and the runoff event parameters (water depth, etc.) must be within the limit ranges of the equipment
2. Equipment malfunction due to routine wear or vandalism is more likely. Without consistent inspection and maintenance, storm events may be measured or sampled incorrectly or not at all

As with any field work, safety is an important concern and should be addressed when conducting monitoring.

8.2 Monitoring Sedimentation Practices

Some sedimentation practices may be too large for synthetic runoff testing (i.e., a sufficient water supply is not available) and therefore may require monitoring (level 3) to achieve the assessment goals. To successfully monitor a stormwater treatment practice, it is necessary to follow appropriate procedures for Water Budget Measurement, Water Sampling Methods, and Analysis of Water and Soils in Chaps. 9, 10, and 11, respectively. In addition, there are some monitoring considerations specific to dry ponds, wet ponds, and underground sedimentation practices provided in the following sections.

8.2.1 Dry Ponds

With monitoring, one can assess a dry pond's peak flow reduction, runoff volume reduction, and pollutant removal efficiency. Measuring and comparing inflow and outflow hydrographs for a dry pond can give an estimate of the reduction in peak flow for a given runoff event and, therefore, an estimate of the hydraulic effectiveness of the stormwater treatment practice. Dry ponds are not typically designed to, and often do not, infiltrate a significant fraction of the runoff volume. The data collected when monitoring a dry pond, however, can be used to determine how much, if any, of the influent runoff volume is infiltrated. Because dry ponds are typically designed to drain within 2 days, the fraction of water lost to evapotranspiration is negligible and can be ignored. Thus, with no other mechanism available for water to leave the pond, the difference between the total volume of influent and effluent is the volume lost to infiltration.

To assess pollutant removal efficiency, influent and effluent samples must be collected and analyzed for the concentration of the target contaminant. Pollutant removal effectiveness is typically reported as the reduction in the total mass load between the influent and effluent locations or the percent difference between the influent event mean concentration (EMC) and effluent EMC. See Chap. 12, for guidance on analyzing data collected from monitoring studies.

8.2.2 Wet Ponds

Monitoring of wet ponds (also known as wet detention basins) is well documented (Wu et al. 1996; Comings et al. 2000; Koob 2002; Mallin et al. 2002). Monitoring of wet ponds is accomplished in the same manner as with dry ponds (see previous section). Short-circuiting within a wet pond can be estimated by monitoring the movement of a naturally occurring conservative tracer, such as chloride, as it moves through a wet pond if a sufficient pulse in concentration has occurred at the inlet. Plotting the inflow and outflow tracer concentrations as a function of time and comparing the two curves can determine if, and to what extent, short-circuiting may be occurring (see Example 7.2).

8.2.3 Underground Sedimentation Devices

Monitoring wet vaults and underground sedimentation devices for hydraulic performance or water quality treatment is not recommended because wet vaults and underground sedimentation devices are typically designed for small subwatersheds in urban areas and are located underground with limited access. Monitoring equipment should be kept in an environmental cabinet for protection from extreme weather, vandalism, and theft. Finding a semipermanent place for the equipment cabinet in an urban area such as a busy parking lot or intersection can be difficult, and if a cabinet were located in such a place it would be an easy target for vandals.

Also, sampling tubes must often run from the sampling location to the cabinet where the samples are stored, and these tubes are sometimes protected by metal or plastic piping. Such tubes and pipes crossing sidewalks and streets in an urban area would not only present a challenge, but would also create a liability. Furthermore, sampling tubes need to exit the underground device, meaning that catch basin covers would need to be removed during the monitoring period or modified to allow passage of the sampling lines. Because underground sedimentation devices are generally small, at least two cabinets and sampling lines (one each for inlet and outlet monitoring) would be in close proximity to each other, further complicating the matter.

In some situations, however, monitoring of an underground sedimentation device may be feasible and beneficial. In these situations, monitoring can be accomplished in a manner similar to that of monitoring dry ponds.

8.3 Monitoring Filtration Practices

Monitoring (level 3) is the most comprehensive method for assessing filtration practices. Monitoring can assess how well a filter reduces runoff peak flow, reduces runoff volume (by infiltration into the surrounding soil), and captures pollutants.

The perforated pipe collection systems that collect stormwater after it passes through a filtration practice are typically 4–8 in. (10–20 cm) in diameter. The small pipe diameter can present a significant challenge to measuring and sampling the effluent. If using a weir to measure flow from a perforated pipe, it is important to design the weir crest (i.e., invert) elevations such that the water level in the perforated pipe will always be below the level of the perforations in the pipe. This is because back pressure in the perforated pipe can prevent filtered water from entering the pipe. Sometimes, the perforated pipe collection system is connected to a catch basin that has other inflows; in these situations, it could be difficult to separate the outflow from the filtration practice from the other flows entering the catch basin. Thus, it is important to sample and measure flow from the perforated pipe system before it combines with any other surface runoff or conduit flow to ensure an accurate comparison between outflow and inflow for the filtration practice.

Infiltration into the native soil may occur in filtration practices that do not have impermeable liners or other barriers such as concrete walls. If infiltration is possible, infiltration rates should be measured or estimated to complete the water budget. The amount and rate of infiltration will depend on the stormwater filter design and saturated hydraulic conductivity of the underlying soils. Discussion and recommendations for estimating such infiltration as part of a water budget are included in Chap. 9.

Evaporation and transpiration (also discussed in Chap. 9) will likely account for an insignificant ($< 5\%$) portion of the water budget because they are slow processes and water typically does not remain in properly functioning filtration practices for more than 48 h. Additionally, vegetation, which can increase infiltration and evapotranspiration, is often limited in filtration practices to ensure adequate filtration by the filter media and to facilitate maintenance of the filter surface.

8.4 Monitoring Infiltration Practices

Some infiltration practices may be too large for synthetic runoff testing (i.e., a sufficient water supply is not available) and therefore may require monitoring (level 3) to achieve the assessment goals. To successfully monitor a stormwater treatment practice, it is necessary to follow appropriate procedures for Water Budget Measurement, Water Sampling Methods, and Analysis of Water and Soils in Chaps. 9, 10, and 11, respectively. In addition, there are some monitoring considerations specific to infiltration basins, trenches, and permeable pavements provided in the following sections.

8.4.1 Infiltration Basins

If the size of an infiltration basin precludes the use of synthetic runoff testing, monitoring can be used to assess the overall effective saturated hydraulic conductivity and infiltration capacity of the basin. In this case, the major difference between synthetic runoff testing and monitoring is that in monitoring, a natural runoff event is used to fill the basin rather than a fire hydrant or water truck. Once the basin is filled and the water surface elevation begins to drop, the procedures are identical and data analysis is the same in both scenarios. For details on this method, see Chap. 7.

8.4.2 Infiltration Trenches

Monitoring can be used to determine the ability of an infiltration trench to infiltrate runoff by determining the time required to infiltrate the WQV. Infiltration trenches function by storing water within the trench and allowing the stored water to infiltrate through the sides and bottom of the trench. If the trench receives the WQV of runoff, monitoring the water surface elevation within the trench will determine the time required to infiltrate this volume, which is the design volume.

Once the trench is filled with stormwater, data collection and analysis are performed in an identical manner to that of synthetic runoff testing. See Chap. 7 for a detailed discussion of the procedure.

Because water in the trench disperses as it infiltrates into the surrounding soil and does not have a well-defined location at which it exits the practice, obtaining representative effluent samples is extremely difficult. Thus, it is difficult to use monitoring to accurately assess the pollutant removal effectiveness of infiltration trenches.

Monitoring an infiltration trench can be made easier by installing sampling locations underneath the trench during construction. Sampling locations should be located below the bottom of the trench at a distance that ensures the samples represent water that has been fully treated by the soil. Sampling locations should not be located such that water not treated by the trench (e.g., groundwater) is sampled.

8.4.3 Permeable Pavements

It is possible to use monitoring to assess the infiltration capability of permeable pavements, although unique challenges usually exist. If monitoring can be accomplished, however, results do not yield a value of saturated hydraulic conductivity for the pavement. Instead, only an indication of the volume or fraction of runoff that can be infiltrated will be obtained.

Monitoring permeable pavements is challenging because of difficulties in accurately measuring inflows to and outflows from the permeable pavement surface. These measurements, as discussed in Chap. 9, must be obtained in order to successfully complete an assessment. Outflows are difficult to measure because runoff from a permeable pavement typically does not occur at one or two locations where the flow rate can be measured. For example, runoff from a permeable parking lot may flow as sheet or shallow concentrated flow out the driveway, into the gutter, and down the street, making flow measurement extremely difficult.

Inflows, which also must be measured over time, can also present a challenge. For example, runoff may be generated from adjacent permeable surfaces (e.g., grassed yards) and impermeable surfaces (e.g., buildings and impermeable pavements). This runoff can flow on to the permeable pavement over a widespread area, again making it difficult to accurately measure the total flow rate. In other situations it can be difficult to isolate the permeable pavement. For example, some parking lots have permeable pavement in the parking stalls but not in the driving lanes. If all inflows and outflows on the permeable pavement surface can be accurately measured, completing a water budget on the pavement will determine the volume of water that has infiltrated. This volume can be compared to the total volume of influent to indicate the fraction of runoff that can be infiltrated for the rainfall event. The volume infiltrated divided by the area available for infiltration and divided by the time over which infiltration occurs results in an average infiltration rate, which is dependent on but not equal to the saturated hydraulic conductivity.

Significant water depth (i.e., 6 in. (15 cm) or more) on top of permeable pavement is typically not obtainable; therefore, water surface elevation measurements are not made during monitoring. As a result, monitoring efforts on permeable pavements do not yield a value for the overall effective saturated hydraulic conductivity.

8.5 Monitoring Biologically Enhanced Practices

For most biologically enhanced practices, monitoring may be possible but not cost-effective or necessary. In fact, only constructed wetlands and swales may require monitoring. Bioretention practices are often relatively small in size, and assessment may be achieved in a much more cost-effective manner through capacity testing (level 2a) or synthetic runoff testing (level 2b). For example, bioretention practices that are located in residential areas on private property (i.e., resident's front yards) can range from 500 square feet (46.5 m^2) down to 20 square feet (1.86 m^2), and may occur on 50% or more of the properties on a given street. Monitoring each bioretention practice in this area would be costly in both equipment and labor. Monitoring may be appropriate, however, to investigate the overall impact of a number of bioretention practices in a subwatershed or drainage basin. Constructed wetlands and swales are large by comparison, and their design typically facilitates monitoring.

8.5.1 Bioretention Practices (Rain Gardens)

If an adequate water supply is available, synthetic runoff testing (level 2b) is preferred to monitoring because monitoring can be cost intensive and result in minimal conclusive data. Monitoring, however, can provide information regarding the watershed that synthetic runoff testing cannot, such as characteristics of watershed runoff coefficients and pollutant loads.

If monitoring is to be performed, the procedures are identical to those for infiltration basins, which are discussed previously in this chapter. Monitoring bioretention practices is easier if the practice has a subsurface pipe collection system to allow for effluent measurement and sampling. Effluent measurement allows for a water budget to be completed on the practice, which allows for the volume reduction to be determined. Effluent sampling allows for pollutant removal rates to be estimated when compared to influent sampling. Monitoring bioretention practices without subsurface collection systems for pollutant capture can be cost-prohibitive and result in minimal conclusive data (Tornes 2005). See Chaps. 9, 10, 11, and 12 for details on water budget measurement, sampling, sample analysis, and data analysis.

8.5.2 Constructed Wetlands and Swales

Numerous studies have been published concerning the assessment of constructed wetlands and swales with monitoring (Maehlum et al. 1995; Kadlec and Knight 1996; Oberts 1999; Barrett, et al. 1998; Carleton et al. 2000; Laber 2000; Yu et al. 2001; Bulc and Slak 2003; Farahbakhshazad and Morrison 2003; Farrell 2003; Deletic and Fletcher 2006). Monitoring is the most comprehensive assessment technique for measuring the hydraulic and pollutant removal effectiveness of a constructed wetland. Runoff volume reduction (by evapotranspiration) can be estimated by comparing the total influent water volume to the total effluent water volume in the water budget for the constructed wetland. It is important to recognize that constructed wetlands typically do not infiltrate stormwater runoff and may receive substantial (> 5%) inflow from direct rainfall due to their large surface areas (see Chap. 9 for more information). Monitoring constructed wetlands for pollutant removal effectiveness requires that the volume and water quality of all stormwater inputs and outputs are measured. Refer to Chap. 9 for guidance on flow measurement, Chap. 10 for guidance on sampling techniques for gathering stormwater samples, and Chap. 11 for analysis techniques and recommendations. Data from monitoring should be analyzed according to methods described in Chap. 12.

8.5.3 Filter Strips

Filter strips typically lack the inlet and outlet flow structures that would allow for discharge measurement or pollutant sampling. In this case, it is not recommended that filter strips be monitored for water quality or hydraulic performance.

8.6 Case Study: Monitoring a Dry Detention Pond with Underdrains

A dry detention pond with underdrains was evaluated using monitoring. It drains a watershed that encompasses the corner of the local public works facility site, consisting of 45 acres (18.2 ha) with impervious area on the site of 10.2 acres (4.1 ha). Future construction of the facilities may occur on the remainder of the site.

The dry detention pond is approximately three acres with a slope of 1% from inlet to outlet. It is designed to provide storage for up to a 100 year—24 h rainfall event. Stormwater runoff is directed through grass waterways to a small pretreatment pond (forebay) before it enters the main detention pond. After entering the detention pond, the stormwater runoff infiltrates through the soil media. A series of rock-filled trenches holding perforated drain tile acts as an underdrain for the pond, into which most of the stormwater runoff drains. As shown with the circle labeled "Dry Infiltration Basin" in Fig. 8.1, the drain tile system consists of an 8-in. (20.3 cm) diameter, 140-ft (42.7 m) long, perforated polyethylene pipe running down the near middle of the basin. At eight locations on the 8-in. (20.3 cm) pipe, a set of two, 8-in. (10.1 cm) diameter, perforated polyethylene underdrain laterals are attached; one to each side of the main pipe. Each lateral extends outward away from the main pipe at a 45° angle for 30 ft (9.1 m). Thus, in addition to the 140-ft (42.7 m) of 8-in. (20.3 cm) diameter central drain tile, there is a total of 480 ft (146.3 m) of the 4-in. (10.1 cm) diameter laterals.

A cross section of the underdrain system is shown in Fig. 8.2. The underdrain pipe was surrounded by a mixture of soil and ASTM C33 fine sand, which was used as the filter media. A filter fabric was used to separate the soil-sand filter media and underdrain pipe from the surrounding existing soil. A layer of 6 in. (15 cm) of native soils (typically tighter clays for local area) was used to bury the filter fabric so it was not exposed at the surface. The underdrains collect the infiltrated storm water and drain it to the outlet structure. The outlet structure is 5 ft (1.5 m) in diameter and receives infiltrated runoff through an 8-in. (20.3 cm) underdrain pipe as shown in Fig. 8.3. The outlet structure acts as an overflow spillway so that runoff in excess of the design storage volume will bypass filtration and be discharged downstream. An 18-in. (46 cm) diameter reinforced concrete pipe collects runoff from the outlet structure and discharges it to the downstream watershed. Native plants were planted on the site, including the grass waterways (ditches) and areas around the parking lot.

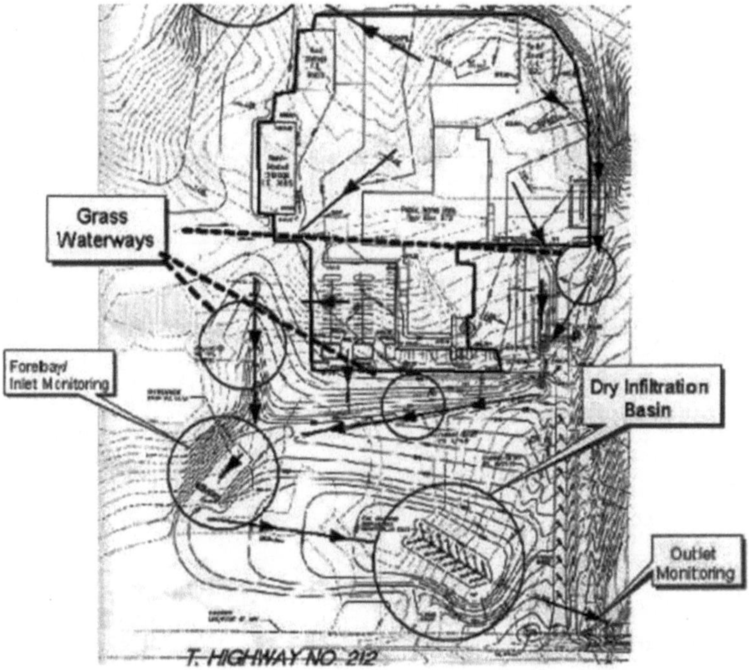

Fig. 8.1 Plan view of dry detention pond

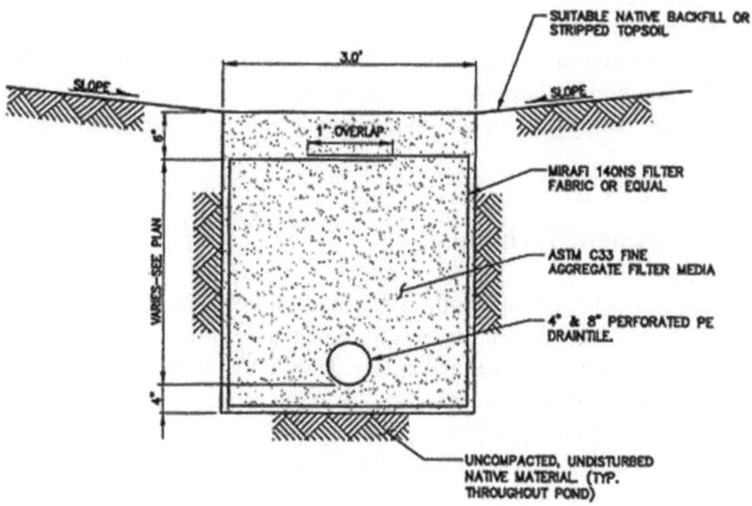

Fig. 8.2 Cross section of pond under-drain system

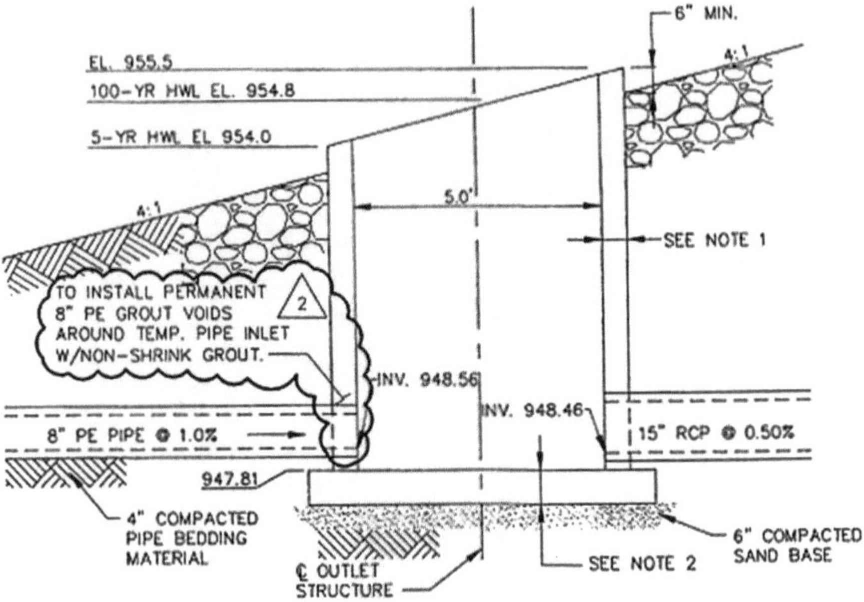

Fig. 8.3 Outlet structure of dry detention pond

8.6.1 Assessment Goals

The goals of this assessment were to (1) assess runoff volume reduction and (2) assess pollutant retention performance of total suspended solids, volatile suspended solids, total phosphorus, and particulate phosphorus. This pond was designed to drain within 48 h after a runoff event by filtering the stormwater through the sand trenches. In addition to filtration, a primary treatment process of dry detention ponds with underdrains is sedimentation, which occurs while the runoff is pooled in the pond.

8.6.2 Assessment Techniques

To meet the assessment goals, both inflow and outflow had to be measured and sampled. The pond was chosen for monitoring because it has one influent and one effluent location and limited overland inflow. Thus, only two flow measurement and sampling stations were needed.

A portable water quality sampler, which contained a complete set of 24, 1 L wedge-shaped bottles, was installed at the inlet of the pond. The unit was programmed to collect flow-weighted samples and to record the depth, velocity, and discharge at 10-min intervals. A tipping bucket rain gauge was also installed near the inlet of the pond to collect data on the total rainfall amount, antecedent dry days, and rainfall intensity for each storm event.

Fig. 8.4 Rectangular weir at the inlet of the pond, which was later modified

Initially, a 5-ft (1.5 m) wide rectangular, sharp-crested weir was installed at the inlet of the pond, as shown in Fig. 8.4, with an ultrasonic flow sensor. The sensor was installed over the water surface just upstream of the weir to measure depth behind the weir. The equipment continuously monitored and recorded the rainfall and water level, which was used to calculate the corresponding flow rate.

Results from preliminary monitoring showed that the rectangular weir did not provide accurate estimates of discharge at the relatively low discharge rates that were most common at the site. Therefore, the 5-ft (1.5 m) wide rectangular weir was modified by cutting a 3-in. (7.6 cm) deep, 90° V-notch into the middle of the rectangular weir such that the result was a sharp-crested compound weir which would more accurately measure low discharges.

At the outlet, another portable automatic sampler was programmed to collect flow-weighted samples. Using a flexible circular spring ring, an Acoustic Doppler Velocimeter (ADV) was installed on the bottom of the outlet culvert. This type of sensor uses Doppler technology to measure average velocities at locations across the flow cross section. A pressure transducer contained within the sensor measured water depths and calculated flow areas based on conduit geometry. The automatic sampler calculated the total discharge by summing the products of all recorded average velocities and their corresponding flow areas.

Initial monitoring revealed that the ADV sensor did not accurately measure flow velocities when water depth was less than approximately 2 in. (5 cm). A 3-in. (7.6 cm) tall plastic circular weir, as shown in Fig. 8.5, was installed to ensure that the area velocity sensor used at the outlet had at least 2 in. (5 cm) or more of water depth needed to accurately measure the velocity profile. The area velocity sensor was located inside of the pipe, 6 in. (15 cm) upstream of the circular weir.

Subsequent monitoring revealed that the circulating flow caused by installing the weir resulted in significant error with the ADV sensor measurement. The sensor appeared to sum the forward and backward velocity caused by the weir, resulting in significantly larger flow velocity than the actual net forward velocity. The depth

Fig. 8.5 Circular weir installed in the outlet pipe of the pond

measurement, however, reported by the probe was correct, and these values were used to calculate the head over the weir. After the weir was properly calibrated, the head measurements could be used to calculate the discharge over the weir and thus the flow in the outlet pipe.

The monitoring systems at both the inlet and outlet of the pond were powered by heavy duty deep-cycle marine batteries and solar powered battery chargers. Although the samplers and data loggers are watertight, corrosion resistant, and can be installed without additional protection, all the monitoring equipment was enclosed in lockable wooden environmental cabinets. A laptop PC with corresponding software was used to retrieve the data from the automatic samplers.

8.6.3 Assessment Results

Data and samples were collected for twelve runoff events over two years. Due to the extra variables and uncertainties introduced when monitoring, this time frame is not uncommon to obtain meaningful results. The results are presented in Tables 8.1, 8.2, and 8.3. The data presented in Tables 8.1 and 8.2 were used to estimate the performance of the pond for volume reduction and pollutant retention as listed in Table 8.3. There was significant infiltration in the pond. Values ranged from 1/3 of the total influent volume at high discharges to greater than 2/3 of the total volume at lower discharges.

Overall, load-based efficiencies are preferred for total load studies. Total load is determined by subtracting the sum of the outflow from the sum of the inflow and dividing by the sum of the inflow (see Chap. 12). The overall load-based efficiencies for the twelve monitored storms were 88% for total suspended solids,

Table 8.1 Rainfall amount, measured, direct, and total influent, measured effluent, and infiltration volume for the pond (1 in. = 2.54 cm, 1 ft^3 = 0.028 m^3)

	Total rainfall (in)	Measured influent volume (ft^3)	Direct rainfall volume (ft^3)	Total influent volume (ft^3)	Measured effluent volume (ft^3)	Total infiltration volume (ft^3)
SE 1	4.1	76,182	44,649	120,831	70,062	50,769 (42.0%)
SE 2	2.23	15,586	24,285	39,871	24,744	15,127 (37.9%)
SE 3	0.7	12,138	7,623	19,761	3,837	15,924 (80.6%)
SE 4	2.25	39,752	24,503	64,255	32,281	31,974 (49.8%)
SE 5	1.58	31,075	17,206	48,281	8,796	39,485 (81.8%)
SE 6	1.39	11,312	15,137	26,449	18,967	7,482 (28.3%)
SE 7	1.67	39,181	18,186	57,367	30,420	26,947 (47.0%)
SE 8	0.41	9,280	4,465	13,745	5,184	8,561 (62.3%)
SE 9	1.16	25,574	12,632	38,206	32,470	5,736 (15.0%)
SE 10	0.4	4,980	4,356	9,336	2,158	7,178 (76.9%)
SE 11	0.51	8,630	5,554	14,184	6,926	7,258 (51.2%)
SE 12	0.18	1,247	1,960	3,207	220	2,987 (93.1%)

81% for volatile suspended solids, 58% for total phosphorus, and 52% for dissolved phosphorus (see Chap. 12). These load-based efficiencies incorporate infiltration as a treatment mechanism and are therefore less comparable between sites.

The average concentration-based retention efficiencies for the twelve storms at dry detention pond with underdrains were 78% for total suspended solids, 64% for total volatile solids, 13% for particulate phosphorus, and 7% for total phosphorus (see Chap. 12). Retention efficiencies for dissolved phosphorus varied significantly and ranged from negative 60% to positive 28%. Dry detention ponds are focused on removing sediment and the associated pollutant concentration, such as particulate phosphorus. The primary retention mechanisms are not designed to retain dissolved phosphorus; thus, dissolved phosphorus retention is minimal.

8.6.4 Conclusions and Recommendations

The dry detention pond with underdrains was selected and monitored from May 2004 to November 2004 and May 2005 to August 2005 to assess its pollutant removal performance. Performance was estimated by comparing the influent and effluent pollutant loads and concentrations. From the results obtained in this study, the following specific conclusions were reached.

The measured concentrations of most parameters in stormwater runoff that entered at the dry detention pond with underdrains were substantially lower than concentrations typically mentioned in other studies throughout the nation, which influenced the pollutant retention efficiency of the pond. The lower values found at this site are thought to be related to pretreatment provided by the small pond near

Table 8.2 Total influent and effluent pollutant load and concentration of TSS, VSS, TP, and DP for the pond (all loads are in kg, all concentration are in mg/L)

	Total suspended solids				Volatile suspended solids			
Storm event	Load in	Load out	Conc. in	Conc. out	Load in	Load out	Conc. in	Conc. out
SE 1	194	19.8	57.6	10	18.2	5.36	5.4	2.7
SE 2	873	76.2	791	109	107	10.9	96.7	15.6
SE 3	10.6	0.98	19.2	9	4.3	0.47	7.8	4.3
SE 4	117	24.9	64.8	27.2	23.8	6.03	13.2	6.6
SE 5	15.6	1.32	11.5	5.3	8.81	0.95	6.5	3.8
SE 6	4.04	1.29	5.6	2.4	2.09	0.75	2.9	1.4
SE 7	29.9	21.7	18.4	25.2	8.45	6.97	5.2	8.1
SE 8	6.32	0.69	16.2	4.7	2	0.29	5.1	2
SE 9	9.66	8.1	8.9	8.8	2.41	1.86	2.2	2
SE 10	2.42	0.5	9.1	8.1	1.21	0.22	4.6	3.5
SE 11	4.28	2.28	10.6	11.6	1.83	1.06	4.6	5.4
SE 12	0.25	0.01	2.7	1.8	0.13	0.01	1.4	0.8
	Total phosphorus				Dissolved phosphorus			
Storm event	Load in	Load out	Conc. in	Conc. out	Load in	Load out	Conc. in	Conc. out
SE 1	0.547	0.175	0.162	0.088	0.209	0.101	0.062	0.051
SE 2	0.273	0.106	0.247	0.151	0.041	0.028	0.037	0.041
SE 3	0.058	0.009	0.105	0.082	0.038	0.006	0.069	0.059
SE 4	0.418	0.19	0.232	0.208	0.254	0.117	0.141	0.128
SE 5	0.359	0.046	0.265	0.183	0.226	0.03	0.167	0.12
SE 6	0.123	0.083	0.171	0.155	0.07	0.048	0.097	0.09
SE 7	0.278	0.194	0.171	0.225	0.175	0.109	0.108	0.127
SE 8	0.078	0.018	0.201	0.125	0.033	0.01	0.084	0.065
SE 9	0.285	0.218	0.263	0.237	0.179	0.15	0.165	0.163
SE 10	0.041	0.008	0.157	0.236	0.026	0.005	0.099	0.08
SE 11	0.065	0.028	0.162	0.142	0.033	0.015	0.082	0.078
SE 12	0.007	0.001	0.077	0.086	0.004	0.0004	0.048	0.077

the inlet and also by the two grassy ditches/swales used to transport stormwater runoff to the detention pond.

The use of a primary device (e.g., V-notch, rectangular or circular weirs, flumes) for flow measurement is strongly recommended, especially in outlet underdrain pipes. These devices are easy to install and can be used to provide continuous flow hydrographs using measurements of water surface level. The study revealed that an AV sensor cannot measure any velocity unless there is at least 2.5–3 in. (6.4–7.6 cm) of water over it, which does not often occur in underdrain outlets.

This research study confirmed that dry detention ponds with underdrains are an effective option for water quality control. The pond provided moderate stormwater treatment and reduced the concentrations of total suspended solids, volatile suspended solids, particulate phosphorus, and total phosphorus, even with small influent concentrations.

Results from the dry detention pond with underdrains indicate that influent pollutant concentrations influenced the pollutant retention efficiencies.

Table 8.3 Load-based and concentration-based removal efficiencies for the pond

	Load-based removal efficiencies				Concentration-based removal efficiencies			
	TSS (%)	VSS (%)	TP (%)	DP (%)	TSS (%)	VSS (%)	TP (%)	DP (%)
SE 1	90%	71%	68%	52%	83%	50%	46%	18%
SE 2	91%	90%	61%	32%	86%	84%	39%	−11%
SE 3	91%	89%	84%	84%	53%	45%	22%	14%
SE 4	79%	75%	55%	54%	58%	50%	10%	9%
SE 5	92%	89%	87%	87%	54%	42%	31%	28%
SE 6	68%	64%	33%	31%	57%	52%	9%	7%
SE 7	27%	18%	30%	38%	−37%	−56%	−32%	−18%
SE 8	89%	85%	77%	70%	71%	61%	38%	23%
SE 9	16%	23%	24%	16%	1%	9%	10%	1%
SE 10	80%	82%	80%	81%	11%	24%	−50%	19%
SE 11	47%	42%	57%	55%	−9%	−17%	12%	5%
SE 12	96%	96%	86%	90%	33%	43%	−12%	−60%

Larger total suspended and volatile solids influent concentrations for Storm Event 2 resulted in greater total suspended and volatile solids retention. Similarly, dissolved phosphorus retention efficiencies were greater for large influent concentrations and less for small influent concentrations. However, the trend between influent pollutant concentrations and retention efficiencies was not consistent for all twelve monitored storms at pond.

The filter underdrain system at the pond exhibited poor hydraulic performance and failed to keep the pond dry between the storm events. The runoff residence time in the pond for the twelve storm events monitored ranged from 2 days to 17 days, with an average of 5 days. The filter system requires continual maintenance to ensure that it is functioning properly. Field maintenance activities to maintain the hydraulic performance of the filter media may include replacing the filter media or replacing only a few inches from the top of the filter media.

8.7 Scheduling Maintenance

While not its primary function, monitoring can be used to schedule maintenance. Monitoring requires frequent visits to the site to check equipment and collect samples, resulting in impromptu visual inspections. During these visits, items needing maintenance can be observed and maintenance can be scheduled.

Scheduling maintenance from monitoring is similar to scheduling maintenance from synthetic runoff testing results. If a stormwater treatment practice is not performing as expected, review the monitoring data to determine if other aspects of the practice are functioning properly such as infiltration, filtration, stormwater storage, and sedimentation. Check the practice for the possibility of short-circuiting, excessive sediment deposition resulting in very little storage volume, and erosion within the practice.

Water Budget Measurement

9

Abstract

Performing a water budget on a stormwater treatment practice is often necessary when assessing a practice. A water budget measures or estimates all the flow rates entering and exiting the practice as a function of time and/or the total water volumes entering and exiting the practice through various pathways. If all pathways are measured accurately, the total volume of water entering the practice less the total volume of water leaving the practice should be equal to the change in water storage within the practice. If the water volume leaving the practice via a single pathway (e.g., infiltration) is not measured, a water budget can be used to estimate this volume. This chapter discusses possible modes of water flow into and out of stormwater practices that must be considered when performing a water budget. These modes include precipitation directly on the practice, infiltration into surrounding soil, evaporation and evapotranspiration, open channel flow, and full conduit flow. Techniques for measuring the flow rate of each mode are presented, discussed in detail (with examples, where beneficial), and compared. Recommendations are made regarding preferred measurement methods based on site conditions.

A water budget for a stormwater treatment practice is the accounting of water that enters, exits, and is stored by the stormwater treatment practice (9.1). The water budget assigns discharge values to each of the processes that affect the fate of water, including input processes (e.g., direct precipitation into the treatment practice, surface runoff, and conduit or open channel flow) and output processes (e.g., infiltration, evapotranspiration, and conduit or open channel flow). The inflows and outflows in the water budget should be determined as accurately as possible. If all but one of the terms can be determined, the value of the unknown inflow or outflow can be determined by solving the water budget. For example, if all the flows

A.J. Erickson et al., *Optimizing Stormwater Treatment Practices: A Handbook of Assessment and Maintenance*, DOI 10.1007/978-1-4614-4624-8_9,

into and out of an infiltration basin are known except the volume that has infiltrated, a water budget can be used to determine the volume of water that has infiltrated:

$$\begin{aligned} \Delta S &= \sum V_{\text{in}} - \sum V_{\text{out}} = S_2 - S_1 \\ &= \left(\sum_{i=1}^{N} Q_i \Delta t_i \right) + P(A_{\text{W}}) - \left(\sum_{k=1}^{Z} Q_k \Delta t_k \right) - V_{\text{ET}} - V_{\text{infiltration}} \end{aligned} \tag{9.1}$$

where

ΔS = change in water volume stored in the stormwater treatment practice
ΣV_{in} = sum of all water volumes that enter the stormwater treatment practice
ΣV_{out} = sum of all water volumes that exit the stormwater treatment practice
S_1 = volume of water stored in the stormwater treatment practice prior to runoff event
S_2 = volume of water stored in the stormwater treatment practice after runoff event
Q_i = influent flow rate data point
i = influent data point number
Δt_i = time duration between data point i and $i + 1$
P = depth of precipitation falling directly into the stormwater practice
A_{W} = surface area of the stormwater treatment practice
Q_k = effluent flow rate data point
k = effluent data point number
Δt_k = time duration between data point k and $k + 1$
V_{ET} = volume of water exported by evapotranspiration
$V_{\text{infiltration}}$ = volume of water exported by infiltration
N = number of influent data points
Z = number of effluent data points

It is important to note that this chapter does not examine all possible methods of water budget measurement. The intent is to discuss the most common methods and provide guidance for method selection. Those interested in other methods of discharge measurement or other water budget parameter estimation techniques should consult discharge measurement (e.g., Bos 1998; Herschy 1995), fluid mechanics (e.g., Franzini and Finnemore 1997), hydrology (e.g., Bedient and Huber 1992), or other similar texts.

9.1 Water Budgets

Water budgets require measurement of all water transport into and out of the stormwater treatment practice, including open channel flow, conduit flow, overland flow, direct precipitation, evaporation, transpiration, infiltration, ground water seepage, and any other influent sources and effluent transport processes. Figure 9.1 is an illustration of water budget processes on a typical stormwater treatment practice, and (9.1) can be used to calculate the mass balance of water from these processes.

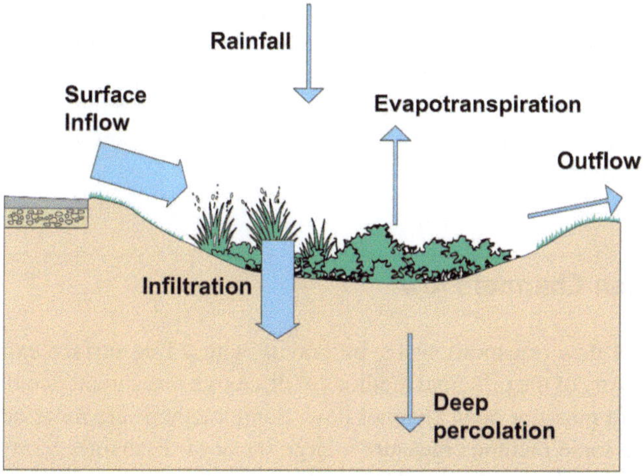

Fig. 9.1 Water budget processes for a typical stormwater treatment practice

Measuring all sources of water transport may not be practical or possible. Some stormwater treatment practices (e.g., infiltration basins) do not have a central effluent location that can be measured easily, and some transport processes (e.g., overland flow) are not easily measured or sampled. This chapter contains sections on water budget measurement techniques for open channel flow, conduit flow, infiltration, and rainfall. More information can be found in "Urban Stormwater BMP Performance Monitoring" (US EPA 2002).

This chapter will not discuss surface flow (overland flow), deep percolation, and evaporation and transpiration. Surface flow is by definition not channelized and therefore can be difficult to measure due to the shallow depth and wide extent of flow. Surface flow can, however, be channelized into a conduit or open channel, facilitating flow measurement, as discussed in the following sections. Also, for most stormwater treatment practices, surface flow from the immediate topography is a small component of the overall water budget compared to the contributing watershed area.

For stormwater treatment, it is often not important to distinguish deep percolation from infiltration because deep percolation is more important on an annual basis than on a storm event basis. The amount of water that comprises deep percolation to the aquifer is the portion of infiltration that does not evaporate or transpire to the plants. Evaporation and transpiration are typically not significant during storm events but can become important for an annual water budget. Evaporation and transpiration can be estimated using several documented techniques (Brutsaert 2005). By incorporating estimates of operation and transpiration into a water budget, one can distinguish deep percolation from infiltration by difference.

Assessment of stormwater treatment practices is significantly simpler and more accurate if the stormwater treatment practice is constructed or retrofitted to minimize modes of water transport into and out of the practice. For example, a detention pond with two or more inlet structures would require multiple discharge measurement and sampling stations. If, however, all inlets were combined into a central influent, only one discharge measurement and sampling station would be required. As a result, assessment costs would be significantly reduced and the process simplified.

9.2 Open Channel Flow

Open channel flow transports water by gravity with a free surface exposed to the atmosphere. Any of the principal methods of discharge measurement outlined below can be used to measure open channel flow. Some methods are more accurate than others, while some methods measure a large range of discharge. Stormwater flow rates are variable; thus, the method for measuring stormwater discharge must be able to measure both small and large values of discharge accurately. For reference, the depth–discharge (also called the stage–discharge) relationship for six discharge measurement techniques is shown in Fig. 9.2.

Methods for estimating discharge in open channels are different for steady and unsteady flow. For flow to be considered steady, all flow properties (velocity, depth, etc.) must remain constant with respect to time. For example, large open channel flow (e.g., large rivers) can be approximated as steady flow for time periods in which the flow changes are not significant. The principal methods of discharge

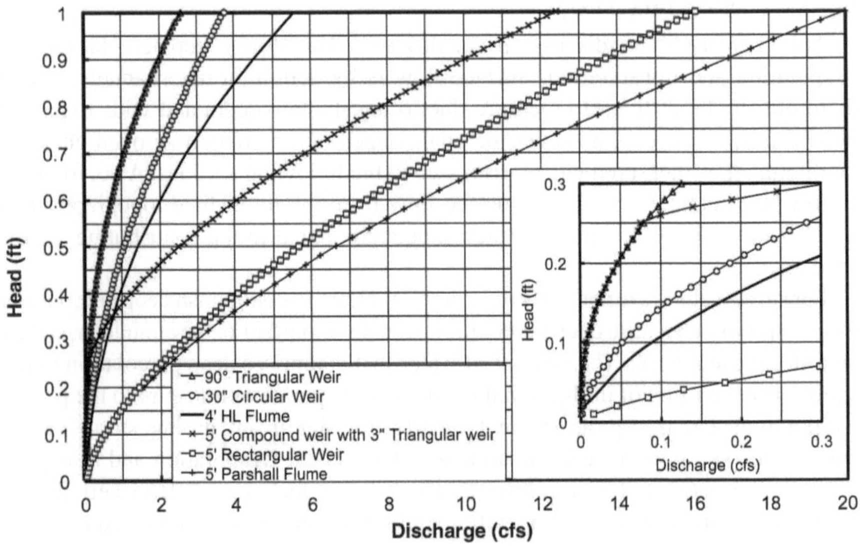

Fig. 9.2 Stage–discharge relationships for common discharge measurement devices (1 foot = 0.305 m; 1 cfs = 0.028 m^3/s)

measurement described below assume steady flow conditions, but in most natural systems, steady flow is only present for short time periods.

To analyze unsteady flow using steady flow concepts, flow data must be collected near-continuously over small time steps. If the time step is small, the flow can be considered steady for that time step and the total volume of flow can be estimated by multiplying the discharge (volume per time) by the time step duration (time) for each data point and summing the products for an entire event (or day, month, year, etc.).

The principal methods to estimate discharge for steady flow are as follows:

- Continuity (flow rate (Q) = Velocity (V) multiplied by area (A)): The average (or area-weighted) flow velocity can be multiplied by the cross-sectional flow area to estimate the discharge.
- Weirs (e.g., V-notch, rectangular, circular, compound): As fluid passes over a weir, it transitions through critical flow (Froude number = 1) to supercritical (Froude number > 1). Discharge at critical flow is solely dependent on the cross section. Discharge can be estimated accurately under critical flow conditions using the depth of water behind the weir and equations corresponding to the type of weir used.
- Flumes (e.g., Parshall, Palmer-Bowlus): Discharge measurement flumes produce a constriction in the flow, causing it to transition to critical flow (Froude number = 1). Similar to weirs, a measurement of the depth of critical flow and relevant flume dimensions can be used to estimate the discharge through the flume.
- Discharge measurement probes (e.g., area–velocity probes, sometimes called area–velocity meters or area–velocity sensors; current meters): Area–velocity (AV) meters use sonic waves to measure the discharge velocity throughout the flow cross section. The velocity values are multiplied by the corresponding cross-sectional area and summed to estimate the total discharge. To ensure accuracy, area–velocity meters require a minimum water depth over the probe as specified by the manufacturer. Most meters do not correctly integrate negative (i.e., flowing upstream) velocities that may occur as a result of turbulence in a backwater profile. Therefore, these meters can produce erroneous data during small discharge conditions and in situations with downstream obstructions in the flow (such as a weir or debris) that may cause negative velocities. Current meters measure velocity at a point in the flow that represents a portion (i.e., area) of the flow cross section. Discharge is then computed from continuity ($Q = \Sigma(V \times A)$) and related to a stage–discharge relationship (i.e., rating curve). For more information on current meters, refer to Chapter 10 in the "Water Measurement Manual" (US Bureau of Reclamation 2001).
- Backwater (water surface) profiles: Backwater profiles for gradually varied flow use discharge, channel geometry, conservation of energy, and estimates of friction losses (usually based on Manning's equation) to calculate the water surface elevation in the channel as a function of distance from a channel location of known depth. When used to estimate discharge, the water depth is measured at

some distance from a control (such as weir or free outfall) and other variables are either calculated or measured. Backwater profile calculations are iterative and are performed with a guessed value of discharge that is adjusted until the calculated depth at the known distance from the control matches the measured depth. For a more complete explanation with examples of backwater profile calculations, see an open channel flow text or manual (e.g., Sturm 2001).

- Manning's equation: Robert Manning developed Manning's equation (9.2) in the nineteenth century to estimate discharge for uniform open channel flow using cross-sectional area, hydraulic radius, energy grade–line slope, and an empirically defined roughness coefficient (n) (Sturm 2001). The potential measurement uncertainty in roughness coefficient, however, is large, and it is recommended that Manning's equation be used only as a last resort to estimate discharge in stormwater applications:

$$Q = \frac{K_n}{n} R^{2/3} S_f^{1/2} A \tag{9.2}$$

where

Q = discharge
K_n = unit conversion factor, $K_n = 1.0$ for SI units, $K_n = 1.49$ for English units
n = Manning's roughness coefficient
R = hydraulic radius, $R = A/P_w$
A = cross-sectional area
P_w = wetted perimeter
S_f = friction slope (i.e., energy grade line)

Two components of Manning's equation make it potentially inaccurate when estimating stormwater discharge. First, the slope of the channel bed (or pipe) is often assumed to approximate the energy grade line; second, the empirically defined roughness coefficient is often estimated from a table of values. For long channels of constant slope, one can often assume that the channel slope approximates the energy grade line, but short channels, transitions, and changes in the flow, which are common in stormwater systems, invalidate this assumption. Additionally, measurement uncertainty is large for short channels of shallow slopes because of human and instrument error. The empirically defined roughness coefficient often must be calibrated for a specific system and the potential measurement uncertainty is large. Again, Manning's equation should be used only as a last resort to estimate discharge in stormwater applications.

Selection of a discharge measurement method is dependent on many factors, including accuracy, cost, range of discharge, and site conditions. For further discussion of individual factors, see Chapter 4 in the "Water Measurement Manual" (US Bureau of Reclamation 2001).

All the discharge measurement principles previously discussed require a measurement of water depth and a known channel (or pipe, etc.) geometry to calculate

Table 9.1 Depth measurement device accuracy (Teledyne Isco Inc. 2006)

Depth measurement device	Range (ft)	Accuracy (ft)	Range (m)	Accuracy (m)
Ultrasonic sensor	<1.0[a]	0.02	<0.31	0.006
	1.0–10[a]	0.03	0.31–3.05	0.009
Pressure transducer	0.1–5.0	0.01	0.03–1.52	0.003
	0.1–7.0	0.03	0.03–2.13	0.009
	0.1–10.0	0.1	0.03–3.05	0.03
Bubbler	0.1–5.0	0.005	0.03–1.52	0.002
	0.1–7.0	0.01	0.03–2.13	0.003
	0.1–10.0	0.035	0.03–3.05	0.011
Area–velocity probe	0.05–5.0	0.01	0.015–1.52	0.003
	0.05–7.0	0.03	0.015–2.13	0.009
	0.05–10.0	0.1	0.015–3.05	0.03

[a]Range for ultrasonic sensor is the actual change in vertical distance between the sensor and the liquid surface. All other ranges are ranges in liquid depth

discharge. In the case of a weir, the water depth is measured behind the weir and weir equations (discussed in detail in the following section) convert depth to an estimated discharge over the weir. In the case of discharge measurement probes, a water depth is needed to determine the wetted perimeter. The principal methods of depth measurement use pressure under hydrostatic conditions and water density. Bubbler probes and pressure transducers, when located under the water surface, measure the pressure of water (i.e., hydrostatic pressure), which corresponds to a specific depth of water. Ultrasonic and Doppler probes, typically positioned above the water surface, locate the water surface using the change in density from air to water because the water surface reflects the acoustic signal back to the probe.

The accuracy of any depth measurement should be verified prior to installation of equipment and re-verified each time the site is visited to ensure that the equipment is calibrated correctly and in good working condition. A graduated ruler (i.e., staff gauge) affixed to a nonmoving structure (such as the weir or a post) can be used to verify the depth visually. If the depth measured by the staff gauge does not correspond with that of the depth measurement device (e.g., bubbler), verify that the staff gauge has not been disturbed and that the depth measurement device is working properly. Most manufacturers provide documentation that describes measurement range and accuracy for their respective depth measurement devices. For example, Isco (Teledyne Isco Inc. 2006) reports the measurement accuracy for the Isco 4200 discharge meters as shown in Table 9.1.

9.2.1 Weirs

V-notch weirs measure small discharge accurately (±1 to 2%, ASTM 2008) because small changes in discharge result in large changes in depth. Therefore, measurement uncertainty associated with the depth measurement has less effect on

Table 9.2 Effect of depth measurement uncertainty of ±0.02 ft (0.006 m) on accuracy of discharge estimation methods, expressed as a percent of discharge

Discharge (cfs) =	0.1 cfs (%)	1 cfs (%)	10 cfs (%)
90° V-notch weir	18%	7%	3%
5′ rectangular weir	>100%	74%	3%
5′ compound weir with 3″, 90° triangular weir	58%	21%	5%
30″ circular weir	61%	18%	6%

the estimated discharge than other weirs. For example, a measurement uncertainty of ±0.02 ft (0.006 m) in a 90° triangular weir with a discharge of 0.1, 1.0, and 10 cfs results in a discharge accuracy of ±18%, ±7%, and ±3%, respectively, as shown in Table 9.2. The discharge equation for triangular weirs is given in (9.3) (Franzini and Finnemore 1997). Examples 9.1 and 9.2 are provided to show how (9.3) is applied in two different situations. The discharge coefficient (C_d) as shown in (9.3) varies from 0.58 to 0.62, is dependent on θ and h, and may be determined graphically or experimentally. However, a value of 0.60 may be assumed with a measurement uncertainty of ±3%:

$$Q = \frac{8}{15} C_d \left[\tan\left(\frac{\theta}{2}\right) \right] \left(\sqrt{2g}\right) h^{5/2} \tag{9.3}$$

where

Q = discharge
C_d = discharge coefficient
θ = angle of the V-notch
g = gravitational acceleration
h = head above the vertex of the weir

Example 9.1: Flow rate calculations

Drew, a recent civil engineering graduate, is verifying the discharge values estimated by a computer program. He inputs a specific set of conditions into the program: 18-in. (0.46 m) conduit, 120° V-notch weir, water depth of 4 in. (0.10 m) that does not exceed the top of the V-notch. He then uses (9.3) to verify the discharge estimated by the program:

$$Q = \frac{8}{15} C_d \left[\tan\left(\frac{\theta}{2}\right) \right] \left(\sqrt{2g}\right) h^{5/2}$$

$$Q = \frac{8}{15} (0.6) \left[\tan\left(\frac{120°}{2}\right) \right] \left(\sqrt{2 \times 32.2 \frac{\text{ft}}{\text{s}^2}}\right) (0.33\ \text{ft})^{5/2}$$

$$Q = 0.32[1.73](8.02)(0.64) = 0.278\ \text{cfs} \left(0.008 \frac{\text{m}^3}{\text{s}}\right)$$

Example 9.2: Flow depth in a V-notch weir

Drew notices that the program can also calculate a water depth based on a specified discharge. He inputs a discharge of 1.5 cfs, specifies a 90° V-notch weir in the computer program, and uses (9.3) to verify the results:

$$Q = \frac{8}{15} C_\mathrm{d} \left[\tan\left(\frac{\theta}{2}\right)\right]\left(\sqrt{2g}\right) h^{5/2}$$

$$1.5\ \mathrm{cfs} = \frac{8}{15}(0.6)\left[\tan\left(\frac{90°}{2}\right)\right]\left(\sqrt{2 \times 32.2\frac{\mathrm{ft}}{\mathrm{s}^2}}\right)(h)^{5/2}$$

$$1.5 = 0.32[1.0](8.02)(h)^{5/2} = 2.57(h)^{5/2}$$

$$\frac{1.5}{2.57} = (h)^{5/2} \rightarrow h = (0.58)^{2/5} = 0.81\ \mathrm{ft}\ (0.247\ \mathrm{m})$$

90° V-notch weirs are limited, however, because a large discharge requires more depth as compared to other weirs and flumes for the same discharge. For example, a 90° V-notch weir requires 0.9 ft of depth to measure 2 cfs, whereas a 30-in. circular weir requires less than 0.3 ft of depth, a Parshall flume requires less than 0.25 ft of depth, and a 5-foot rectangular weir requires less than 0.05 ft of depth, as shown in Fig. 9.2. Rectangular weirs require less depth for the same discharge than all the other measurement devices shown in Fig. 9.2. Rectangular weirs do not accurately measure small discharge, however, because small changes in depth result in large changes in discharge. Therefore, measurement errors associated with the depth measurement have a significant effect on the discharge estimation, as shown in Table 9.2. As mentioned above, the optimal method for measuring stormwater discharge must be able to measure small discharge accurately while also having the capacity to measure large discharge.

A compound weir and circular weir (Addison 1941) both measure small discharge while also having the capacity to measure large discharge. As shown in Table 9.2 and Fig. 9.2, the compound weir composed of a 3-in. 90° V-notch section and a 5-foot rectangular section (see Fig. 9.3) measures small discharge as accurately as a 90° V-notch weir but also measures large discharge without large head requirements (e.g., 14 cfs with less than 1.0 ft of head over the weir).

A schematic of a V-notch compound weir is shown in Fig. 9.3. Using the water surface elevation and the weir dimensions, (9.4) can be used to estimate the discharge for a compound weir with a 3-in., 90° V-notch (Hussain et al. 2006), as performed in Example 9.3. A circular weir also measures both small and large

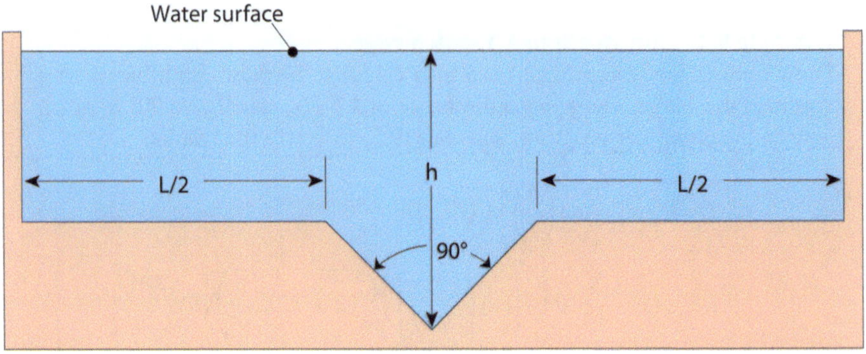

Fig. 9.3 Schematic of V-notch compound weir

discharge but is less accurate at large discharge than the other methods listed in Table 9.2:

$$Q = \frac{8}{15} C_{d1} \left(\sqrt{2g}\right) h_1^{5/2} - \frac{8}{15} C_{d2} \left(\sqrt{2g}\right) (h_1 - h_2)^{5/2} + \frac{2}{3} C_{d3} L \left(\sqrt{2g}\right) (h_1 - h_2)^{3/2} \quad (9.4)$$

where

Q = discharge (cfs)
C_{d1} = coefficient of discharge for the V-notch = 0.57
g = gravitational constant (32.2 ft/s^2)
h_1 = total head above the vertex of the V-notch (ft)
C_{d2} = coefficient of discharge for the overlapping portion of the V-notch and the rectangular weirs = 0.55
h_2 = depth of the V-notch portion (ft) = 0.25 ft
C_{d3} = coefficient of discharge for the rectangular weir = 0.64
L = combined length of the horizontal sections (ft)

Example 9.3: Compound weir flow calculation

Drew, the new civil engineer, is verifying the discharge values estimated by a computer program. He chooses a V-notch compound weir with 2-foot (0.61 m) horizontal sections on each side and a 3-in. (7.62 cm), 90° V-notch and specifies a water depth of 15 in. (0.381 m) above the vertex of the V-notch. He then verifies the results using (9.4) as follows:

$$Q = \frac{8}{15} C_{d1} \left(\sqrt{2g}\right) h_1^{5/2} - \frac{8}{15} C_{d2} \left(\sqrt{2g}\right) (h_1 - h_2)^{5/2} + \frac{2}{3} C_{d3} L \left(\sqrt{2g}\right) (h_1 - h_2)^{3/2}$$

$$Q = \frac{8}{15}(0.57)(8.02)(1.25\text{ ft})^{5/2} - \frac{8}{15}(0.55)(8.02)(1.25\text{–}0.25\text{ ft})^{5/2} + \frac{2}{3} \times (0.64)(2\text{ ft} + 2\text{ ft})(8.02)(1.25\text{ ft} - 0.25\text{ ft})^{3/2}$$

(continued)

Example 9.3: (continued)

$$Q = 2.44(1.25\ \text{ft})^{5/2} - 2.35(1\ \text{ft})^{5/2} + 13.7(1\ \text{ft})^{3/2}$$

$$Q = 4.26 - 2.35 + 13.7 = 15.6\ \text{cfs}\left(0.442\frac{\text{m}^3}{\text{s}}\right)$$

When using a weir to estimate discharge, it is very important to ensure that all flow enters by traveling over the weir and not around the weir or under the weir. It must also be noted that

- The weir (or some other barrier) should be extended into the ground (sometimes three or more feet) to minimize groundwater seepage under the weir.
- To ensure critical flow over the crest of the weir, it is important to maintain a "free outfall" over the weir. As long as the flow conditions downstream of the weir do not affect the flow over the weir, a free outfall is maintained.
- Weirs will back up the flow in the channel or conduit, which may alter the locations of flow entrance or exit for the stormwater treatment practice.
- The weir itself requires inspection and any necessary maintenance at least once a month to ensure that water does not leak or scour under the weir, that it is free of debris that may collect on the upstream side and disturb the water surface, and that it is in proper, working condition.

9.2.2 Flumes

A Parshall flume (Parshall 1936) may also be used to estimate discharge in open channels. Parshall flumes are rectangular sections that constrict flow to create critical flow through a specific section of the flume. The discharge may be estimated by measuring the water surface elevation just upstream of the critical section and converting it to discharge using a calibration curve, which is most often provided by the manufacturer. Parshall flumes are readily available in widths from 2 to 120 in. (5.08–304.8 cm).

H, HS, and HL flumes (Gwinn and Parsons 1976) combine the sediment movement capabilities of a flume with the accuracy of a weir. The cross section of H flumes, which is initially rectangular, converges at the downstream end with the top side walls sloped downward.

There are three types of H flumes, categorized by size: the smallest size is the HS flume, the intermediate is the H flume, and the largest is the HL flume. Many manufacturers sell preconstructed H flumes with rating curves that provide the relationship between water level and discharge with ranges from 0.085 cfs for HS flumes to 117 cfs for HL flumes (0.002–3.313 m^3/s, respectively).

Compared to weirs, flumes are different in the following ways:

1. Flumes do not create a pool upstream of the flume
2. Flumes are less prone to collecting debris

3. Flumes obstruct the movement of sediment less than weirs
4. Flumes require more space and effort to install
5. There is, in general, a smaller measurement range of discharge when using a flume as compared to a weir in the equivalent space

9.2.3 Recommendations for Open Channel Flow

Open channel flow in stormwater applications is most often unsteady, and discharge magnitude is often varied. Compound weirs, as shown in Fig. 9.2, provide a combination of accurate small discharge estimation and capacity to measure large discharge; therefore, it is recommended that compound weirs be used whenever possible. In channels with a large sediment load, weirs may create excessive deposition that will eventually affect the accuracy of the weir. In such cases, a properly sized H-flume (open channel flow) or Parshall flume may be used to measure open channel flow.

9.3 Flow in Conduits

Conduits can transport two types of flow: pressurized conduit flow and open channel conduit flow. Pressurized conduit flow is defined as the transport of water in closed conduits (e.g., pipes) that are flowing full. Flow occurs because there is a longitudinal pressure difference along the conduit. Open channel conduit flow is the transport of water by gravity with a free surface open to atmospheric pressure in which the channel determines the size, shape, and slope of the conduit.

9.3.1 Closed Conduit Flow

Stormwater conduits are designed for a specific capacity (i.e., maximum discharge) that depends on the upstream conditions and downstream controls. Conduits flowing full operate at, or near, that capacity. Measuring discharge in a full-flowing conduit with a weir or flume is not recommended because weirs and flumes reduce the capacity of the conduit and the relationship between discharge and the water surface elevation is not well established without a critical depth. Area–velocity probes, however, can measure discharge without causing significant obstruction in conditions that provide adequate depth over the probe. The following are advantages and disadvantages to using probes to measure full-flowing conduit discharge in lieu of weirs or flumes:

Advantages:
- Probes create less flow obstruction than weirs or flumes
- Probes can accurately measure depth or discharge in full-flowing conditions
- Probes are usually easier to install

Disadvantages:

- Probes cannot accurately measure small discharge associated with small storm events or at least a portion of the rising and falling limbs of hydrographs
- Probes sometimes require calibration, which may be difficult for certain site conditions
- Probes require additional cost and maintenance

A discharge measurement probe is usually attached to a flexible metal ring, which, when compressed, can be slid into the conduit to the desired location. When the compression is released, the ring expands against the inside of the conduit where friction holds the ring and probe in place. The probe is connected to a data logger that records such information as depth and velocity, which is then converted to a cross-sectional area of discharge, and then to a measurement of discharge. Additional equipment, such as tipping bucket rain gauges, can be connected to the data logger, as well.

Area–velocity probes must be located at the bottom of the conduit and oriented so they face the oncoming discharge directly. They also require a minimum water depth (usually 1–2 in., ~2.5–5 cm) in order to obtain accurate measurements. Stormwater pipe systems can have supercritical flow that produces large discharge values with minimal water depth. Significant errors can occur when using probes to measure these discharge conditions when the depth does not exceed the minimum suggested by the manufacturer.

Two common brands of area–velocity probes are Isco and Campbell Scientific. Campbell Scientific produces a velocity sensor that must be combined with a depth measurement and area conversion computation to estimate discharge. Campbell Scientific equipment is capable of connecting with equipment from other manufacturers but requires computer code written by the user to communicate with the equipment. Isco equipment does not require code, but it cannot be used in combination with equipment from other manufacturers. Most discharge measurement probes require connection to a data logger to record measurements with respect to time.

Area–velocity probes should be used only when an insignificant portion of the runoff event will occur at depths below those required for accurate measurements. Otherwise, a large portion of the total runoff volume may not be measured accurately.

9.3.2 Partially Full Conduits

Conduits flowing partially full are a specific instance of open channel flow in which the channel is simply the size, shape, and slope of the conduit. Therefore, a V-notch compound weir (Fig. 9.3), a circular weir (Addison 1941, Fig. 9.4), or a V-notch weir (Fig. 9.5) may be used to measure the discharge. To ensure accurate discharge measurements in a conduit with a weir, the weirs and probes must receive regular inspection and maintenance to remain free of sediment and debris that may accumulate behind the probe or weir.

All weirs should be constructed so that the bottom of the weir fits the contour of the conduit and can be sealed with a waterproof sealer such as polyurethane. If a

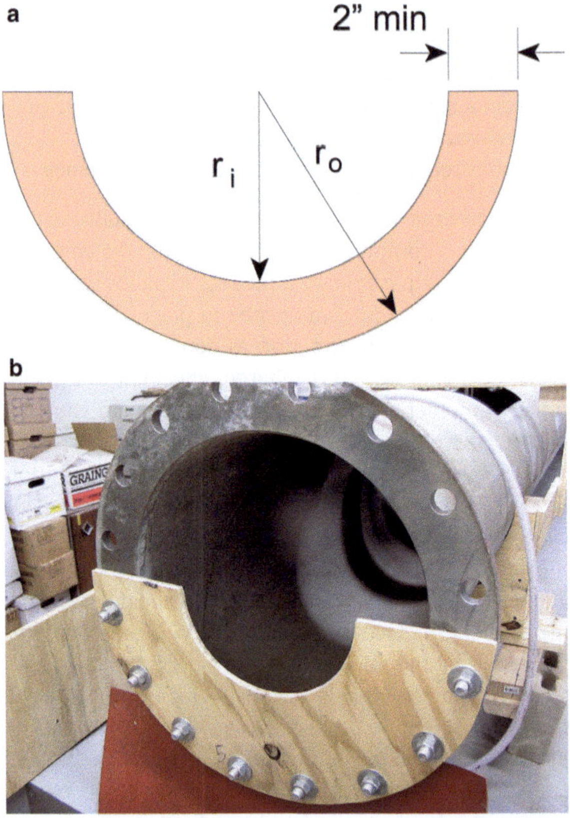

Fig. 9.4 Circular weir schematic (*top*) and test setup photo (*bottom*)

circular weir is to be used in a noncircular conduit, it is important that the crest of the weir remains circular (unless a calibration curve is determined for a specialized weir). For a circular weir, depth can be converted to an estimated discharge using (9.5). Example 9.4 demonstrates the use of (9.5) for discharge estimation using a circular weir:

$$Q = C_\mathrm{d}\left[10.12\left(\frac{h}{d}\right)^{1.975} - 2.66\left(\frac{h}{d}\right)^{3.78}\right](d)^{5/2} \tag{9.5}$$

where

Q = discharge (L/s)
d = diameter of circular orifice (dm)
h = height over the weir (dm)
C_d = coefficient of discharge as given by:

$$C_\mathrm{d} = 0.555 + \frac{1}{110(h/d)} + 0.041\left(\frac{h}{d}\right) \tag{9.5a}$$

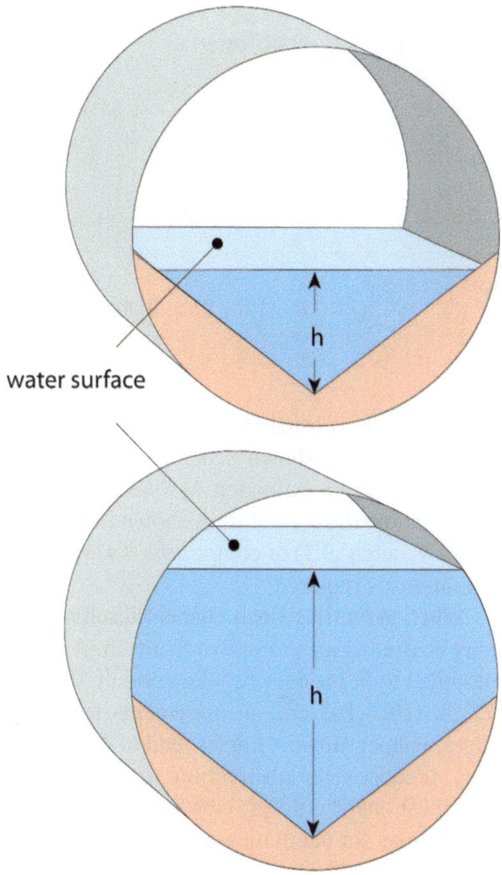

Fig. 9.5 Schematic of V-notch weir inside a circular conduit in normal flow conditions (*top*) and overflow conditions (*bottom*)

Example 9.4: Discharge over a circular weir

Drew, the new engineer, is verifying the discharge values estimated by a computer program. His final test of the computer program's accuracy is for a circular weir. He chooses a 14-in. (0.36 m) inside diameter circular weir placed in an 18-in. (0.46 m) pipe and a 2-in. (0.051 m) water depth over the weir. He then verifies the results of the computer program by calculating the discharge using (9.5):

$$C_{\mathrm{d}} = 0.555 + \frac{1}{110(0.51\ \mathrm{dm}/3.56\ \mathrm{m})} + 0.041\left(\frac{0.51\ \mathrm{dm}}{3.56\ \mathrm{dm}}\right)$$

(continued)

Example 9.4: (continued)

$$C_{\mathrm{d}} = 0.555 + \frac{1}{15.71} + 0.0059 = 0.624$$

$$Q = 0.624\left[10.12\left(\frac{0.51\ \mathrm{dm}}{3.56\ \mathrm{dm}}\right)^{1.975} - 2.66\left(\frac{0.51\ \mathrm{dm}}{3.56\ \mathrm{dm}}\right)^{3.78}\right](3.56)^{5/2}$$

$$Q = 0.624[0.217 - 0.0017](23.85) = 3.2\ \frac{L}{s}\ (0.113\ \mathrm{cfs})$$

A V-notch weir may be used as an alternative to a circular weir for partially full conduit flow, as shown schematically in Fig. 9.5. For normal flow conditions, discharge can be estimated by (9.3), which applies to any V-notch weir section. Note that overflow conditions shown in Fig. 9.5 will not be estimated accurately by either V-notch (9.3) or compound (9.4) weir equations. A calibration to the specific conditions is required.

When measuring open channel discharge in a conduit, a Palmer-Bowlus flume may be used as an alternative to a weir. A Palmer-Bowlus flume is a Parshall flume modified to fit inside a circular conduit. Commonly available sizes range from 4 to 72 in. (10.2–183 cm) in increments similar to those of commercially available pipes. Palmer-Bowlus flumes tend to collect less debris compared to weirs because they produce less obstruction to the flow through the conduit. Manufacturer specifications should provide calibration or rating curves along with installation instructions for depth measurement equipment.

9.3.2.1 Recommendations for Partially Full Conduits

For conduits with small discharge (i.e., not sufficient to provide adequate depth over a probe), it is recommended that a weir be used for discharge measurement. Circular weirs provide a good combination of small discharge accuracy and large discharge capacity and are recommended for open channel conduit flow when there is adequate capacity to pass the design flow. The combination of a circular weir and a pressure sensor has proven to be effective for conduit flow over a wide range of discharges. The pressure measurement can be used to indicate depth over the weir. Pressure sensors in combination with circular weirs in large discharge conditions (nearly full-flowing conduit) may require an individual calibration.

If a pressure sensor is to be used in conjunction with a weir, it is important to remember that a minimum depth above the probe is typically required to obtain accurate measurements. It is recommended that the weir height be set to achieve this minimum depth or more.

9.4 Infiltration

Various stormwater treatment practices use infiltration as a primary or supportive process for stormwater treatment. When developing a water budget for a specific practice, it is important to consider infiltration and determine whether infiltration will represent a significant fraction of the total water outflow. For example, a dry pond may use sedimentation as the primary treatment process, but if the structure does not have an impermeable liner, it may also infiltrate a significant portion of the stormwater entering the pond. Neglecting infiltration may result in significant errors when performing a water budget analysis.

Infiltrometers and permeameters can be used to measure soil infiltration properties, such as saturated hydraulic conductivity (K_s), at a specific location within a stormwater treatment practice as part of capacity testing (Chap. 6). To obtain estimates of the overall effective K_s of a practice, synthetic runoff testing (Chap. 7) or monitoring (Chap. 8) may be used.

9.4.1 Infiltration Measurement Devices

Capacity testing uses a series of point measurements to determine the filtration or infiltration capacity on the media surface at various locations within a treatment practice. Infiltration measurements can be divided into two groups: constant head and falling head. Constant head devices measure infiltration rate until it approaches steady state, at which time it is assumed that the ground is saturated, and infiltration rate is equal to $K_{s.}$ Falling head devices measure head level in the device at various times as it falls. The Green-Ampt assumptions are applied (Klute 1986), and soil moisture is measured or estimated to estimate $K_{s.}$ The advantage of the falling head technique is that it requires less water and is quicker. Falling head devices may not, however, detect the presence of confining layers below the depth in which their fixed, typically small, volume of water infiltrates. The advantage of the constant head technique is that one does not need to measure soil moisture. While constant head devices use more water volume than falling head devices, determining the effect of confining layers in the soil may be challenging because these measurements were intended for homogeneous soils.

Various devices are available for simple estimations of infiltration at a specific location within a stormwater treatment practice, including the single ring infiltrometer, double ring infiltrometer, Philip-Dunne permeameter, Guelph permeameter, and tension infiltrometer. Many infiltration measurement devices also require soil moisture to be measured, procedures for which are discussed in Chap. 11. Infiltration can also be estimated by numerical methods such as the Horton and Green-Ampt models. Some of the available devices for measuring saturated hydraulic conductivity (K_s) are discussed in the following sections.

Fig. 9.6 Photograph of a double ring infiltrometer (St. Paul, MN)

9.4.1.1 Double Ring Infiltrometer

A double ring infiltrometer (Fig. 9.6) is made of two concentric tubes, typically of thin metal or hard plastic, that are both continuously filled with water such that a constant water level is maintained as water infiltrates into the soil (ASTM 2005). The rate at which water is added to the center tube is measured to determine the infiltration rate. The accuracy of this device is only moderate relative to air-entry and borehole permeameter methods (ASTM 2010a). One limitation with the calculations used to determine K_s with the double ring infiltrometer is that the infiltrating water tends to flow outward due to the soil's resistance to flow, and this is not considered in the calculations. While the flow through the outer ring is intended to promote one-dimensional (i.e., vertical) flow from the inner ring into the media, some lateral flow still occurs. With the assumed calculation method, the measured K_s will tend to be biased toward large values (i.e., overestimated). It is also difficult to determine when steady state has been sufficiently approached, which will also tend to bias K_s toward large values (Fig. 9.7).

A typical plot of the infiltration rate vs. time for a double ring infiltrometer is shown in Fig. 9.7. After a certain period of time, the infiltration rate will approach a constant value (i.e., steady state). The rate of infiltration as it approaches steady state is assumed to be equivalent to the saturated hydraulic conductivity (K_s). Steady state conditions are dependent on the soil type and initial moisture content and typically will require between 20 min and 4 h.

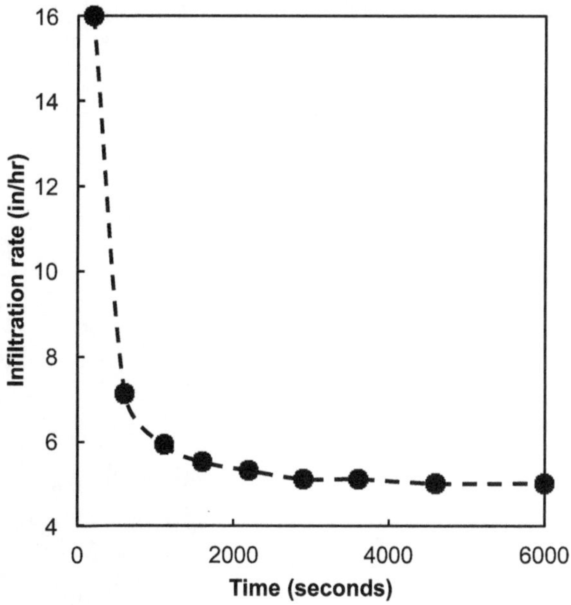

Fig. 9.7 Infiltration rates obtained from a double ring infiltrometer experiment (1 in./h = 7.06×10^{-4} cm/s)

9.4.1.2 Single Ring Infiltrometer

The single ring infiltrometer is similar to the double ring infiltrometer, except with only one ring. The simpler construction allows it to be used as either a constant head (similar to the double ring infiltrometer) or a falling head infiltrometer (see Modified Phillip-Dunne infiltrometer). A standard design is a 30-cm (11.81 in.) diameter, 20-cm (7.87 in.) tall ring, which is driven approximately 5 cm (1.97 in.) into the soil and filled with water (Klute 1986). By measuring the flow of water to the ring as a function of time (constant head method), the saturated hydraulic conductivity (K_s) is determined (Klute 1986). The soil surrounding the ring may also be flooded to encourage vertical flow of water into the soil. Similar to the double ring infiltrometer, lateral flow will tend to bias estimates of K_s toward large values (i.e., overestimated) and it is difficult to determine when steady state has been sufficiently approached, which will also tend to bias K_s toward large values.

9.4.1.3 Guelph Permeameter and Tension Infiltrometer

The Guelph permeameter (GP) is another tool for measuring soil-water properties (Bagarello et al. 2004). The GP estimates field-saturated hydraulic conductivity (K_s), matric flux potential, and soil sorptivity based on constant head calculations (SoilMoisture Equipment Corp. 1986) and is available for purchase from SoilMoisture Equipment Corp., Santa Barbara, CA, http://www.soilmoisture.com.

The Guelph permeameter is a constant head well permeameter consisting of a mariotte reservoir (McCarthy 1934) that maintains a constant water level inside an augured hole that is typically 4 in. (10.16 cm) deep, cored into the unsaturated soil. This permeameter requires steady discharge from two different water levels (heads)

Fig. 9.8 Illustration of the Guelph tension infiltrometer

in the augured hole. Steady state discharges are measured at two different water pressure heads. Generally, the water pressure heads are a 2-in. (5.08 cm) and a 4-in. (10.16 cm) head, as recommended by the manufacturer. The measured discharges are used with the change in volumetric water content ($\Delta\theta$) to determine field-saturated hydraulic conductivity (K_s), matric flux potential (ϕ_m), and sorptivity (S) (SoilMoisture Equipment Corp. 1986).

A limitation of the Guelph permeameter is that it is applied to a borehole and not to the soil surface. Therefore, the effect of the top layer of the stormwater treatment practice surface is not reflected in the results. An alternative test to the conventional Guelph permeameter is a tension infiltrometer. For unsaturated soil conditions, a tension infiltrometer can be added to the Guelph permeameter.

The tension infiltrometer consists of a 4-in. (10.16 cm) or 8-in. (20.21 cm) diameter porous disc connected to a Mariotte bottle. An illustration of the tension infiltrometer is presented in Fig. 9.8. The procedures for using the method are described by Reynolds and Elrick (1991). In applying the method, the porous disc is placed in contact with the soil surface. This usually requires that vegetation and debris be removed from the surface and that the surface be flat. In many cases, it is also desirable to place a thin layer of fine sand onto the soil surface to provide good contact between the disc and the soil.

Once the disc is in place on the soil surface, the steady state discharge for infiltration into the soil is measured for two applied water pressures. The tension infiltrometer can facilitate the measurement of unsaturated hydraulic conductivity for various applied tensions, but only the saturated hydraulic conductivity (K_s) value is typically desired for stormwater treatment practices. To estimate this value, the pressures need to be slightly negative (i.e., tension) and it is recommended that successive pressures of −5 cm (−1.97 in.) and −1 cm (−0.394 in.) be used. At each of these pressures the corresponding steady state discharge is measured. The steady state discharge and change in volumetric moisture content ($\Delta\theta$) are used in equations derived by Reynolds and Elrick (1991) to find the desired soil properties.

9.4.1.4 Mini Disk Tension Infiltrometer

The Mini disk Tension infiltrometer (available for purchase through Decagon Devices, Pullman, WA, http://www.decagon.com/) is a small, falling head tension infiltrometer that uses a method developed by Zhang (1997) to determine the saturated hydraulic conductivity (K_s) and sorptivity of a soil. The Mini disk Tension infiltrometer has a base diameter of 4.5 cm and an infiltration volume around 90 mL. The sintered steel disc is placed directly on the smooth surface of the soil and the volume within the device is recorded at a regular time interval until the water reservoir is empty. Relationships to determine K_s are given in Zhang (1997) and typically in the manufacturer's literature.

9.4.1.5 Philip-Dunne Permeameter

The Philip-Dunne permeameter (Munoz-Carpena et al. 2002) is a single ring, falling head device that estimates saturated hydraulic conductivity (K_s). It is made of a plastic or metal tube that is inserted between 5 and 15 cm (1.97–5.91 in.) into the ground. In the standard Philip-Dunne permeameter procedure, a tube is inserted into the bottom of an auger hole of the same radius. The initial moisture content of the soil is measured, the tube is filled with water, and the observer measures the time required for the water level in the tube to reach the halfway mark on the tube as well as the time required for the tube to empty completely. After the experiment, the final water content is measured. Generally, the porosity of the soil can be used as the final water content because the soil should be saturated in the vicinity of the auger hole. The radius of the tube, the two measured times, and the measured initial and final water contents are used to estimate the hydraulic properties of the soil. The equations for performing this analysis are given in Philip (1993).

9.4.1.6 Modified Philip-Dunne Infiltrometer

The Modified Philip-Dunne (MPD) infiltrometer (available for purchase through St. Anthony Falls Laboratory, Minneapolis, MN, http://www.safl.umn.edu/) is a single ring, falling head device that measures the saturated hydraulic conductivity (K_s) of surface soil. It is a falling head device and suitable for infiltration practices because measurements can be performed relatively quickly, which allows for accurate representation of the large spatial variability in infiltration rates that commonly occurs within stormwater treatment practices. It is suitable for assessment of the practice to determine required maintenance because accumulation of fine particles on the soil surface is typically the limiting factor affecting infiltration rates in these

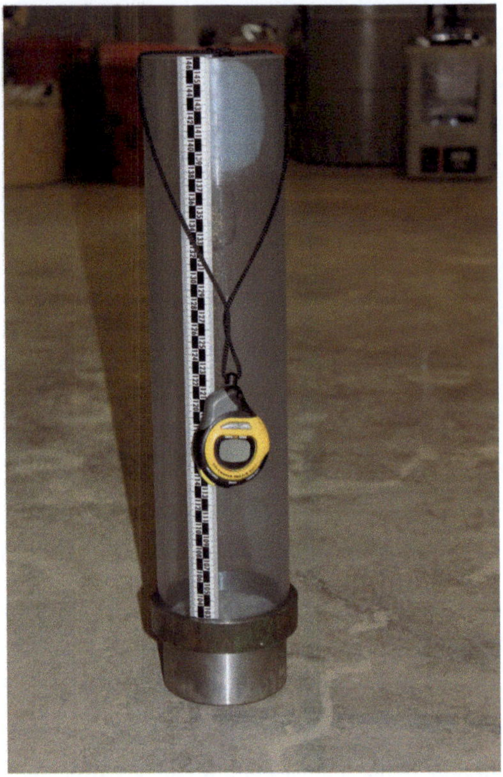

Fig. 9.9 Photo of a Modified Philip-Dunne (MPD) infiltrometer

practices. The MPD infiltrometer, shown in Fig. 9.9, is an open ended 50 cm long, clear plastic cylinder with walls 2 mm thick, and a 10 cm inner diameter inserted into a machined metal base. The bottom edge of the metal base is beveled from the outside to ease the process of inserting the device 5 cm into the soil surface. A metric measuring tape is adhered to the outside wall of the tube so the water level inside the tube can be quickly recorded.

Using a spreadsheet program (Ahmed and Gulliver 2011) and the initial and final moisture content of the soil, the saturated hydraulic conductivity (K_s) of the soil can be determined. Because K_s values typically have a large variability (Warrick and Nielsen 1980) it is useful to collect many samples to understand how K_s varies throughout the practice. Typically multiple MPD infiltrometers are used at up to 24 locations at a time, allowing for up to 60 measurements per day. As shown in Fig. 9.10, the theory used to calculate K_s assumes a capped spherical shape of the wetting front.

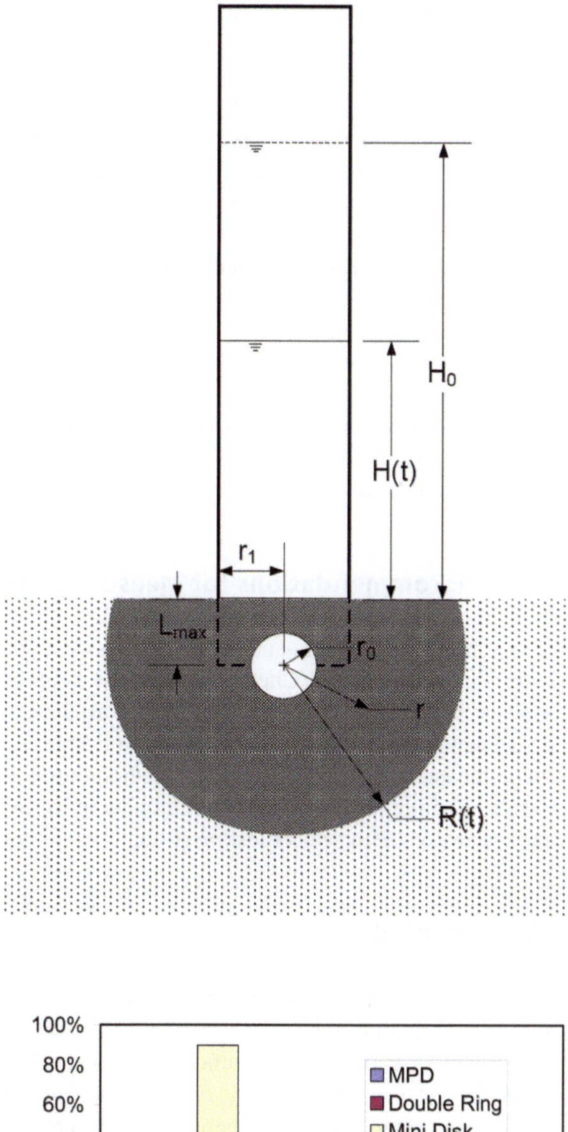

Fig. 9.10 Schematic of flow beneath a Modified Philip-Dunne (MPD) infiltrometer

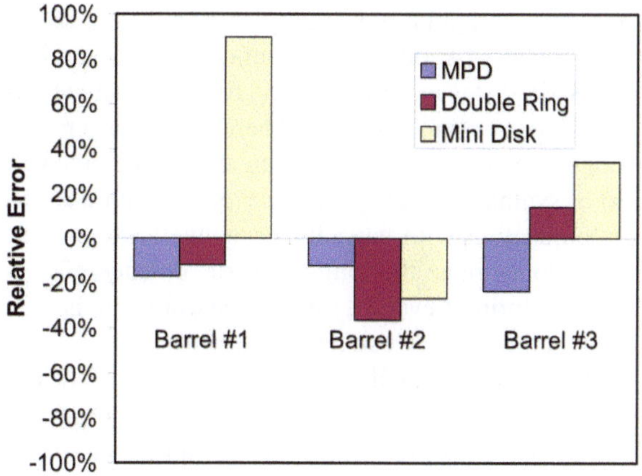

Fig. 9.11 Relative error of saturated hydraulic conductivity (K_s) for devices compared to the reference falling head tests in each barrel (Nestingen 2007)

9.4.2 Comparison of Field Infiltrometers

Nestingen (2007) compared the performance of the double ring infiltrometer, Mini disk Tension infiltrometer, and the Modified Philip-Dunne (MPD) infiltrometer with a falling head test on the entire media in controlled laboratory studies of three particle size distributions that mimicked field conditions yet minimized the potential differences in setup due to compaction and layering of the soil. A 55-gal (208 L) barrel with a diameter of 1.8 ft (56 cm) and height of 3 ft (91 cm) was used for the comparison tests. The falling head tests on the barrels indicated saturated hydraulic conductivities of 0.035 cm/s, 0.0078 cm/s, and 0.0057 cm/s for barrels 1, 2, and 3, respectively.

The Minidisk was not sufficiently accurate in the large saturated hydraulic conductivity (K_s) media of barrel No. 1, but otherwise all three infiltrometers performed acceptably ($\pm 40\%$ error), considering the orders of magnitude spatial variation in most field soils (see Fig. 9.11). The double ring and MPD infiltrometers were judged to be of a roughly equivalent accuracy.

9.4.3 Recommendations for Measuring Infiltration

The various infiltrometers and permeameters described previously may be used to perform capacity testing (Chap. 6) to estimate saturated hydraulic conductivity (K_s) values at specific locations throughout a stormwater treatment practice. Table 9.3 lists advantages and disadvantages for all of the previously mentioned infiltrometers and permeameters. The most accurate methods currently available for measuring the overall effective saturated hydraulic conductivity (K_s) in a stormwater treatment practice are synthetic runoff testing (Chap. 7) and monitoring (Chap. 8).

9.5 Rainfall

Rainfall measurement is an important aspect of any stormwater treatment practice assessment program, especially for monitoring. Rainfall data are collected in many locations (e.g., airports), but rainfall amounts and intensities can vary significantly over short distances, especially in regions where atmospheric convection cells (e.g., thunderstorms) are common. Therefore, to ensure an accurate and complete water budget, on-site rainfall measurement is recommended. For capacity testing (level 2a) or synthetic runoff testing (level 2b), rainfall measurement results can be used in conjunction with capacity or synthetic runoff testing results and a watershed model to estimate the long-term performance of a stormwater treatment practice. For monitoring (level 3), rainfall measurement is required for an accurate estimate of the rainfall water budget component.

Rainfall data provide an accurate account of the amount of rain that falls directly on the stormwater treatment practice and its drainage area. Rain falling on a stormwater treatment practice is not measured by the influent discharge

Table 9.3 Advantages and disadvantages for infiltrometers and permeameters

Device	Type of measurement	Advantages	Disadvantages
Single ring	Constant head, variable flow rate	Simple, small, measurement of the soil surface	May be significantly biased by lateral flow in soils, potentially large volume of water and significant length of time required for each measurement to reach steady state, requires measurement of varying flow rate to maintain constant water level in ring
Double ring	Constant head, variable flow rate	Common technique, measurement of the soil surface	May be significantly biased by lateral flow in soils, potentially large volume of water and significant length of time required for each measurement to reach steady state, requires measurement of varying flow rate to maintain constant water level in both rings
Guelph permeameter	Constant head, variable flow rate	Built for accuracy in research, Well-documented procedures	Measurement in a borehole and not to the soil surface, potentially large volume of water and significant length of time required for each measurement to reach steady state, requires measurement of varying flow rate to maintain two different, yet constant, water levels
Guelph permeameter with tension infiltrometer	Constant head, variable flow rate	Measurement of the soil surface, can measure unsaturated hydraulic conductivity	(Potentially) large volume of water and significant length of time required for each measurement to reach steady state, requires measurement of varying flow rate to maintain two different, yet constant, water levels
Mini Disk Tension Infiltrometer	Constant head	Small, measurement of the soil surface	May be difficult to achieve adequate contact with soil surface
Philip-Dunne	Falling head	Small, measurement of the soil surface, constant volume of water for each test, possible to perform many measurements per day	May be biased by lateral flow in soils, need to measure or estimate soil moisture before and after the tests
Modified Philip-Dunne	Falling head	Small, measurement of the soil surface, constant volume of water for each test, possible to perform many measurements per day, lateral flow is incorporated into K_s estimates, most accurate representation of lateral flow in the soil	Need to measure or estimate soil moisture before and after the tests

measurement device (e.g., weir, probe) but may constitute a significant portion of water entering the stormwater treatment practice, depending on watershed and stormwater treatment practice characteristics.

Several tools are available for rainfall measurement, ranging from the simple depth measurement rain gauge to the more advanced tipping bucket rain gauges that record depth and intensity with a data logger. Depth rain gauges require prompt recording of the rainfall depth to avoid any loss due to evaporation or spillage, and tipping bucket rain gauges may require calibration. All rainfall measurement equipment should be installed according to the manufacturer's instructions and maintained regularly to ensure accurate measurements.

Compared to depth rain gauges, tipping bucket rain gauges are a more accurate measurement of incremental rainfall because measurements are recorded near continuously with a data logger. Prompt inspection and recording of depth measurements can make depth rain gauges an accurate method of total rainfall measurement, as long as the depth rain gauge is readily accessible.

9.5.1 Recommendations for Measuring Rainfall

It is recommended that rainfall be measured at each location in which monitoring (level 3) is conducted. For small drainage areas, a single rain gauge is sufficient, but larger watersheds will require multiple gauges. Manufacturer recommendations or hydrologic texts (e.g., Bedient and Huber 1992) and manuals can provide additional guidance on spacing and placement of rainfall gauges.

Accurate rainfall measurement can be achieved in several ways. Depth rain gauges can measure total rainfall for individual storm events inexpensively but require prompt inspection and measurement recording to be accurate. To ensure timely measurements of rainfall depth and intensity, it is recommended that a tipping bucket rain gauge with a data logger be used.

Water Sampling Methods 10

Abstract

In order to determine influent and effluent contaminant loads or concentrations and treatment practice performance, assessment efforts often include stormwater runoff sampling. Depending on the water quality parameter of interest, sampling can be done in situ, by grab samples, or by automatic sampling devices. Samples can also be collected on a time-weighted basis (equal time between samples) or on a flow-weighted basis (equal volume of flow passing the sampling site between samples). This chapter discusses available sampling methods and when and how to implement a particular method, and discusses the number of sampled events required to achieve a desired confidence interval. Also included is a discussion of sample storage and handling is also included.

The effectiveness of a stormwater treatment practice at capturing a pollutant or pollutants can be assessed by comparing the amount of pollutant that enters the stormwater treatment practice to either the amount of pollutant that exits the stormwater treatment practice or to the amount that is retained. Pollutant quantities are measured in mass or concentration of pollutant, and these measurements can be collected using one of four methods. First, pollutants can be measured and recorded in situ, or in place, using pollutant sensors or probes placed directly in the stormwater runoff to collect near-continuous measurements with respect to time (in situ sampling). The measurements are later downloaded to a computer by an individual on site or via cell phone connection. Second, stormwater samples can be collected manually and analyzed on site with sensors, probes, or by other analytical methods (on-site sampling). Third, a sample can be collected manually in the field and transported back to a laboratory for analysis ("grab" sampling). Fourth, stormwater runoff can be collected with an automatic sampler, retrieved at a later time, and analyzed in a laboratory (automatic sampling). Some advantages and disadvantages of each method are given in Table 10.1. For more information on sampling methods, consult Standard Methods (APHA 1998b), "Urban Stormwater

A.J. Erickson et al., *Optimizing Stormwater Treatment Practices: A Handbook of Assessment and Maintenance*, DOI 10.1007/978-1-4614-4624-8_10,

Table 10.1 Comparison of in situ, on-site, grab, and automatic sampling methods

	Sampling approach			
Characteristic	In situ	On-site	Grab	Automatic
Sample of stormwater collected	No	Yes	Yes	Yes
Personnel required to collect sample	No	Yes	Yes	No
Sample transported	No	No	Yes	Yes
Relatively large setup costs	Yes	No	No	Yes
Possibility of equipment damage or theft	Yes	No	No	Yes
Parameters or pollutants that can be measured				
Suspended solids	No	No	Yes	Yes
Pathogens (i.e., coliforms)	No	No	Yes	No
Nutrients				
Phosphate	Yes	Yes	Yes	Yes
Nitrate	Yes	Yes	Yes	Yes
Ammonia	Yes	Yes	Yes	Yes
Specific organic chemicals[a]	No	No	Yes	Yes
Oxygen demand	No	No	Yes	No
Heavy Metals	No	No	Yes	Yes
Water quality indicators				
Dissolved oxygen	Yes	Yes	No	No
Temperature	Yes	Yes	No	No
pH	Yes	Yes	No	No
Conductivity	Yes	Yes	Yes	Yes
Turbidity (a surrogate for suspended solids)	Yes	Yes	Yes	Yes
Organic carbon	No	No	Yes	Yes

[a]Examples include petroleum hydrocarbons (e.g., benzene), pesticides, chlorinated solvents

BMP Performance Monitoring" (US EPA 2002), or "Wastewater sampling for process and quality control (Manual of practice)" (WEF 1996).

One advantage of in situ sampling is that data can be collected frequently, in small time steps, with the results available remotely (e.g., cellular phone connection) once the sampling equipment is installed. Another advantage of in situ sampling is that it can be used to measure some of the water quality parameters that are likely to change during sample storage or transport, such as pH and dissolved oxygen. Although personnel are not required to collect samples or perform the chemical analyses, someone must periodically (e.g., weekly) visit the site to maintain and recalibrate the equipment. Unfortunately, a different probe or sensor is required for each pollutant being measured, and not all pollutants can be measured with sensors or probes. For example, nutrient measurement technology is currently cumbersome, with potential improvements that are currently research topics (Arai et al. 2009). There are available, however, in situ bundles that include several common probes and sensors used in water quality assessment.

On-site sampling can also be used to measure water quality parameters that are likely to change during transport. Setup costs for on-site sampling are minimal because it does not require that any equipment remain in the field. Nevertheless, on-site sampling requires individuals to collect samples and perform the analysis on site.

Grab sampling works well for parameters that cannot be accurately or quickly measured in the field or in situ (e.g., phosphorus). Grab sampling also does not require any equipment to remain in the field, where it would be susceptible to damage from the weather or vandalism. Portable pumps and tubing may be used to collect samples from locations that are difficult to access, such as the bottom of an underground sedimentation device or the center of a wet pond. The primary advantage of grab sampling is that setup costs are small. Nevertheless, flow measurement equipment must be installed because pollutant removal efficiency and effluent pollutant loads cannot be determined without discharge measurements. The disadvantages of grab sampling include (1) inconvenience and cost of sending a crew to the site to collect samples during a storm event and (2) lack of an ability to perform flow-weighted sampling.

Automatic sampling requires someone to set up the sample collection system, periodically retrieve the samples from the sampler, and transport the samples to a laboratory for analysis. The time spent in the field for automatic sampling after sampler installation is minimal because samples collected automatically from a storm event can be retrieved and the automatic sampler reset within a few minutes. Automatic samplers are commonly used for stormwater monitoring operations because of the ability to accurately sample nutrients and metals. As will be discussed later, however, the accuracy of automatic samplers rapidly decreases when sampling suspended solids larger than 88 μm. This inaccuracy can affect the particulate or total phosphorus concentration because suspended solids can adsorb a significant amount of phosphorus.

Choosing from in situ, on-site, grab, and automatic sampling will depend on budget constraints, personnel availability, and the goals of the assessment program. The three levels of assessment, listed in order of increasing complexity are visual inspection (level 1), capacity testing (level 2a), synthetic runoff testing (level 2b), and monitoring (level 3). Visual inspection is the only level of assessment that does not require sampling. Capacity testing for saturated hydraulic conductivity (K_s) determination often requires samples for soil moisture measurements at each location (see Chap. 11). Some stormwater treatment practices for which synthetic runoff testing is applicable may require sampling of the influent or effluent synthetic runoff, or both. In these cases, the sampling methods for synthetic runoff testing are the same as the sampling methods for monitoring, which are discussed in the rest of this chapter.

The following five key questions should be considered when incorporating sampling into an assessment program:

1. How many storm events should be sampled to make statistically accurate estimates of performance?
2. How many samples should be collected per storm event?
3. When multiple samples are collected per storm event, should they be collected based on discharge amount, elapsed time, or a user-defined basis?

4. When multiple samples are collected per storm event, should they be collected in individual bottles (discrete samples) or combined into a single bottle (composite samples)?
5. Should stormwater runoff be sampled in situ, on-site, manually (i.e., grab), or automatically?

The next several sections provide discussion and recommendations for each of the above criteria, all of which should be thoroughly considered before sampling is included in any assessment program.

10.1 Representative Samples

Regardless of the type of samples collected (in situ, on-site, grab, or automatic), it is imperative that representative samples are measured or collected. A representative sample is a sample in which the measured parameter (e.g., phosphorus) is the same in the sample as in source from which the sample was measured or collected. In many cases, samples are only representative for a very short period of time and a small, specific location.

To make conclusions in an assessment program, it may be necessary to make assumptions about the dynamics of a system and to what degree a collected sample is representative. When planning a sampling program to include representative samples, the following should be considered:

1. How will sample contamination be prevented?
2. Does the measured parameter (e.g., phosphorus) change significantly in time or space?
3. Do conditions other than the measured parameter (e.g., discharge) change significantly in time or space?
4. Is the system poorly mixed or very large, such that a sample in one location is not representative of the entire system?

In order to measure or collect representative samples, it may be necessary to use a specific (or more than one) sampling method. Several examples for choosing sampling methods to ensure representative samples include:

- Measuring dissolved oxygen with on-site, grab, or automatic sampling in some situations requires careful sample collection, storage, handling, and analysis to prevent contamination. It may be more cost-effective to measure dissolved oxygen in situ.
- Some systems change quickly and capturing representative samples in these systems may require measuring or collecting several samples in a short period of time. On-site and grab sampling may be limited by the capacity of the personnel measuring or collecting the samples; therefore, in situ or automatic sampling may ensure more (temporally) representative samples.
- Some systems vary or change drastically in space, and capturing representative samples in these systems may require measuring or collecting samples in several

locations. It may be cost-prohibitive to install in situ or automatic samplers in several locations; therefore, it may be more cost-effective to measure or collect on-site or grab samples to ensure more (spatially) representative samples.

- Some systems vary or change drastically in time and space. Capturing representative samples in these systems may require measuring or collecting samples repeatedly in several locations simultaneously. On-site and grab sampling may be limited by the capacity of the personnel measuring or collecting the samples; therefore, in situ or automatic sampling may ensure representative samples. It may, however, be cost-prohibitive to install in situ or automatic samplers in several locations. In these situations, the best solution may be to choose a different (often simpler) assessment method or study site.

10.2 Number of Storm Events

The most important sampling consideration in an assessment program is the number of storm events to be sampled. The number of storm events sampled and the variance in the results from those storm events will determine assessment uncertainty. Assessment uncertainty must to be minimized so that comparisons to other stormwater treatment practices, comparisons to past assessments, predictions of future performance for Total Maximum Daily Load (TMDL) calculations, and maintenance scheduling are accurate and reliable. For example, suppose the event mean concentration (EMC) for a specific pollutant during a storm was reduced from an influent value of 100 mg/L to an effluent value of 40 mg/L in a stormwater treatment practice. It cannot be assumed that the stormwater treatment practice reduces the EMC by 60% for all storm events. Several storm events, representing a range of conditions (i.e., flow rate and pollutant concentration), need to be sampled and analyzed before predictions of treatment practice performance can be made. The rest of this section describes a process that can be used to select an appropriate range of assessment uncertainty, and subsequently determine the number of storm events that should be sampled.

To simplify the statistical analysis related to determining the number of storm events that should be sampled, several assumptions can be made. One assumption is that the percent removal data are normally distributed about a mean value and that one storm event does not influence other storm events. Another assumption is that there is no storm event bias (systematic uncertainty) in percent removal. Finally, the number of storms required will likely be fewer than 30 and the actual variance in the data is unknown. From these assumptions, the Student (Gosset 1908) *t*-distribution is used. The Student (Gosset 1908) *t*-distribution is a probability distribution used to estimate the mean of a normally distributed population from a sample of the population and is more accurate for small ($n < 30$) sample sizes than the similar *z*-distribution. For more information on distributions, consult a statistics text (e.g., MacBerthouex and Brown 1996; Moore and McCabe 2003).

The 95% confidence interval is recommended to adequately represent uncertainty in average pollutant removal efficiency because it indicates that there is a

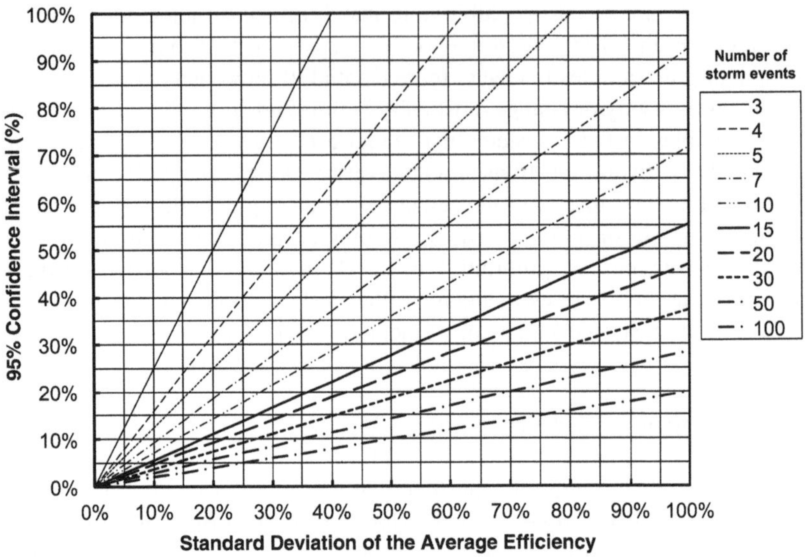

Fig. 10.1 Relationship between number of storm events and standard deviation for a 95% confidence interval

95% probability that the actual average performance will be within the confidence interval. For example, a stormwater treatment practice with an average pollutant capture rate of 72 ± 17% confidence interval ($\alpha = 0.05$) will have a 95% (19 out of 20) probability that the actual average pollutant capture rate is between 55 and 89%. The range of the confidence interval (in this case, ±17% for $\alpha = 0.05$) is dependent on the standard deviation and the number of monitored storm events. The relationship between standard deviation, number of storm events, and 95% confidence interval is shown in Fig. 10.1.

The process for determining the number of storm events can be performed in three steps, as illustrated in Example 10.1. This process should be performed during development of an assessment program to estimate the cost and effort associated with sampling multiple storm events based on the estimated uncertainty. As assessment results are gathered, this process should be performed again using actual assessment data to determine the actual uncertainty:

1. Compute the standard deviation of the percent removal values for storm events that have been sampled. If there are no storm event data, select a standard deviation; typical standard deviations for percent removal of stormwater treatment practices range from 20 to 40% (Weiss et al. 2005).
2. Select the desired range of the 95% confidence interval for the mean removal over all storms (10–15% is recommended).
3. Using the standard deviation (step 1) and the confidence interval (step 2), estimate the number of storm events required to achieve the desired range for the 95% confidence interval of the mean removal over all storms from Fig. 10.1.

Example 10.1: Determining the number of storm events required
Gina, an engineer in training (EIT) at a local consulting firm, is developing an assessment program that includes monitoring (level 3). She is tasked with determining how many storms will be required to attain 95% confidence that the average total suspended solids (TSSs) removal is within ±15%. From previous monitoring data, Gina finds that the stormwater treatment practice is expected to remove 72% (standard deviation = 27%) of TSS from any given storm. She then uses this information (standard deviation = 27%, 95% confidence interval = 15%) and Fig. 10.1 to determine that roughly 15 storm events are required, as shown in Fig. E10.1.

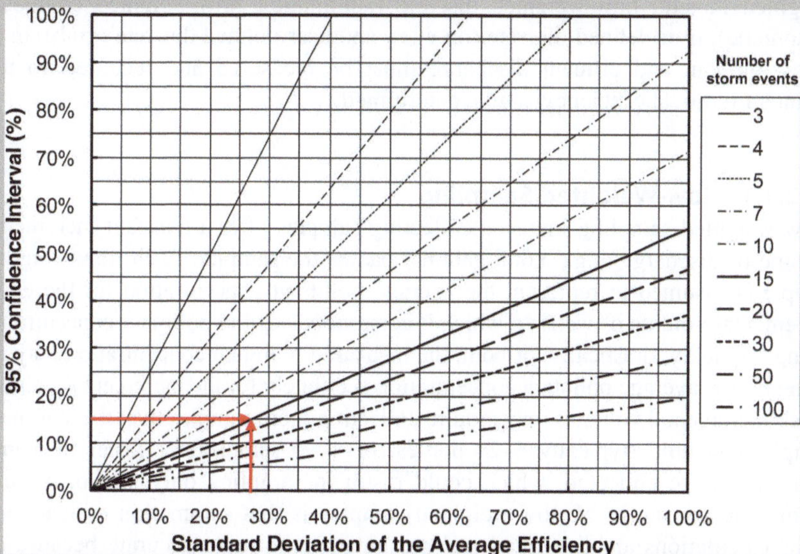

Fig. E10.1 Determining the number of storms required using Fig. 10.1

10.3 Samples Per Storm Events

The US EPA (2002) recommends that multiple samples be collected throughout a storm event to incorporate changes in concentration and discharge and therefore accurately represent the storm event. Choosing an appropriate number of samples per storm event will depend upon the basis on which the samples will be collected: discharge, time, or grab samples.

10.3.1 Flow-Weighted, Time-Weighted, and User-Defined Sampling

The frequency with which samples are typically collected can be defined by three approaches: (1) flow weighted, (2) time weighted, and (3) user defined. Any of these sampling approaches can be used to collect in situ, on-site, grab, or automatic samples. There are also two methods of sample storage: discrete and composite. Sample storage is not required for in situ samples and most in situ samples are time weighted. Samples are typically described by the method of collection and storage (e.g., flow-weighted discrete samples) but could be in situ, on-site, grab, automatic, or some combination thereof. The most common sampling programs are time-weighted in situ, flow-weighted discrete (automatic), flow-weighted composite (automatic), user-defined discrete (on-site), and user-defined discrete (grab). In all cases, influent and effluent discharge must be measured and recorded so that pollutant removal efficiency can be determined.

10.3.1.1 Flow-Weighted Sampling

Flow-weighted sampling involves collecting samples after a constant incremental volume of discharge (e.g., 5,000 gallons) passes the sampler. Each flow-weighted sample is assumed to represent the average pollutant concentration for the entire incremental volume of water to which it corresponds. If the pollutant concentration changes quickly, drastically, or both, the measured pollutant concentration may not represent the average pollutant concentration accurately for the incremental volume. Small incremental volumes may require collecting more samples than the automatic sampler can hold (typically 4–24 bottles, or 4–96 samples) or faster than grab samples can be collected, which could result in sampling only part of a storm event. The advantage of flow-weighted samples is that summation of loads and EMC calculations are simplified and presumed to be more accurate because the discharge volume is constant for each representative sample. The most common flow-weighted samples are discrete or composite samples that are collected automatically. The relationship between sampling accuracy and the number of samples collected is shown in Example 10.2.

Example 10.2: Error associated with number of samples

Gina, the EIT at a local consulting firm, has been contracted to assess the effectiveness of a dry pond that treats runoff from a Public Works facility. Preliminary monitoring determined the inflow and outflow hydrographs and pollutographs for total phosphorus (TP), as shown in Fig. E10.2. (1 cfs = 0.028 m^3/s)

(continued)

Example 10.2: (continued)

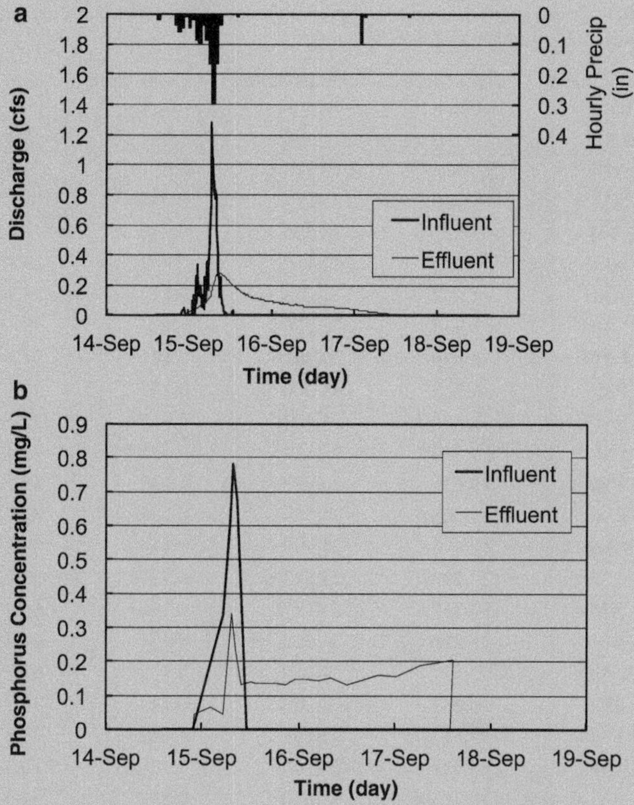

Fig. E10.2 Hydrograph (**a**) and pollutograph (**b**) for example storm event

Gina uses the hydrograph and pollutograph data to determine the error associated with the number of samples collected, assuming the sampled concentrations are correct. First she considers the effluent data, in which 23 individual samples were collected, and calculates the pollutant load, as shown in Table E10.1:

(continued)

Example 10.2: (continued)

Table E10.1 Storm event discharge, sample volume, pollutant concentration, and sum of pollutant mass (23 samples) (1 ft3 = 28.3 L)

Sample collected at (mm/dd hh:mm)	Discharge volume (ft3)	Incremental volume (L)	Concentration (mg/L)	Sum of mass load (g)
9/14 10:00 PM	0	0	0.047	0.0
9/15 2:22 AM	675	19,110	0.067	1.3
9/15 5:23 AM	1,566	25,233	0.045	2.4
9/15 7:16 AM	**2,573**	**28,506**	**0.342**	**12.2**
9/15 8:30 AM	3,710	32,197	0.232	19.6
9/15 9:39 AM	4,832	31,774	0.132	23.9
9/15 10:49 AM	5,934	31,226	0.136	28.1
9/15 12:04 PM	**6,976**	**29,481**	**0.141**	**32.3**
9/15 1:25 PM	7,965	28,006	0.137	36.1
9/15 2:54 PM	8,862	25,407	0.137	39.6
9/15 4:33 PM	9,749	25,128	0.138	43.1
9/15 6:28 PM	**10,631**	**24,973**	**0.138**	**46.5**
9/15 8:36 PM	11,505	24,754	0.131	49.7
9/15 11:03 PM	12,416	25,799	0.149	53.6
9/16 1:43 AM	13,293	24,813	0.149	57.3
9/16 4:46 AM	**14,167**	**24,757**	**0.146**	**60.9**
9/16 8:02 AM	15,018	24,102	0.151	64.5
9/16 11:49 AM	15,877	24,329	0.132	67.7
9/16 3:26 PM	16,682	22,792	0.144	71.0
9/16 7:40 PM	**17,527**	**23,931**	**0.162**	**74.9**
9/17 12:22 AM	18,267	20,948	0.156	78.2
9/17 6:23 AM	18,785	14,670	0.190	81.0
9/17 2:21 PM	**18,866**	**2,287**	**0.204**	**81.4**

Based on the calculations, the storm produced 18,866 ft^3 (534 m^3) of effluent discharge with 81.4 g (0.179 lb) of phosphorus load. The cost, however, to analyze 23 samples for each storm event could be expensive. Gina estimates the total load if there had been only six equally distributed samples collected during this same storm event in Table E10.2 (bold text from Table E10.1) (1 cubic foot = 28.3 L).

Table E10.2 Storm event discharge, sample volume, pollutant concentration, and sum of pollutant mass (six samples) (1 ft3 = 28.3 L)

Sample collected at (mm/dd hh:mm)	Discharge volume (ft3)	Incremental volume (L)	Concentration (mg/L)	Sum of mass load (g)
9/15 7:16 AM	2,573	72,849	0.342	24.9
9/15 12:04 PM	6,976	124,678	0.141	42.5
9/15 6:28 PM	10,631	103,514	0.138	56.8
9/16 4:46 AM	14,167	100,123	0.146	71.4
9/16 7:40 PM	17,527	95,154	0.162	86.9
9/17 2:21 PM	18,866	37,905	0.204	**94.6**

(continued)

Example 10.2: (continued)

Gina determines that if only six samples had been collected during the storm event, the pollutant load calculated using the same method above would be 94.6 g (0.209 lb), which is 16.2% more than the estimate resulting from 23 samples. If, however, the automatic sampler was programmed to collect 4 subsamples in each sample bottle, the same 23 bottles above would be collected in 6 composite samples and would result in a total phosphorus effluent load calculation of 78.7 g (3.3% error).

The number of samples collected depends on the influent discharge of each storm event and the incremental volume. Selecting the optimum volume increment depends on the size of the watershed, land cover, soil type, slopes, and expected rainfall intensity and discharge volume of the storm events. Due to the unpredictability of rainfall, the selection of a flow increment will always involve some uncertainty. An approach for selecting the incremental sampling volume is provided in Example 10.3.

Example 10.3: Determining the incremental volume for automatic sampling

Gina, the consulting EIT, realizes that storm event volumes will vary and therefore not every storm will produce exactly 24 samples. She therefore does an analysis of the variation of storm events to determine what incremental volume the samplers should be set at to capture the most storm events. Based on Gina uses the watershed area (A), runoff coefficient (C), and the previous season's rainfall (P), to estimate the inflow volume for each storm using $V = P \times C \times A$. The estimated inflow volumes for 12 storms are shown in Fig. E10.3.

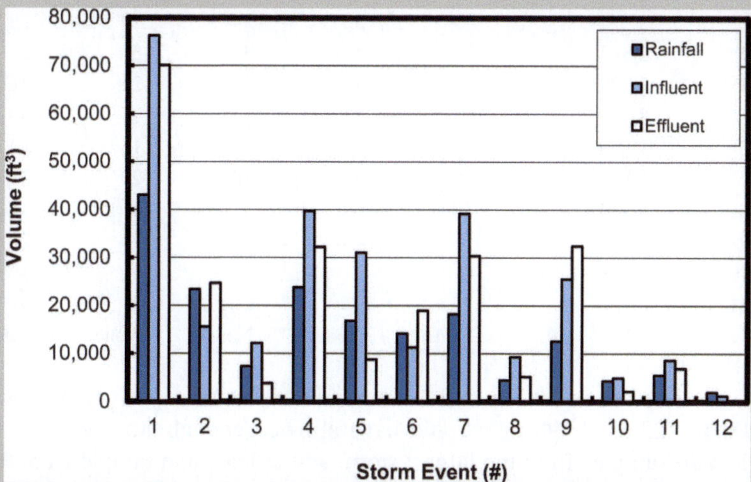

Fig. E10.3 Influent, rainfall, and effluent discharge volume for example storm event

(continued)

Example 10.3: (continued)

Based on the influent volumes, Gina calculates the relationship between incremental volume and number of samples for each storm event. She determines the number of samples by dividing the runoff volume by the incremental volume. She would like to capture as many storm events as possible, including the large storms, so the automatic samplers will be programmed to collect four small samples into each of the 24 sample bottles, allowing for 96 total samples.

Table E10.3 Number of samples required for each storm influent and effluent discharge volumes as a function of incremental volume (SE = Storm Event) (1 ft3 = 28.3 L)

		Incremental volume (ft^3)				
Influent	Volume (ft^3)	400	800	1,200	1,600	2,000
SE 1	76,182	>96	95	63	47	38
SE 2	15,586	38	19	12	9	7
SE 3	12,138	30	15	10	7	6
SE 4	39,752	>96	49	33	24	19
SE 5	31,075	77	38	25	19	15
SE 6	11,312	28	14	9	7	5
SE 7	39,181	>96	48	32	24	19
SE 8	9,280	23	11	7	5	4
SE 9	25,574	63	31	21	15	12
SE 10	4,980	12	6	4	3	2
SE 11	8,630	21	10	7	5	4
SE 12	1,247	3	1	1	None	None
		Incremental volume (ft^3)				
Effluent	Volume (ft^3)	500	750	1,000	1,250	1,500
SE 1	70,062	>96	93	70	56	46
SE 2	24,744	49	32	24	19	16
SE 3	3,837	7	5	3	3	2
SE 4	32,281	64	43	32	25	21
SE 5	8,796	17	11	8	7	5
SE 6	18,967	37	25	18	15	12
SE 7	30,420	60	40	30	24	20
SE 8	5,184	10	6	5	4	3
SE 9	32,470	64	43	32	25	21
SE 10	2,158	4	2	2	1	1
SE 11	6,926	13	9	6	5	4
SE 12	220	None	None	None	None	None

Gina determines that an inflow incremental volume between 800 and 1,200 ft^3 (22.65–33.98 m^3) would have allowed enough storage space to collect all samples from the largest storm and at least one sample from the smallest storm from the previous year. Similarly, an effluent incremental volume between 750 and 1,000 ft^3 (21.24–28.32 m^3) allows ample storage

(continued)

Example 10.3: (continued)
for the largest storm and several samples from the smallest storms, excluding the smallest storm from the previous year.

Gina realizes that this procedure should be revised and adjusted before each rainy season and sometimes during rainy seasons to ensure that the most storm events are sampled. To increase the accuracy of this procedure, rainfall data from more than one preceding year could be used to determine the appropriate incremental volume.

10.3.1.2 Time-Weighted Sampling

Time-weighted samples are collected at a user-specified, constant time interval (e.g., 30 min). Because the discharge of natural storm events is not constant, time-weighted samples do not represent constant volumes of flow with respect to time. Total discharge volume for each time interval must be calculated before calculation of summation of loads or event mean concentration (EMC). The time-weighted approach is common for in situ, on-site, and grab sampling.

The calculations for time-weighted samples can be more complicated than those for flow-weighted samples because each sample must be weighted by the corresponding discharge volume. In these cases, discharge volume for each time interval must be calculated by integrating the discharge vs. time curve.

Selection of the optimal time increment will depend on the duration of a "typical" storm event and the variation in storm event duration. It is also important to consider the amount of time required to collect in situ, on-site, and grab samples or the maximum number of samples the automatic sampler can collect, if applicable. Due to the unpredictability of rainfall events, however, the selection of a time increment will always involve some uncertainty.

10.3.1.3 User-Defined Sampling

User-defined samples are collected on a basis determined by the user which is commonly chosen based on the hydrology of the system being assessed. For example, some sampling programs may collect a specified number of grab samples during the rising and falling limbs of a storm event. The discharge and time increment between these samples will vary between samples and for each storm event. Similar to time-weighted sampling, total discharge volume for each interval between samples must be calculated before calculation of summation of loads or event mean concentration (EMC). User-defined sampling is most common for manually collected samples (i.e., on-site or grab samples).

10.3.2 Discrete and Composite Samples

Once it is determined how samples will be collected (flow-weighted, time-weighted, or user-defined), the next step is to determine whether to collect discrete (i.e., separate) samples or a composite sample(s). Discrete samples are collected in individual containers and the contents of each container are analyzed separately. Composite samples are collected in a single container and analyzed as a single sample representative of the entire sampling period.

Discrete samples can be collected manually for on-site or grab sample analysis, or with automatic samplers equipped with multiple sample containers. Most often, discrete sampling is only necessary when a record of temporal variation in pollutant concentration throughout a storm event (e.g., minimum, maximum) is desired. The main disadvantage of discrete sampling is that multiple samples must be analyzed for pollutant concentration for each storm event, which can increase the costs of an assessment program significantly.

Composite sampling combines all collected samples into one large storage container and should only be used in conjunction with flow-weighted sampling. Time-weighted composite samples cannot be used to determine pollutant loads or event mean concentration (EMC) because each time-weighted subsample does not represent equivalent volumes of discharge. Thus, if time-weighted sampling is used, samples should not be stored as composite samples.

Discrete and composite sampling can be used with on-site, grab, or automatic samples, but most automatic sampling equipment is designed specifically for one method or the other. Thus, in order to ensure compatibility between an assessment program and sampling equipment, the goals and details of the assessment program should be developed before purchasing sampling equipment.

It is important to note that flow-weighted samples can be collected either as discrete or composite samples because the volume increment is the same for each sample. Each sample added to a composite sample represents the same volume increment of stormwater and is therefore equally representative. Therefore, chemical analysis is considerably cheaper for flow-weighted composite samples compared to flow-weighted discrete samples, but only the event mean concentration (EMC) can be determined. If discrete samples are collected, the EMC and the concentration as a function of time over the runoff event can be determined.

Time-weighted samples, however, can only be collected as discrete samples because each sample represents a different volume of stormwater. It may be important to consider the parameters used by stormwater models (e.g., XP SWMM, and WinSLAMM, among others) when developing a sampling program because some models input sampling parameters (such as discrete samples) directly. Unless the goal is to measure pollutant removal performance as it changes with time throughout the runoff event, flow-weighted, composite sampling is recommended because of the cost savings of analyzing only one sample per storm event.

10.4 In Situ, On-Site, and Grab, and Automatic Sampling

Some pollutants can be measured in situ, on-site, or by analysis of grab or automatic samples. In situ, on-site, and grab sampling for assessment of stormwater treatment practices are cost-effective for some parameters that may be of interest. For example, capacity testing (level 2a) of a stormwater treatment practice for saturated hydraulic conductivity (K_s) requires measurement of soil moisture content. Soil moisture can be measured either by using a field soil moisture probe (in situ sampling) or by collecting a soil sample and analyzing it in the laboratory (grab sampling). Another example includes synthetic runoff testing of a wet pond for hydraulic performance using a conservative tracer. Rather than using grab or automatic sampling, a conductivity probe could be used in situ to measure salinity when sodium chloride (NaCl) is used as the conservative tracer. In this case, in situ sampling is simpler and cheaper than grab or automatic sampling, and therefore recommended.

In situ and on-site sampling for stormwater assessment are often limited by the availability of probes for many pollutants of concern. In addition, in situ probes may become fouled when they are not maintained as recommended by the manufacturer's instructions, which can produce erroneous measurements. It is also important to recognize may affect changes that occur over time in the stormwater treatment practice system, such as sediment collection in an inlet pipe or structure, that may affect in situ measurements. Some in situ probes such as pressure transducers or dissolved oxygen probes may require recalibration as conditions change. The following sections describe in situ, on-site, grab, and automatic sampling techniques as they apply to various stormwater pollutants.

10.4.1 Temperature

The temperature of stormwater runoff may be of interest depending on assessment goals and downstream conditions (e.g., temperature-sensitive trout streams). Unlike most water quality parameters such as phosphorus and suspended solids, temperature can be easily measured with in situ or on-site techniques. One method is to collect a stormwater sample and measure the temperature on-site with a thermometer immediately after the sample is collected. Another method is to use a probe or sensor in situ to collect near-continuous temperature data with a data storage device. There are two types of data storage devices that are used for in situ temperature measurement: devices that are integrated with temperature probes and devices that are externally attached to them (often called data loggers). Temperature must be measured either in situ or on-site because water temperature can change during transportation or storage.

For near-continuous in situ sampling using a data storage device, the probe or thermocouple must be submerged during a runoff event. The device will continually measure and record temperature at a user-specified time interval until the data storage capacity is exhausted. Most devices can be set such that the oldest data are overwritten with new data when storage capacity is exceeded. For data storage devices that are integrated with the probe, data are usually downloaded directly to a computer through a data transfer cable or infrared connection. For data storage devices connected to an external thermocouple, data are typically accessible via modem, cellular connection, or direct download (via serial cable) from the data storage device.

Some advantages of integrated and external data storage devices include:

Integrated data storage device advantages include

- Less expensive than data logger and thermocouple
- Data can be downloaded using infrared wireless connection
- Does not require protective cabinet to store data storage equipment

External data storage device advantages include

- Less expensive if a data logger is already in use
- Temporally synchronized with other measurements stored in the data logger (e.g., discharge, rainfall)
- Typically more storage capacity than an integrated device
- Thermocouples respond more quickly to temperature changes
- Data retrieval does not require disturbance of the thermocouple
- Data can be downloaded via modem or cellular connection

The US EPA (2002) notes that some pressure transducers have built-in thermometers so that water depth values can be corrected for temperature. Probes are available for different temperature ranges, depths, and prices.

Prior to monitoring, the temperature probes should be calibrated against a NIST (National Institute of Standards and Technology) traceable thermometer or against 0 °C (32 °F) temperature by placing the probe in a mix of ice and water. Probes should be placed in shaded areas of the sewer pipes or channels whenever possible, to avoid solar heating of the probe. It is recommended the probe be placed inside a PVC pipe anchored to the sewer to protect it from debris, as shown in Fig. 10.2.

To measure the water temperature in a stream or creek, the probe should be installed at least a few inches above the streambed and attached to stakes that are inserted securely into the streambed. The probe should not be installed directly on, or buried in, the sediment bed, because the sediment is often a different temperature than the stream water due to groundwater inputs. Most streams are well-mixed water bodies, and the temperature near or above the sediment surface is representative of the entire water column temperature. For shallow streams (less than 8 in. deep), the temperature probe should be installed in a shaded area of the stream channel to avoid direct solar radiation affecting the temperature measurement of the probe.

A measured difference in stream temperature between upstream and downstream locations may be due to atmospheric heating or surface inflow, not necessarily due to stormwater inflows. During hot summer days, solar radiation can heat the stream such that the water temperature at the downstream location becomes warmer than at the upstream location. The temperature difference varies diurnally and depends upon the solar radiation received, the distance between the upstream and downstream measurement locations, stream discharge, and stream geometry. During storm events and for several hours after, inflow of surface runoff directly into the stream may have a significant impact on the temperature difference between two locations. The thermal impact of surface inflow may be identified as transient change in the temperature difference (see Fig. 10.3).

Fig. 10.2 Installation of a temperature probe in a sewer pipe

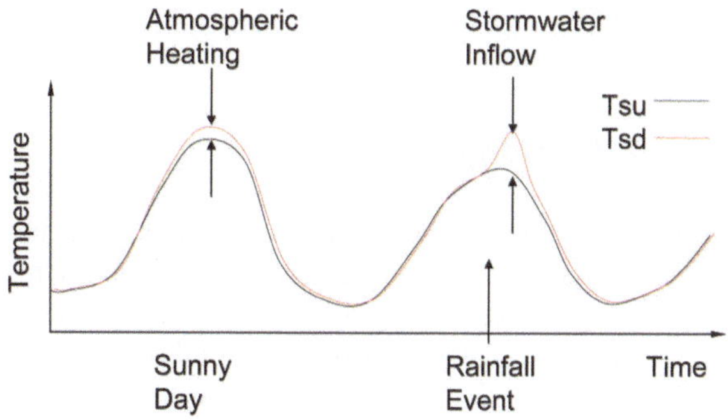

Fig. 10.3 Typical characteristics of stream temperature impacts due to atmospheric heating and stormwater inflows

10.4.2 pH or Hydrogen Ions

The acidity or basicity of water is indicated by pH, which is a function of the molar concentration of hydrogen ions in solution ($[H^+]$), $pH = -\log_{10} [H^+]$. Thus, for a water of pH 8, the hydrogen ion concentration is 10^{-8} moles/L. Acidic waters have relatively large hydrogen ion concentrations and therefore small pH values (< 7). Alkaline waters have relatively small hydrogen ion concentrations and large pH values (> 7) and neutral waters have pH values of approximately 7. USA federal and state regulations suggest that pH values remain between approximately 6.5 and 8.5 to ensure the quality of water for recreational use, aquatic life, and drinking water (MPCA 2003; US EPA 2004).

In situ or on-site sampling should be used to measure pH values. Grab samples for pH with subsequent analysis in a laboratory are permissible if the samples are transported on ice and analyzed within 2 h of collection; automatic sampling for pH measurement is not recommended. Probes should be calibrated weekly, after every 25 samples (US EPA 1997), or as recommended by the manufacturer to ensure accurate and consistent results.

10.4.3 Conductivity

Conductivity is an indirect measure of the ion concentration in water and is often measured with a probe or meter using in situ or on-site sampling techniques. Conductivity is often used as a surrogate for total dissolved solids (TDS) or salinity. Large concentrations of TDS can be toxic to aquatic life and can reduce habitat.

Conductivity is most often measured in situ or on-site but can also be measured using grab or automatic sampling techniques. Most "bundled" probes (multiple probes in one device) include a conductivity probe for in situ sampling. Grab or automatic samples transported to an analytical lab for conductivity measurements must be analyzed within 28 days of collection and should be kept on ice or refrigerated.

10.4.4 Turbidity

Turbidity is a measure of water clarity and can be measured in situ with a turbidity meter. Large turbidity values can block sunlight required for photosynthesis by aquatic vegetation and subsequently reduce aquatic life and diversity. Turbidity can be used as a surrogate for suspended solids concentration but requires calibration at each location and for different seasons to ensure accuracy (Stefan et al. 1983). Refer to the turbidity section in Standard Methods (e.g., APHA 1998b) for details about correlating turbidity and suspended solids.

Turbidity is most often measured in situ or on-site, but samples collected manually (i.e., grab) or by automatic samplers can also be transported to an analytical lab for analysis. Some "bundled" probes include a turbidity meter. Grab or automatic samples transported to an analytical lab for turbidity measurements must be analyzed within 24 h of collection and should be kept in the dark and on ice or refrigerated.

10.4.5 Dissolved Oxygen

Dissolved oxygen (DO) is the amount of oxygen dissolved in water. DO is necessary for the survival of aerobic aquatic organisms such as fish and invertebrates. Minimum dissolved oxygen concentrations for surface waters typically range from 5 to 7 mg/L depending on the use, class, and temperature of the water body, though values are rarely less than 5 mg/L (MPCA 2003; WDNR 2004; MDEQ 2006).

DO should be measured using in situ or on-site sampling techniques. Grab and automatic sampling for DO measurement is not recommended because DO concentrations can change during transport or storage. For on-site sampling, DO should be measured immediately after sample collection. Most DO probes require weekly cleaning and recalibration to ensure accuracy. Luminescent DO measurement techniques are available, but their accuracy and stability have not been thoroughly field tested.

10.4.6 Nutrients

Nutrients (e.g., phosphorus and nitrogen) support aquatic vegetation and organisms. Excess nutrients, however, can cause nuisance algae blooms that generate negative aesthetic and eutrophic conditions in receiving lakes and rivers (US EPA 1999a). In temperate fresh waters, dissolved phosphorus is typically the limiting nutrient (Schindler 1977; Aldridge and Ganf 2003).

Until recently, it was necessary to collect water samples by automatic or grab sampling techniques for subsequent analysis in a laboratory to determine nutrient levels. In situ probes, however, are now available for nitrate, ammonia, and phosphate. Nevertheless, the accuracy and stability of these probes have not been thoroughly demonstrated in the field.

10.5 Additional Considerations for Automatic Sampling

10.5.1 Automatic Sampling Equipment

Stormwater sampling equipment is designed to collect samples either manually when triggered by the user or automatically when predefined criteria are met, with the aid of data loggers and computer software. Available equipment is summarized herein and discussed in greater detail in "Urban Stormwater BMP Performance Monitoring" (US EPA 2002). Grab sampling equipment is also discussed in Stenstrom and Strecker (1993).

Automatic samplers, which collect and store water samples until they can be retrieved, are recommended for sampling suspended solids, phosphorus, nitrogen, salts, metals, and other pollutants that do not change or degrade rapidly. For pollutants that may undergo rapid transformation, such as temperature, fecal coliforms, and organic chemicals, it may not be possible to retrieve and analyze the samples before transformation compromises the sample integrity. For such pollutants, in situ or on-site measurement, grab samples, or rapid retrieval of automatically collected samples followed by prompt analysis are recommended to ensure accurate representation of the pollutant concentration. Alternatively, sample refrigeration or chemical preservatives can be used to reduce the rate at which pollutant transformation occurs. Consult an analytical methods manual (e.g., APHA 1998b) to determine if refrigeration or preservatives will reduce transformation of pollutants and whether addition of preservatives will interfere with analysis of other pollutants of concern.

Automatic samplers do not require anyone to be present for sample collection; they can be programmed to begin sampling when a user-specified rainfall amount or intensity occurs (electronic rain gauge required), after a predefined depth or quantity of flow occurs, or after some combination of conditions is met. They can also be programmed to collect varying sample sizes, collect samples at user-specified time intervals (i.e., time-weighted) or flow volume increments (i.e., flow-weighted), or collect samples over an entire runoff event that lasts 2 days or more.

Some automatic samplers are powered from an external 120-V AC power source. Many locations, however, do not have an external power source and therefore most monitoring applications use automatic samplers that are powered by one or more deep-cycle, 12-V battery. Solar panels are also available to recharge the batteries, provided that adequate sunlight is available and the solar panel is free from obstructions (e.g., snow and leaves, among others). Another option is to use an additional battery as backup to the power supply.

Automatic samplers are available to collect discrete samples or composite samples. The sampling portion of an assessment program should be planned before sampling equipment is purchased to ensure the appropriate equipment is available and does not exceed the budget of the program. Automatic samplers with refrigerated sample storage compartments can be used to preserve the integrity of samples that degrade. For example, sample storage for dissolved phosphorus determination recommends refrigeration to reduce the transformation of dissolved phosphorus to particulate phosphorus, or vice versa (APHA 1998a). Refrigeration units, however, require an AC power supply.

While samples must always be manually retrieved from the storage unit for analysis, some samplers and data loggers have modem or cellular connections that allow measurement data such as flow rate, water depth, and rainfall intensity to be retrieved without physically visiting the sampling location. Some systems also allow users to remotely determine whether samples have been collected.

10.5.2 Equipment Placement and Maintenance

Placement of sampling equipment is site specific and depends on a number of factors, including equipment type, amount of equipment, availability of protective cabinets, and type and design of stormwater treatment practice. As described in Chap. 9, influent (or effluent) flow measurement and sampling is simplified if all

Fig. 10.4 Pressure probe for flow measurement and sampling tube for pneumatic sample collection placed at a weir to measure stormwater inflow

stormwater inflow (or outflow) is routed to a single location. Placing sampling equipment near flow measurement equipment is advantageous because sampling is typically triggered by flow measurement equipment and all instrumentation can be housed in the same enclosure. An example of flow measurement and sampling equipment in the same location is shown in Fig. 10.4, and a protective cabinet housing automatic sampling equipment is shown in Fig. 10.5.

Automatic sampling equipment that remains in the field for long periods of time should be maintained at weekly intervals. Sampling equipment maintenance will vary, but manufacturer's recommendations are typically provided and should be followed. Additional sampling bottles are available for purchase, and it is recommended to have at least two sets of sample bottles (one for the sampler and one for transporting samples to the analytical lab). More sets of sample bottles may be required, depending on the frequency of storm events and the processing time of the analytical lab.

It is important to recognize that some pollutants adsorb to the surface of collection bottles (organic compounds), degrade over time (coliforms), or may volatilize (dissolved gases). These confounding processes can be minimized by choosing sample bottle material properly (e.g., plastic or glass) and cleaning sample bottles appropriately. Consult Standard Methods (e.g., APHA 1998b) or the analytical lab performing the water quality analysis to determine whether the pollutants of interest for the assessment program will adsorb or degrade and which bottle material or preservation technique is recommended. If analyte degradation is a

Fig. 10.5 Protective cabinet housing automatic sampling equipment

concern, then sample preservation (e.g., refrigeration), collection followed by rapid analysis, or both may be necessary.

Care and cleaning of sampling equipment and bottles will prolong proper functionality and reduce analytical error. Sample bottles should be cleaned according to Standard Methods (e.g., APHA 1998b). Depending on the pollutant, special procedures may be required to prepare the bottles for sampling. For example, sample bottles for metals or phosphorus should be acid washed, and sample bottles for coliforms should be sterilized (e.g., autoclaved). Refer to the analytical procedure for pollutants of interest, Chap. 11, or the analytical lab that will process the samples for more information.

10.5.3 Winter Sampling in Cold Climates

Stormwater treatment practices may function differently during the winter than during the summer. For example, a layer of ice in a wet pond can reduce the effective volume of the pond and cause short-circuiting, which will reduce hydraulic residence times and lower sediment removal rates. Some of the largest concentrations of pollutants in stormwater are found in late winter/early spring runoff (i.e., snowmelt). Unfortunately, winter runoff and snowmelt events are not commonly monitored, most likely due to the inherent challenges imposed by the weather.

One winter challenge that must be overcome is the formation of ice in and around sampling lines and bubbler lines that are used for water depth measurement at weirs (see Sect. 9.2 in Chap. 9). Ice formation in sampling lines can prevent samples from being collected. Ice formation over bubbler tubes will result in erroneously large pressure readings and inaccurate depth measurements. In addition, if an automatic sampler is installed to collect samples when the water depth exceeds a certain value, then a false pressure reading could trigger a sampling sequence when insufficient water depth is available. Because of this possibility, a pressure transducer is recommended to measure water depth. Caution must be exercised, however, because the flexible diaphragm inside a pressure transducer can be damaged by ice formation. In one monitoring attempt in Minnesota, USA, the bubble tube developed ice over the discharge end, which prevented air from being pushed out of the tube. Although the resistance to air flow and large pressure that developed was due to the ice, the monitoring equipment registered an inaccurately large value of water depth and attempted to collect water samples when no water was present.

It is possible to maintain a charge on the batteries used to power the sampling and flow monitoring equipment during winter months with solar panels (Hussain et al. 2006). Solar panels should be faced toward the south and angled steeply (near vertical) to capture the most sunlight and to remain free of snow accumulation. Because of the potential problems of winter sampling, grab sampling is advised in conjunction with automatic sampling to ensure that appropriate samples are collected.

10.5.4 Automatic Sampling of Water Containing Suspended Solids

The accuracy of automatic sampling of water that contains suspended solids has been documented (Reed 1981), and research conducted at the University of Minnesota's St. Anthony Falls Laboratory to investigate the limits of sampling suspended solids and particulates and to improve sampling methods for automatic samplers has shown that large errors can exist when using automatic samplers to collect water for analysis of suspended solids.

Research conducted on an ISCO 3700 automatic sampler at the St. Anthony Falls Laboratory has shown that samples collected by automatic samplers may not accurately represent the suspended solids concentration in stormwater runoff (Gettel et al. 2011). A sediment feeder was installed at the upstream end of an 18-in. (45.7-cm) diameter pipe and sediment and water were fed into the pipe. Suspended solids were sampled 34.8 ft (10.6 m) downstream of the feed point. Discharge through the pipe was measured using a V-notch weir downstream of the pipe. The tests were conducted using five sediment size distributions.

1. Silts and clays with a median diameter of 25 μm and a maximum diameter of 88 μm
2. Silts of size 44–88 μm
3. Sands with size range of 125–180 μm
4. Sands with size range of 180–250 μm
5. Sands with size range of 250–355 μm

Table 10.2 Experimental results for sampled and feed suspended sediment concentration for conventional sampling methods when sampling intake without strainer (tube only) was mounted at the bottom of pipe and facing upstream or downstream of the flow

Particle size (μm)	Water discharge (cms)	Water surface slope (%)	Mean feed conc. (mg/L)	Mean sampled conc. (mg/L)[a]	Mean relative conc. (%)[b]	95% Confidence interval (± around mean %)
Sampling tube only facing downstream						
Silt (D_{50} = 20 μm)	0.095	0.90%	199.5	253.4 (8)	126.9%	12.7%
44–88	0.095	1.55%	122.8	187.2 (21)	148.6%	7.8%
125–180	0.092	N/A	210.1	1049.9 (13)	499.7%	59.4%
180–250	0.091	0.65%	220.4	1396.1 (14)	1396.0%	142.4%
250–355	0.092	0.65%	201.6	2420.9 (7)	1200.8%	107.7%
Sampling tube only facing upstream						
Silt (D_{50} = 20 μm)	0.089	1.00%	143.1	154.9 (19)	106.4%	3.6%
44–88	0.089	1.55%	221.5	524.0 (16)	246.9%	5.3%
125–180	0.084	N/A	247.6	2093.5 (21)	1011.1%	11.5%
180–250	0.092	0.65%	135.1	5405.6 (21)	2907.0%	10.3%
250–355	0.088	0.45%	218.5	14826.8 (20)	6583.5%	7.3%

Depth of flow was set at 0.23 m (Gettel et al. 2011)
[a]The value in parentheses indicates the number of samples per test
[b]Relative concentration is based on the sample-weighted average concentration

Four automatic intake configurations were tested:

1. Sampling tube oriented parallel to the flow and facing upstream
2. Sampling tube oriented parallel to the flow and facing downstream
3. A commercially available intake manifold attached to the end of the sampling tube and to the bottom of the pipe
4. A commercially available intake manifold attached to the end of a sampling tube and only the tube attached to the side of the pipe so that the manifold was able to move freely in the flow

The results given in Tables 10.2 and 10.3 show that the automatic sampler overestimated the concentration of the suspended sediment by up to 6,600% for large particle sizes. Silts and clays are typically sampled more accurately. The configuration in which the manifold was allowed to move freely in the flow yielded the most accurate results. For TSS concentration, if the size distribution is not too large, automatic sampling with the manifold free to move laterally will collect samples with concentrations within 25% of the actual concentration. The sampling of large sand particles in this configuration, however, resulted in errors of approximately 200%. This indicates that size distributions will be skewed toward larger sizes by the use of an automatic sampler.

Table 10.3 Experimental results for sampled and feed suspended sediment concentration for conventional sampling methods when the sampling intake with strainer position is fixed or flexible

Particle size (μm)	Water discharge (cms)	Water surface slope (%)	Mean feed conc. (mg/L)	Mean sampled conc. (mg/L)[a]	Mean relative conc. (%)[b]	95% Confidence interval (± around mean %)
Sampling intake strainer in fixed position						
Silt (D_{50} = 20 μm)	0.095	0.84%	140.4	143.9 (13)	101.0%	9.0%
44–88	0.089	0.90%	107.8	137.1 (21)	127.2%	5.0%
125–180	0.089	0.84%	211.2	541.0 (20)	258.7%	5.6%
180–250	0.089	0.77%	225.6	543.0 (19)	303.0%	11.9%
250–355	0.089	0.77%	216.9	338.4 (23)	169.3%	8.6%
Sampling intake strainer in flexible position						
Silt (D_{50} = 20 μm)	0.089	0.71%	142.5	142.7 (12)	100.1%	4.0%
44–88	0.090	1.48%	113.6	124.2 (20)	109.3%	2.7%
125–180	0.088	1.42%	219.8	438.6 (9)	199.6%	18.5%
180–250	0.087	1.36%	230.2	471.4 (14)	204.8%	18.8%
250–355	0.089	1.42%	211.7	277.5 (14)	131.0%	22.6%

Depth of flow was set at 0.23 m (Gettel et al. 2011)
[a]The value in parentheses indicates the number of samples per test
[b]Relative concentration is based on the sample-weighted average concentration

Solids suspension is a function of flow characteristics and particle size, density, and shape. Sampling suspended solids concentration is strongly influenced by the location of the intake within the depth of the flow. For a typical stormwater conduit, concentrations larger than the mean concentration are found at lower relative depths for most particle sizes (> 10 μm). Intakes of automatic samplers are typically placed at the base of conduits, which can result in suspended solids concentrations containing larger particles being overestimated.

As depicted by the dotted line in Fig. 10.6, if a sampler intake is located at 10% of the total depth ($y/d = 0.1$), the resulting sampled concentration for 250-μm sand particles will be approximately 2.1 times the mean concentration for the given flow condition. Similarly, at that same relative depth and flow condition, 100-μm fine sand/silt and 11-μm clay particles are sampled at approximate concentrations of 1.3 and 1.0 times the mean concentration, respectively. For this conduit, only clay particles can be sampled accurately. Suggestions have been made to place the sampler intakes at a depth above the bed, but in this configuration the automatic sampler is unable to collect samples below the intake depth.

Developed from equations given in Rouse (1937), Fig. 10.7 represents a limiting particle size for a measured flow condition to ensure a sample concentration within 20% of the mean. Figure 10.7 assumes that the flow is fully developed, i.e., does not

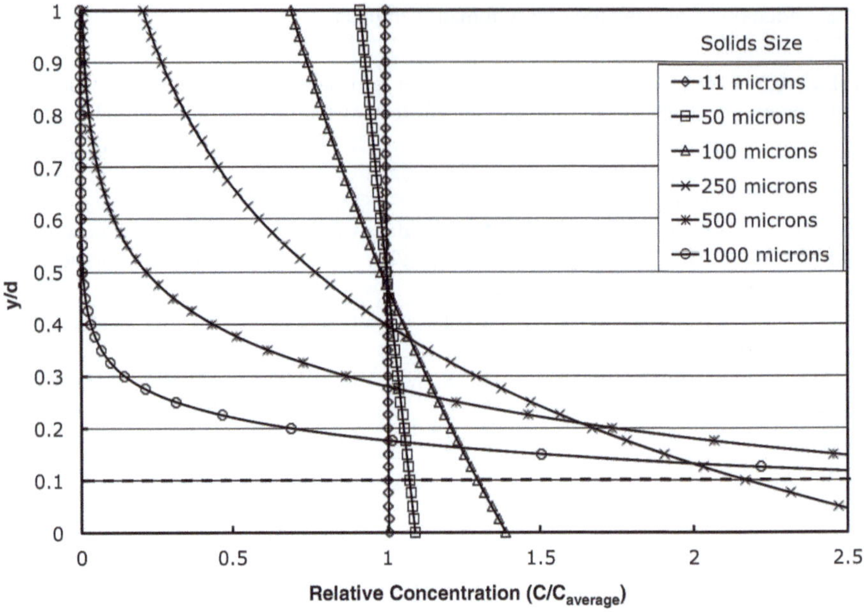

Fig. 10.6 Suspended solids concentration in a given flow condition (slope = 0.02) as a function of depth (Rouse 1937) where C = actual concentration, C_{average} = mean concentration, y = distance up from the bed, and d = depth of flow. Uniform flow in a wide open channel with particle density of sand is assumed

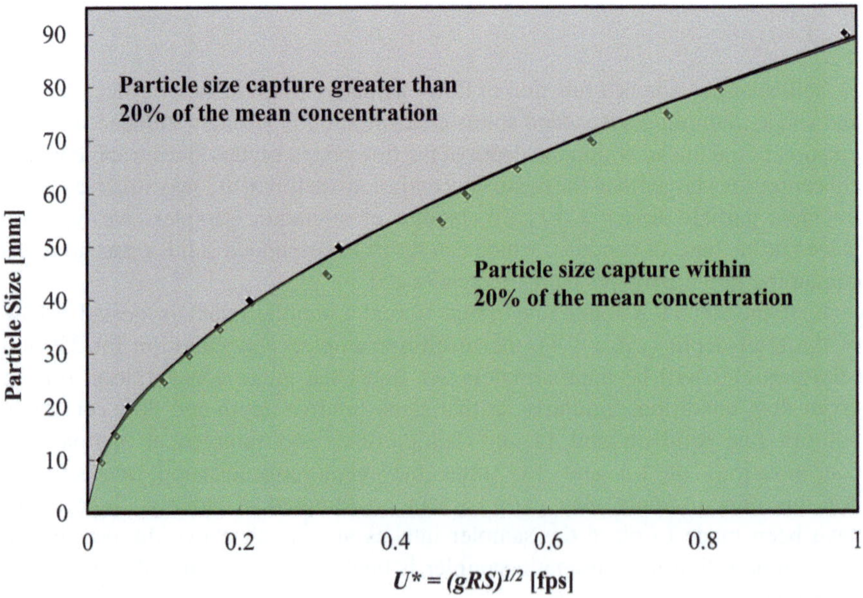

Fig. 10.7 Particle size capture (density of sand) is considered within 20% of the mean concentration under a measured flow condition typical of stormwater culverts. U^* = shear velocity (the square root of wall shear divided by liquid density), g = gravitational acceleration, R = hydraulic radius (A/P, where A = cross-sectional area and P = wetted perimeter), and S = water surface slope, assumed equal to the culvert slope

depend upon upstream entrance conditions. When sampling solids concentrations with automatic samplers, the sample is within 20% of the mean when the maximum particle size is at or below the limiting line depicted in Fig. 10.7.

Example 10.4: Accuracy of automatic sampling of water containing suspended solids

Gina, the consulting EIT, wants to determine the solid size of sand density particles that will be captured within 20% of the mean concentration of that particle size in an 18-in. (46 cm) inside diameter culvert is oriented at a 2% slope and with 6 in. (15 cm) of water depth. To use Fig. 10.7, Gina needs to calculate the shear velocity of the flow, which is the shear stress on the culvert wall divided by the density of stormwater. The hydraulic radius is also needed for the calculation, but Gina can find it in books on fluid mechanics. In this case, Gina determines that $R = 0.31$ ft (9.4 cm). Then she can determine the shear velocity using (E10.1):

$$U^* = \sqrt{gRS} = \sqrt{32.2\frac{\text{ft}}{\text{s}^2} \times 0.31\ \text{ft} \times 0.02\frac{\text{ft}}{\text{ft}}} = 0.45\frac{\text{ft}}{\text{s}} \qquad \text{(E10.1)}$$

where

U^* = shear velocity (ft/s)
g = gravitational acceleration (ft/s^2)
R = hydraulic radius
$R = A/P_w$ (ft)
S = energy grade line slope, assumed to be equal to pipe slope (ft/ft)

Now Gina can use Fig. 10.7 to determine that particles less than or equal to 60 μm in equivalent diameter (i.e., silts and clays) will be measured within 20% of their true mean concentration. From Fig. 10.7, Gina knows that sand-like particles greater than 60 μm, such as fine sand and larger, will not be measured within 20% because of their vertical distribution in the flowing stormwater.

A second challenge is the velocity with which the sample is drawn. Automatic samplers are equipped with pumps to draw samples, which create velocities different from localized streamflow velocities at the intake. When the intake velocity is equal to the streamline velocity (i.e., localized streamflow velocity), the sampled suspended solids concentration equals the mean suspended solids concentration. This is referred to as isokinetic sampling. With varying flow velocities and fixed intake velocities, automatic samplers rarely sample isokinetically.

Research on non-isokinetic samplers (FISP 1941) found significant errors for particle sizes greater than 60 μm silt. Errors associated with non-isokinetic

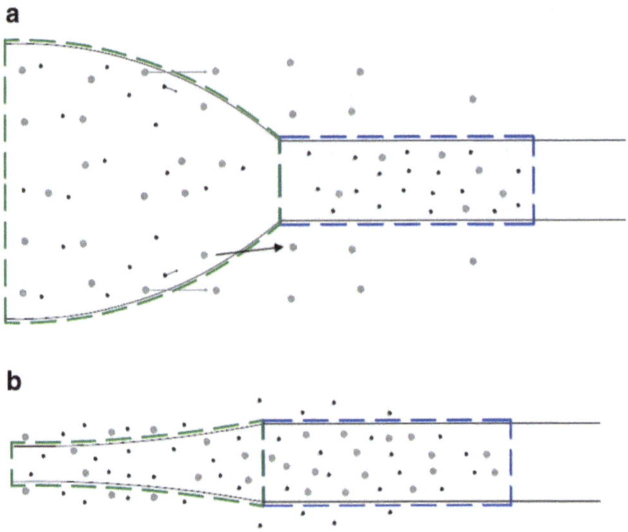

Fig. 10.8 Examples of non-isokinetic sampling (*arrows* indicate larger particles crossing streamlines). (**a**) The intake velocity is greater than the flow velocity; (**b**) the intake velocity is less than the flow velocity

sampling are due to inertial effects of the particles. The larger particles have a significant mass, which corresponds to inertial forces that can result in particles not following curved flow streamlines coming into a non-isokinetic sampling port. Dividing flow streamlines are indicated in Fig. 10.8a, b as illustrations of non-isokinetic sampling. Figure 10.8a is an example of when the intake velocity is greater than the flow velocity. Figure 10.8b is an example of when the intake velocity is less than the flow velocity. The green dashed line is an initial capture control volume upstream of the intake, and the blue dashed line is the corresponding capture control volume of the intake. Both figures contain two particle sizes, one significantly larger than the other. The small particles have minimal inertial forces and have less of a tendency to cross streamlines. The larger particles have enough inertia to move in a horizontal direction and can cross the streamlines. For the case in Fig. 10.8a, a portion of the larger particles leaves the flow streamlines and is not captured by the intake, resulting in a measured concentration smaller than the true mean. When the flow velocity is greater than the intake velocity, as in Fig. 10.8b, the larger particles cross into the streamlines, resulting in a larger measured concentration than the true mean.

For most stormwater conditions, the sampling of fine sand and sand will not be sufficiently accurate. This does create additional challenges in sampling chemicals that are attached to particles, such as particulate phosphorus and many metals and organic chemicals. Currently, the only means of ensuring accurate solids sampling is to capture all of the solids over a known length of time and discharge (Sansalone et al. 1998). Then, suspended solids concentration may be computed from (10.1), noting that each side of the equation must have equivalent units:

$$C = M \times \frac{t}{Q} \tag{10.1}$$

where

C = solids concentration (mg/L)
M = mass of solids collected (mg)
t = time of collection (s)
Q = stormwater discharge (L)

Designing sedimentation practices from inaccurate sampling of suspended solids can lead to practices that are significantly larger than required and therefore more expensive to construct and maintain.

10.6 Sample Handling

Proper sample handing is essential for representative samples. Some constituents undergo rapid reaction, such as degradation (e.g., BOD), degassing (e.g., oxygen), adsorption to the walls of bottles (many metals and organics), and coagulation (e.g., fine suspended sediment). Without proper handling, sample contamination can occur and result in inaccurate results for some analyses, most notably phosphorus and some metals. It is also important that sample handing procedures are documented and that personnel collecting samples are properly trained.

Prior to sampling, sample bottles, filtration apparatuses, filters, and other equipment must be cleaned properly. Bottles used for collection and storage of samples containing nutrient or metals often need to be cleaned with special detergents and acid rinses. Details are provided in Standard Methods (APHA 1998b).

Some analyses require that samples undergo some process prior or during storage, such as filtration and preservation. An extensive list of sample handing requirements is presented in Table 1060.1 in Standard Methods (APHA 1998b). For example, samples to be analyzed for dissolved constituents should be filtered within a few hours of collection.

Some dissolved gases (e.g., dissolved oxygen and total dissolved gas) are readily measured in situ, using field instruments. If laboratory analysis is necessary, samples must be analyzed within a few minutes or collected in sample bottles that are filled completely with water (i.e., no gas bubbles) and sealed tightly to avoid contamination by gas exchange.

Many types of samples require preservation, such as refrigeration, acidification, or reaction to form stable samples for storage. Even with preservation, acceptable holding times vary from a few hours to a few months. Analysis requirements for storage containers, cleaning, filtration, and preservation vary significantly;

therefore, it is possible for several samples to be collected at one time, or for samples to be split into many subsamples, for each type of analysis.

10.7 Recommendations for Water Sampling Methods

Sampling methods will vary based on the goals and budget of the assessment program. In the case of synthetic runoff testing or monitoring, the number of storm events sampled and the number of samples collected during storm events (synthetic or natural) will also vary depending on the assessment goals. For most assessment programs that use synthetic runoff testing or monitoring to assess pollutant removal effectiveness, however, it is generally recommended that:

1. In situ pollutant sensors be used whenever possible
2. Grab samples or automatic samples be collected promptly for pollutants or characteristics that change rapidly (e.g., temperature, bacteria, DO)
3. Flow-weighted composite samples be collected by automatic samplers, unless there is a specific need to measure pollutant concentration over time
4. Proper sample collection and handling techniques be followed to ensure representative samples and avoid contamination

Analysis of Water and Soils

11

Abstract

The water or soil samples of any stormwater treatment practice assessment effort must be analyzed in order to provide useful information. Depending on the characteristic to be determined, one or more analytical method may be available or required. This chapter introduces and discusses the most common soil and water parameters used in stormwater management and offers guidance to help the user select the most appropriate analytical method and incorporate precision and bias through a quality assurance/control program.

Except for visual inspection (level 1), each level of assessment may require the collection of samples. Analysis of these samples will determine soil or water properties such as soil moisture and pollutant concentration. The goal of this section is to identify specific parameters to be measured and to outline the analytical process that occurs after samples have been collected. A key guide for specific methods for sample collection and analysis of water is Standard Methods (APHA 1998a) and for analysis of soils is Klute (1986). A compilation of EPA methods is available on the Internet (Nelson 2003). Finally, the American Society for Testing and Materials (ASTM) publishes individual methods, which are also available online.

11.1 Selecting Analytical Methods

After assessment goals and approach have been identified and a sampling program developed, analysis methods will need to be selected. There are often several different analysis methods for measuring a given constituent in water or soil. In some cases, regulatory requirements may specify the analysis method and constituent (e.g., total phosphorus at a certified laboratory), but most assessment goals and programs allow for several methods of analysis that could satisfy the assessment goals. The following questions can be used to select an appropriate analysis method:

A.J. Erickson et al., *Optimizing Stormwater Treatment Practices: A Handbook of Assessment and Maintenance*, DOI 10.1007/978-1-4614-4624-8_11,

1. **Is a specific analysis method required by regulatory or other restrictions**? National Pollution Discharge Elimination System (NPDES) permits, Total Maximum Daily Load (TMDL) programs, or other regulatory requirements may require specific analysis methods, analysis by certified laboratories, or both. These restrictions should be listed when developing an assessment program and considered when selecting analytical methods.
2. **Will analytical results need to be compared with results from other assessment programs**? If so, sample collection, preservation, and analytical methods must be as similar as possible to minimize bias. This is particularly important for samples that must be compared to other samples within the same assessment program or at the same location. For accurate and unbiased comparison of several locations within the same assessment program, a consistent quality control program for all analysis methods may be needed.
3. **What is the quantification range of interest**? Often there are several analytical methods available for a given constituent, with varying limits of quantification (i.e., detection limits or the smallest concentration that can be measured). It is important to select an analysis method with limits of quantification that include the expected range for the constituent of concern. Also, the potential for contamination increases as the limits of quantification decrease, so additional care may be needed in sampling.
4. **Will measuring total concentration (dissolved + particulate-bound pollutants) satisfy the assessment goals, or is it necessary to measure both dissolved and particulate forms of a pollutant separately**? For example, in assessing phosphorus capture, it may be necessary to analyze both particulate and dissolved forms in order to develop an understanding of the capture and transformation mechanisms.
5. **Can multiple constituents be measured with a single analytical method**? Some analytical methods measure multiple constituents without the need for separate samples or additional analysis. For example, inductively coupled plasma (ICP) emission can measure several different constituents from one analytical injection, and ion chromatography (IC) can measure several different ions from one injection. These methods are often more expensive than analysis of a single constituent, but can be considerably cheaper than analyzing several constituents separately. For many parameters, field test kits are also available and can provide fast, inexpensive analysis without sending samples to an analytical laboratory.

The next sections discuss each of these questions and make recommendations.

11.2 Constituents in Water

Many constituents in stormwater runoff may be of concern for an assessment program. Several are discussed in the following sections. Within each group, the appropriate analysis may depend on the assessment goals, available analytical equipment, analysis cost, and potential for in situ measurement.

11.2.1 Suspended Solids

Analysis for total suspended solids (TSS) is based on filtration of a subsample, which is withdrawn from a larger sample bottle through a glass fiber filter (APHA 1998b). This technique may be inaccurate when samples contain a significant amount of sand-sized particles ($> 0.062\ \mu m$) because these particles settle quickly and it is therefore challenging to obtain a representative subsample (Gray et al. 2000, Selbig et al. 2007). Gray et al. (2000) claim that the TSS analysis method is "fundamentally unreliable" for the analysis of natural water samples and recommend measuring suspended solids concentration (SSC).

ASTM (2007b) includes three analysis methods for determining SSC. For samples with large suspended concentrations, the wet-sieving method (method C) is recommended. This method involves wet sieving the entire sample through a 62-μm (0.002 in.) sieve, followed by filtration through a glass fiber filter. In addition to providing better measurement of SSC, this method also provides some information on the distribution of particle size diameter.

For detailed analysis of particle retention, more detailed information on the particle size distribution in stormwater can be obtained by sieving followed by analysis using the hydrometer method (ASTM 2007b), which yields information on silt and clay-sized particles. Analysis of volatile suspended solids (VSS) can be used to estimate the contribution of organic matter to TSS.

11.2.2 Salinity-Related Variables

Salinity is generally defined as total dissolved solids (TDS), which is the mass concentration (mg/L) of all ions in solution. In practice, eight ions comprise nearly all of TDS (APHA 1998b), as given in (11.1):

$$\mathrm{TDS} \cong \mathrm{Ca^{2+}} + \mathrm{Mg^{2+}} + \mathrm{Na^{+}} + \mathrm{K^{+}} + \mathrm{Cl^{-}} + \mathrm{SO_4^{2-}} + \mathrm{HCO_3^{-}} + \mathrm{CO_3^{2-}} \qquad (11.1)$$

It is often unnecessary to measure all major ions for a stormwater assessment program. Gravimetric analysis for measuring TDS is tedious; therefore, it is common to estimate TDS from measurements of specific conductance (SC), which can be measured in situ. A regression relationship between TDS and SC must be developed from at least 25 samples collected over a range of TDS and SC values (e.g., see Fig. 11.1). Once the relationship is determined, SC is measured directly, and the SC–TDS regression relationship is used to estimate TDS. The SC–TDS relationship may not be valid if the ionic composition of the water changes significantly, as may occur with road salt deicer application in cold climates. The SC–TDS regression should be verified regularly with multiple regressions developed if there are significant differences between different time periods (e.g., winter months vs. summer months).

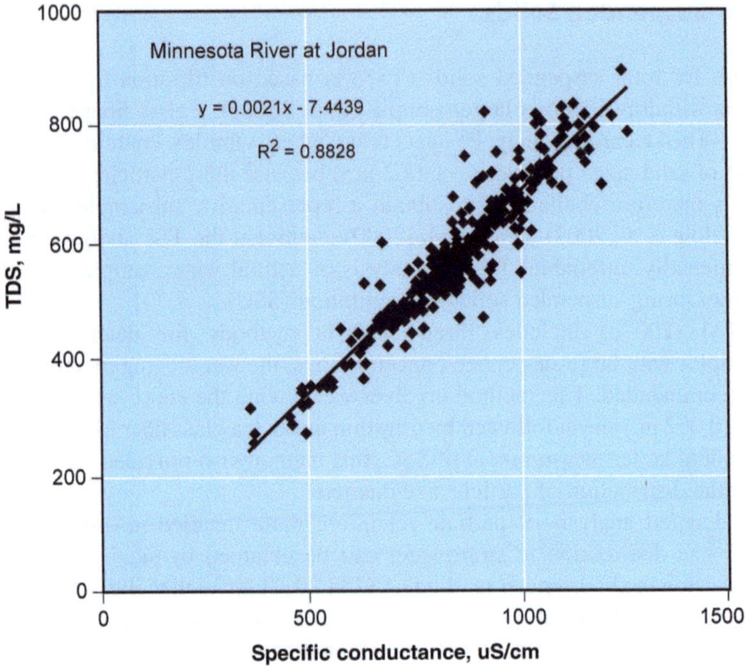

Fig. 11.1 An example of the relationship between specific conductance and total dissolved solids (TDS). Data are from water samples of the Minnesota River at Jordan, MN, USA. Source: USGS water quality database

11.2.3 Natural Organic Matter

In addition to VSS, two common metrics of organic matter in water include biochemical oxygen demand (BOD) and chemical oxygen demand (COD). BOD is a measure of readily decomposable organic matter, generally measured over a period of 5 days (BOD_5). Measurements of BOD are needed when an assessment goal is to determine the reduction of oxygen-depleting material by a stormwater treatment practice. COD measurements are often less expensive, and it is common practice in wastewater to estimate BOD from COD because the relationship between BOD and COD is approximately constant at ~2:1 (Metcalf and Eddy 1991). However, this is not advisable for stormwater because the ratio varies significantly (Maestre and Pitt 2005).

11.2.4 Phosphorus

Phosphorus (P) exists in many forms in the environment: as phosphate ions, as polyphosphates, as a component of RNA and DNA, and in phospholipids. Analysis

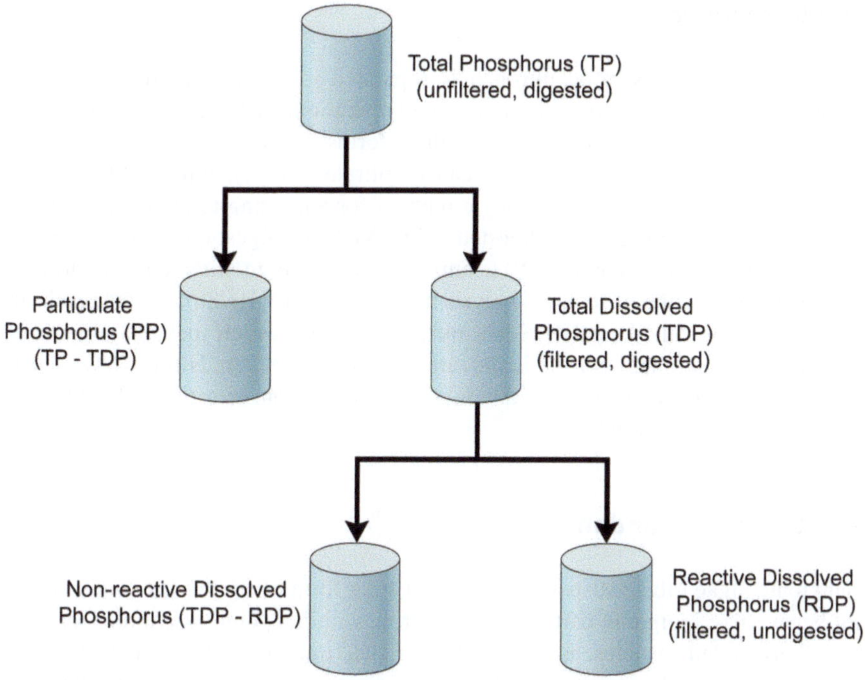

Fig. 11.2 Analytical methods for determination of the multiple forms of phosphorus in stormwater

of the exact chemical form of P requires specialized analytical techniques that are generally used only in research projects. For stormwater assessment, total P is commonly separated into particulate and dissolved by filtering through a 0.45-μm filter. Sometimes, dissolved P is separated into total dissolved and reactive dissolved (also called soluble reactive), although it is more common to measure total P and reactive dissolved P due to the additional cost of measuring total dissolved P. Distinction between these forms of P is based on operational definitions of analysis, including filtration and type of digestion (Fig. 11.2).

Total P is determined by using one of several strong acid digestion techniques on an unfiltered sample and measuring P concentration, typically by a colorimetric method. Dissolved reactive P is determined by measuring the phosphorus concentration of a filtered sample (filtrate). Filtered samples may be digested to determine total dissolved P. Particulate phosphorus and nonreactive dissolved phosphorus are calculated by difference, as shown in Fig. 11.2. Knowledge of the different forms of P is important because P removal processes are dependent upon the form. For example, sedimentation or filtration can typically capture particulate P but not dissolved P. It is also possible that transformation of P occurs within stormwater treatment practices, which can only be determined if both particulate and dissolved phosphorus fractions are measured.

11.2.5 Nitrogen

Similar to phosphorus, total nitrogen is separated in several common forms of nitrogen (N), which are typically determined operationally by filtration, digestion, and chemical analysis. The most common forms of N in stormwater include particulate N, dissolved organic N (DON), nitrate (NO_3^-), nitrite (NO_2^-), and ammonium (NH_4^+). There is no single method for analyzing total N; therefore, it is common for samples to be digested using Kjeldahl digestion, which converts organic N (particulate N and DON) to ammonium. The ammonium concentration is then measured by one of several analysis methods (APHA 1998b) and the resulting concentration is called total Kjeldahl nitrogen (TKN), which includes organic N plus the ammonium that existed in the sample before digestion. Nitrate (NO_3^-) and nitrite (NO_2^-) can be measured individually and total N can be estimated as the sum of TKN + (NO_3^-) + (NO_2^-).

11.2.6 Algae Abundance

Planktonic algae (algae suspended in water) are a form of suspended solids but do not behave like inorganic particles. Algae grow within a pond, essentially forming "new" suspended particles, as opposed to particles from the contributing watershed. In addition, algae do not settle in the same manner as inorganic particles because they have smaller density and different shapes. Large concentrations of algae, particularly blue-green algae, can become a nuisance and health hazard (primarily to pets) for homeowners living near a wet pond or downstream receiving water bodies.

Chlorophyll concentration provides an accurate measure of total algal abundance, and an estimate of blue-green algae. For most stormwater applications, chlorophyll samples are collected and then filtered. The chlorophyll on the filter is then extracted into acetone or another solvent and measured by spectrophotometry or fluorometry (APHA 1998b). When samples cannot be analyzed immediately, filters are frozen for preservation. Chlorophyll can also be measured in situ using fluorescence-based monitors. This enables real-time continuous measurement, which could be useful for assessment programs that focus on algae.

As a rule of thumb, chlorophyll comprises approximately 1% of algae (dry weight basis). Therefore, a sample with 0.1 mg chlorophyll/L (a hypereutrophic pond) is equivalent to a TSS concentration of approximately 10 mg/L.

11.2.7 Metals

Urban stormwater often contains metals at environmentally significant levels, including cadmium, zinc, copper, and lead (see Chap. 3). Metal species in water are most commonly measured as "total" concentration (unfiltered samples) and "dissolved" concentration (filtered samples, generally through a 0.45-μm filter). Metals must be in the dissolved state to be biologically available, so measuring both dissolved and total metal concentrations is recommended. Samples for "total" metals analysis are digested with strong acids and/or oxidants to dissolve metals that are particles or bound to particles. Digestion techniques vary with respect to completeness in releasing metals from solution, so the description of "total" is operationally defined by the type of digestion used (APHA 1998b). The most rigorous digestions may involve hazardous materials (such as hydrofluoric acid or perchloric acid), so unless this level of digestion is required, milder, safer digestion procedures are generally used. The analytical laboratory involved in the assessment project should be consulted regarding digestion procedures.

Once in soluble form, metals are measured by atomic adsorption spectrometry or ICP emission spectrometry. Atomic adsorption measures one element at a time but generally has lower detection limits, while ICP can be used to measure several different metals simultaneously.

Metals readily bind to soils but do not degrade; therefore, accumulation of metals over time is expected in infiltration and filtration practices. Using typical values of stormwater loading and soil capacity estimates, Davis et al. (2003) estimated that, after 20 years, concentrations of cadmium, lead, and zinc would reach or exceed levels permitted by US EPA biosolids land application regulations. Legret et al. (1999) conducted studies and applied mathematical models to determine that the increase in soil metal concentrations would be slight after 50 years. Studies have found that metals are typically retained on the soil particles within the first 10–40 cm (5–15 in.), with cadmium having the most downward mobility (Barraud et al. 1999, Dierkes and Gieger 1999).

Some metals such as zinc and copper are also micronutrients used by plants and may be accumulated into plant biomass as the plant grows. Plant species accumulate metals at different rates and have different metal tolerances (Sun and Davis 2007). Metals that are not micronutrients may adsorb to plant roots, but they are typically not assimilated into plant material. There may therefore be a need to measure changes in the metal content of soils over time. Further discussion, including analysis methods, is included in the following section.

11.3 Soils

This section addresses the measurement of soil properties and soil constituents that are important to stormwater treatment practices. Before soil properties and constituents can be determined, a soil sample must be collected using a standardized method such as ASTM D6640-01 (2010b). There may be several methods available for each analysis, but only the most commonly used methods are discussed. Detailed description of the following analysis can be found in Methods of Soil Analysis Part 1 and Part 2 (Klute 1986, Black 1965).

11.3.1 Soil Properties

Soils are an integral component of a variety of stormwater treatment practices and provide numerous functions for the treatment of stormwater runoff. Efficient treatment of stormwater runoff by soil processes requires properly functioning and stable soils. Therefore, measurement and understanding of soil properties are important for the overall assessment of stormwater treatment practices. Soil is composed of three phases: the soil matrix (solid phase), the soil solution (liquid phase), and the soil atmosphere (gaseous phase). The volume and mass relationships among these phases, along with some basic parameters, are useful to characterize the physical characteristics of the soil. The processes and the physical properties of the soil vary with surface location and soil depth; therefore, soil analysis will typically require samples from several locations spatially at the soil surface and vertically within the soil profile.

11.3.1.1 Bulk Density

Bulk density is the ratio of the mass of solids to the total soil volume. It can be used to estimate degree of compaction and is needed to calculate soil moisture content and porosity. The bulk density of soil is influenced by soil structure due to its looseness or degree of compaction and by its swelling and shrinking characteristics (Hillel 1998). Soil compaction in stormwater treatment practices reduces infiltration by reducing the pore space available for water transmission. Soils in stormwater treatment practices can become compacted during construction. Post-construction soil compaction typically does not occur unless heavy machinery is used for maintenance/redevelopment of the practice or surrounding areas.

Bulk density is typically measured using the core method, which involves drying and weighing a soil sample of a known volume (Klute 1986). Bulk density can also be measured or estimated using digital or analog soil penetrometers and the sand cone test (ASTM 2007a), among others.

11.3.1.2 Soil Texture

Many of the physical and chemical properties of soil are affected by soil texture (Pepper et al. 1996). Soil texture is described by classifications that are determined by the particle size distribution of sand, silt, and clay within the soil. The particle size distribution of soil can be measured in the laboratory according to standards of the US Department of Agriculture, the American Society for Testing and Materials, and the International Soil Science Society.

Pretreatment of the soil is required prior to particle size distribution analysis to improve the separation and dispersion of aggregates (Klute 1986). After pretreatment of the soil sample, the sand fractions are measured using mesh sieves of various-sized openings. The fraction of silts and clays can be determined using the pipet or hydrometer methods. The pipet method is a sedimentation analysis that relies on the relationship of settling velocity and particle diameter (Klute 1986). The hydrometer method is similar to the pipet method but makes use of a calibrated

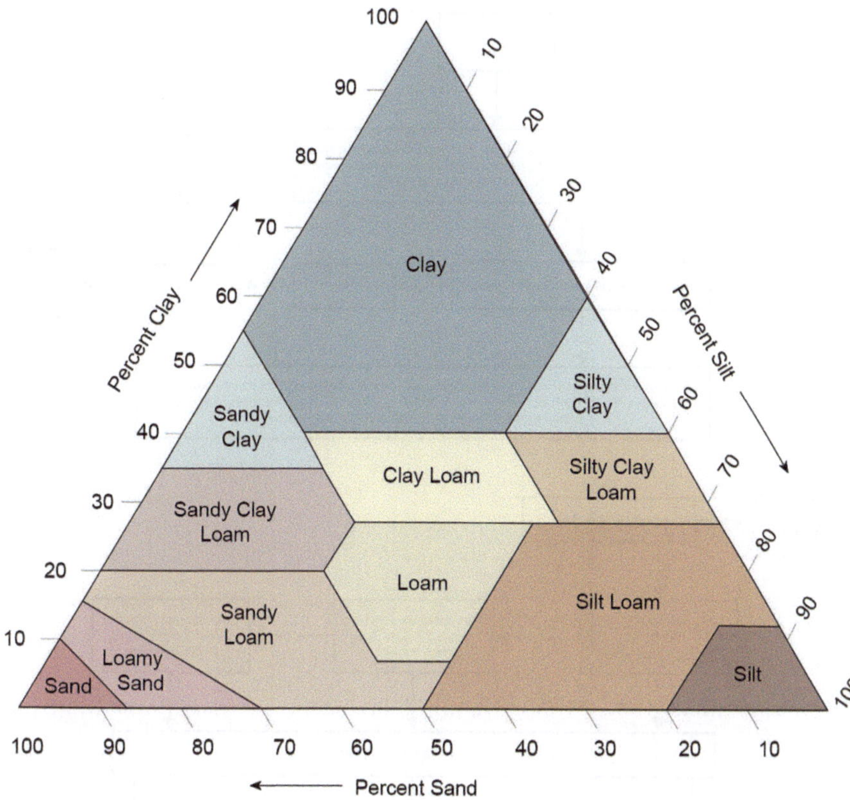

Fig. 11.3 The percentage of sand, silt, and clay in the basic textural classes. (© United States Department of Agriculture, Natural Resources Conservation Service, Reprinted with permission)

hydrometer for multiple measurements of the suspended sediment over time. Given the percent sand, silt, and clay of a soil, the soil texture classification can then be determined from the USDA textural triangle, as shown in Fig. 11.3. Alternatively, a field procedure for approximating of soil texture (Thien 1979) is provided in Fig. 11.4.

Soil texture analysis should also include identification of hydric soils. Hydric soils are formed when anaerobic conditions develop in the root zone due to prolonged saturation during the growing season (Richardson and Vepraskas 2001) and can be identified by the gray color of the soil and the presence of mottles. The gray color indicates a process of "gleying," which includes the chemical reduction of iron or manganese. Mottles are small areas that differ in color (gray, red, yellow, brown, or black) from the soil matrix because of water saturation and chemical reduction. Reddish mottles, for example, are due to the accumulation of iron oxides in root channels or large pore spaces, and black mottles indicate the accumulation of manganese oxides (Richardson and Vepraskas 2001). Hydric soils are evidence

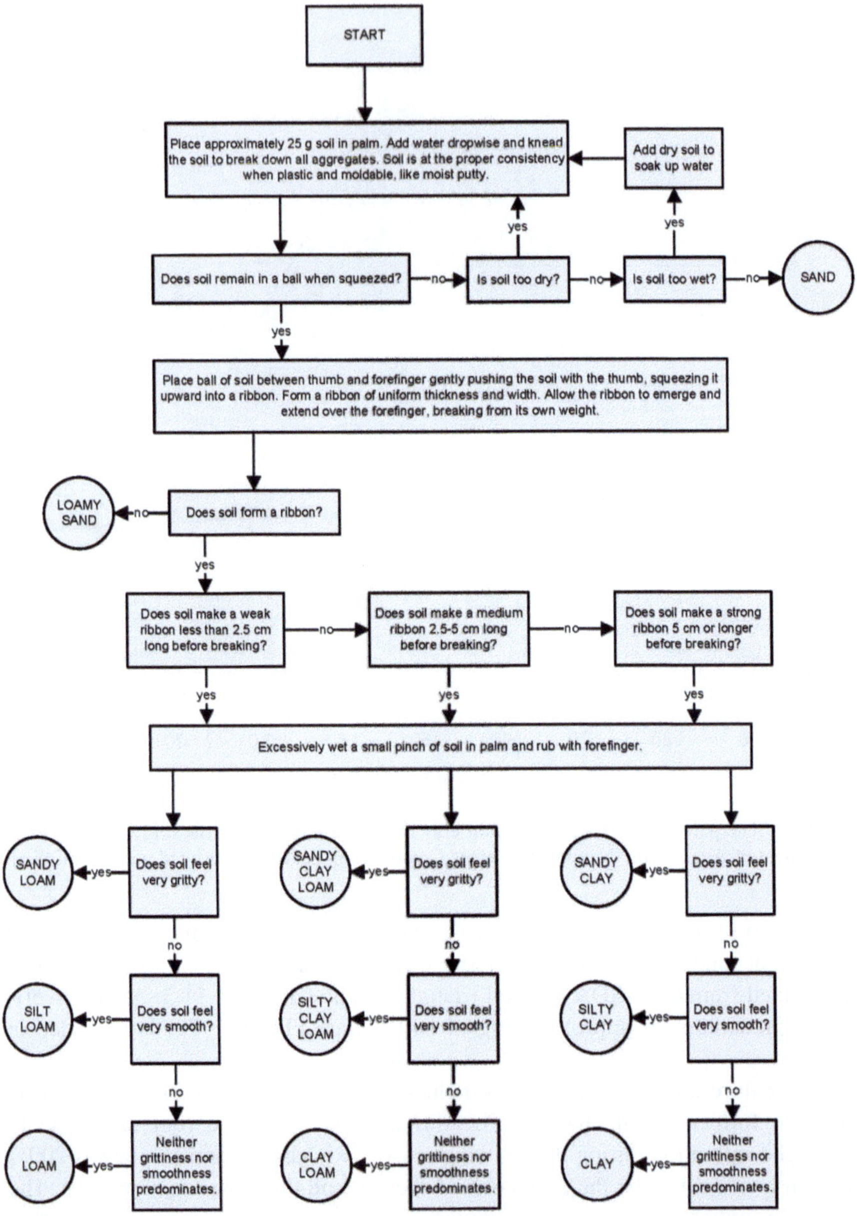

Fig. 11.4 Field method for soil texturing (From Thien (1979), with permission, © American Society of Agronomy)

of prolonged water saturation, indicating that stormwater runoff is not infiltrating at sufficient rates. For more information on identifying hydric soils, refer to "Wetland Soils: Genesis, Hydrology, Landscapes, and Classification" (Richardson and Vepraskas 2001).

11.3.1.3 Porosity

The pore spaces in the soil matrix vary in amount, size, shape, tortuosity, and continuity and are an important physical property of the soil, especially with regard to the retention and transport of solutions, gases, and heat (Klute 1986). For stormwater treatment practices, porosity is important because it is a measure of the soil's capacity to infiltrate water. Porosity is the volume fraction of pores, where most soils are within a porosity range from 0.3 to 0.6 (Hillel 1998). Porosity (f) can be calculated using (11.2) if the particle density (ρ_s) and the dry bulk density (ρ_b) are known. A typical mineral soil has a particle density equal to that of sand (2.65 g/ cm^3). Porosity can also be measured directly (Klute 1986) by saturating a known volume of soil, weighing it, and then drying it at 105 °C (221 °F). The difference in weight is the pore volume (the bulk density of water = 1.0 g/cm^3). Therefore, the porosity is equal to the pore volume divided by the original soil volume:

$$f = 1 - \frac{\rho_b}{\rho_s} \tag{11.2}$$

where

f = porosity (percent)
ρ_b = dry bulk density (gm/cm^3)
ρ_s = particle density (gm/cm^3)

11.3.1.4 Penetrability

Compaction of soils can occur due to normal activities, can be induced by machinery, or both. Compacted soils exhibit small total porosity due to reduced volume and continuity of large pores, which restricts aeration and impedes root penetration, infiltration, and drainage (Hillel 1998). Compaction can be estimated with penetrometers, which measure the ease with which an object can be driven into the soil. Penetration resistance measured by penetrometers is influenced by several soil factors, including water content, bulk density, compressibility, soil strength, and soil structure (Klute 1986).

11.3.1.5 Water Content

The amount of water in the soil influences numerous soil properties, governs the air content and gas exchange of the soil, affects plant growth, influences microbial activity, and dictates the chemical state of the soil (Hillel 1998). The measurement of water content may also be necessary for other assessment or analysis methods, such as determination of saturated hydraulic conductivity (K_s) of the soil with an infiltrometer or permeameter (see Sect. 9.4.1 in Chap. 9). Prolonged saturation of the soil promotes the formation of hydric soils, while soil dryness can inhibit plant growth.

There are both direct and indirect methods for measuring water content. The traditional gravimetric technique involves weighing a fresh soil sample, drying the sample in an oven or microwave, and reweighing the sample to determine the amount of water removed. Indirect methods rely on the relationship between water content and certain physical and physical–chemical properties of the soil (Klute

1986), such as electrical resistance, capacitance, neutron scattering, gamma-ray absorption, and time-domain reflectometry. There are several probes available for manually or automatically measuring water content in the field, with varying degrees of accuracy and calibration requirements. Alternatively, visual (and tactile) assessment of the soil moisture can be made and the soil described as dry, moist, saturated, or inundated.

11.3.1.6 Saturated Hydraulic Conductivity

The saturated hydraulic conductivity (K_s) of soil is a measure of its ability to transmit water and is used in Darcy's Law to calculate flow or infiltration rates (Klute 1986). The terms "permeability" and "hydraulic conductivity" are sometimes used synonymously; however, permeability is an exclusive property of the soil matrix, while saturated hydraulic conductivity (K_s) is also a function of the fluid properties (Hillel 1998). See Chap. 9 for a discussion of some available devices used to measure K_s.

11.3.1.7 Cation Exchange Capacity

The cation exchange capacity of soil is a major sorption mechanism for pollutants and is due primarily to the negative charge associated with clay particles and organic particles. Positively charged ions (cations) such as metals are attracted to the negatively charged sites on the clay particles, which influences the mobility of cations. As plants and microorganisms utilize cations in the soil solution, exchanges are made from the negatively charged particles (soil colloids) to the soil solution (Pepper et al. 1996) to balance the charge. The cation exchange capacity is the sum of the exchangeable cations in the soil and is usually expressed as milliequivalents of positive charge per 100 g of soil (mEq (+) 100 g^{-1}). Common methods for measuring cation exchange capacity include saturating soil with a particular cation and then measuring the absorbed cations (Black 1965).

11.3.1.8 Soil pH

Soils with large concentrations of organic matter in areas with large rainfall tend to be acidic (pH < 5.5). The pH of the soil can influence the degree of ionization of compounds, especially metals, which affects their solubility and may be critical to the transport of pollutants through the soil (Pepper et al. 1996). The measurement of soil pH is split into two methodic groups: colorimetric methods that utilize dyes or acid–base indicators and electrometric methods that utilize an electrode to measure the hydrogen ions (Black 1965).

11.3.1.9 Other Soil Properties

The analysis of other soil properties, such as water potential, evaporation rate, temperature, and air permeability, may also be desirable for assessment. For detailed standard procedures, see Methods of Soil Analysis (Klute 1986).

11.3.2 Soil Constituents

Stormwater runoff carries various types of pollutants with it as it is conveyed. Measuring the type of constituents and their concentration in the soils of a stormwater treatment practice can be a useful assessment tool for understanding the soil's capacity to retain those constituents. Retention (typically via sorption) is one of the major processes influencing the transport of pollutants in soil (Pepper et al. 1996). The retention and transformation of pollutants in the soil can prevent water quality degradation in lakes, streams, and rivers. To ensure that pollutants are retained and that the soil has not reached its capacity, analysis of the soils within the practice may be necessary. The mobility of pollutants and the physical properties influencing their transport vary spatially; therefore, samples from several locations may be required to adequately characterize the pollutants in the soil. This section will discuss some of the key pollutants found in soils of stormwater treatment practices.

11.3.2.1 Organic Matter

Plant residues incorporated into the soil surface are degraded by microorganisms into organic matter that is utilized by plants and microbes for metabolism and also incorporated into macromolecules (Pepper et al. 1996). Organic matter affects physical properties of soil such as bulk density, porosity, and infiltration rate. The humic and nonhumic substances in organic matter contribute to the pH-dependent cation exchange capacity of the soil and the chelation of metals. There are two approaches for determining organic matter content: loss on ignition and carbon content. Loss on ignition is a measure of organic matter volatilization and is determined by the difference between the dry weight of a combusted and noncombusted sample. Measuring organic carbon (OC) content is commonly done by measuring the production of CO_2 during high-temperature combustion. Organic matter is then estimated using a ratio of carbon to organic matter (C:OM) which is often assumed to be approximately 0.5.

11.3.2.2 Salinity (Including Chloride)

Soil salinity refers to the concentration of soluble salts within the soil. Salinity can harm plants by interfering with water uptake or through direct toxicity of ions associated with salinity (especially chloride). The accumulation of salts in soils also indirectly affects soil properties such as swelling, porosity, water retention, and saturated hydraulic conductivity (Hillel 1998). A major source of salinity in stormwater treatment practices is road salt.

A common method for measuring salinity is by electrical conductivity, which is typically expressed as millimho (mmho) per centimeter (Black 1965). According to the US Department of Agriculture *Handbook 60* (Richards 1954), a saline soil has an electrical conductivity exceeding 4 mmho/cm at 25 °C (77 °F) (Hillel 1998).

11.3.2.3 Phosphorus

Phosphorus is retained in soils by adsorption and chemical precipitation. The accumulation of phosphorus in soils may need to be measured in stormwater treatment practices to assess whether the soils are saturated with phosphorus. The most commonly used phosphorus adsorption metrics are the Bray extraction method (for noncalcareous soils) and the Olsen method (for calcareous soils) (AES 1988).

11.3.2.4 Nitrogen

Total nitrogen is separated in several common forms: particulate N, nitrate (NO_3^-), nitrite (NO_2^-), and ammonium (NH_4^+). Plants typically use nitrogen as ammonium (NH_4^+) or nitrate (NO_3^-), which are generated directly from dissolving salts or indirectly through processes such as nitrogen fixation (conversion of atmospheric nitrogen to ammonia) and nitrification (oxidation of ammonia and ammonium to form nitrate) (Pepper et al. 1996). Nitrate is very soluble and has the potential to contaminate groundwater. Nitrate in runoff is also the primary contributor to hypoxia (dead) zones in coastal areas of the oceans. A biological process called denitrification converts nitrate into nitrogen gas (N_2) and is discussed in Chap. 3. There are several methods of nitrogen analysis, and the appropriate technique should be chosen depending on the form(s) of concern.

11.3.2.5 Metals

Metals commonly found in stormwater are lead, zinc, copper, and cadmium. There is the potential for the accumulation of metals in stormwater treatment practices, especially in the soils of infiltration practices. "Total" metal content is determined by rigorous hot-acid digestion of soil samples with one of several strong acids (often nitric), generally with a catalyst. "Extractable" metals are measured using extractions with weaker acids, usually at room temperature. A detailed discussion of metals analyses is found in Standard Methods (APHA 1998b).

11.3.2.6 Microbial Populations

Soil contains billions of living organisms that are essential to biochemical transformations and the overall health of the soil. These organisms can also capture and convert pollutants from stormwater runoff that may filter or infiltrate through the soil profile. The major groups of organisms found in soils include bacteria, actinomycetes, fungi, algae, viruses, and macrofauna (Pepper et al. 1996). The abundance of microorganisms in the soil, and therefore pollutant biodegradation, is dependent on oxygen and nutrient availability, organic matter content, pH, redox potential, temperature, and soil moisture texture (Pepper et al. 1996). Due to the variability in the type of microorganisms present in the soil and the special requirements for each species, it is challenging to measure the entire biological community in the soil. The most-probable-number (MPN) method is an estimate of the population density that avoids direct measurement of actual colonies (Black 1965). There are, however, additional techniques for measuring specific microorganisms (Black 1965).

11.4 Quality Assurance Program

Taylor (1987) defines quality assurance as the "system of activities whose purpose is to provide to the producer or user of a product or a service the assurance that it meets defined standards of quality with a stated level of confidence." Quality assurance has two components: quality control and quality assessment. Quality control is the process of minimizing errors in sample handling and analysis. Quality assessment is the quantification of errors and comparison of errors to acceptability standards. For stormwater assessment and maintenance, quality assurance is important for all assessment processes (e.g., sample collection, sample analysis, flow measurement, and pollutant load calculations, among others) to ensure proper conclusions about performance and proper selection and scheduling of maintenance.

Many assessment programs will require sample collection (see Chap. 10) and analysis. Some assessment program managers may lack access to the proper equipment or expertise to conduct sample analysis and therefore choose to send samples to commercial or government laboratories for analysis. In some cases, regulatory agencies require the use of "certified" analysis laboratories to ensure strict adherence to standardized methods such as Standard Methods (APHA 1998b), the American Society for Testing and Materials (ASTM), and US EPA approved methods (Nelson 2003).

Commercial or government laboratories often make use of quality assurance programs, but these programs assure only the quality of the analytical result for the sample(s) submitted to the laboratory. The laboratory cannot assure that samples are representative or were collected and handled properly (see Sect. 10.6 in Chap. 10). Assessment programs that do not use commercial or government laboratories for analysis must also include quality assurance for sample analysis as well as sample collection and handling. Therefore, the assessment program manager is responsible for developing, implementing, and enforcing a quality assurance (QA) program. The quality assurance program must start with sample collection and proceed through sample processing, lab analysis, and validation of results.

11.4.1 Quality Control

Quality control is the process of minimizing errors in sample collection, handling, and analysis. Minimizing errors in sample collection, handling, and analysis can be achieved by choosing standard recognized methods and performing those methods consistently. The following three steps can be used to provide quality control for sample collection, handling, and analysis: sample procedure documentation, personnel training, and compliance.

Sample procedure documentation includes information on sample collection, handling, and analysis. Documentation of sample procedures should include sufficient detail such that the sampling and analysis program can be replicated by an individual unfamiliar with the assessment program (i.e., third party). This will ensure that the program can be continued by a third party, if necessary, or can be replicated at a different location to facilitate comparison between two or more sites. Documentation at this level of detail also ensures that the program is consistent regardless of personnel changes or length of time.

Sample procedures should be documented before commencing an assessment program to ensure measurements throughout the program are comparable. Any changes to the procedures should be noted in the documentation along with the time frame for which the changes are effective. Documentation of sample procedures is commonly compiled into a guidance manual that could include the following:

- Field sampling methods: sampling locations, methods, measurements, equipment, etc.
- Preparation of samples (precleaning of bottles, filtration, storage, etc.).
- Labeling and chain-of-custody procedures.
- Analytical methods, including any modifications of test procedures from standard conditions.
- Safety protocols.

Personnel should be properly trained on sampling procedures to ensure consistent and quality controlled sampling. Use the documentation of sampling procedures to ensure consistent training and to allow personnel to become familiar with the documentation.

Assessment programs should be regularly checked for compliance with the sampling procedures documentation. Compliance will ensure consistent implementation of the sampling procedures and allow for comparison of assessment results between different sites, different time periods, or both. Regular compliance checks can often quickly identify errors in sampling procedures, equipment, or analysis techniques. If identified quickly and corrected, errors due to non-compliance usually result in minimal loss of data.

11.4.2 Quality Assessment

Quality assessment is the quantification of errors and comparison of errors to acceptability standards. Error in sample collection, handling, and analysis is caused by two components: bias and precision. Bias is systematic error that results in sample values that are consistently the same or different than the "true" value, typically by a constant amount. Precision is a measure of the similarity between repeated values and results in sample values that are consistently the same or different from each other. Both precision and bias decrease (i.e., become imprecise and biased) as measurements and analysis results approach the detection limit. Figure 11.5 is an illustration of bias and precision and the following describe how bias and precision are represented by each of the colored clusters:

- The *green* values are both unbiased and precise. They are close to the true value (center) and relatively similar. This is an optimal sampling procedure.

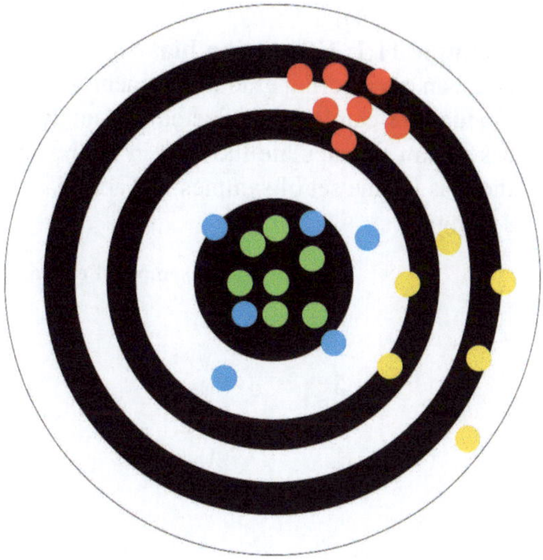

Fig. 11.5 Schematic illustrating sample accuracy and precision

- The *blue* values are unbiased but imprecise. They are close to the true value, but there is substantial variability in the values.
- The *red* values are precise, but biased. They are similar values but consistently different from the true value.
- The *yellow* values are both imprecise and biased. They are consistently different from the true value and there is substantial variability in the values.

11.4.3 Bias

Bias can be calculated using (11.3). For water or soils analyses, bias can be introduced in the analytical system (e.g., by contaminated reagents) or in sampling (e.g., by contaminated filters or sample bottles). Analysis bias is generally determined by analyzing samples mixed to standard concentrations (also called "check samples" or "known values"). A common technique is to include check samples regularly and frequently throughout the analysis process (e.g., every 10th sample). An example of how bias can be calculated is given in Example 11.1:

$$b = \frac{(C_{\mathrm{m}} - C_{\mathrm{T}}) \times 100}{C_{\mathrm{T}}} \tag{11.3}$$

where

b = bias (%)
C_{T} = true (known) concentration of QC standard (commercially prepared solution made with pure chemicals)
C_{m} = average measured value of QC standard

Example 11.1: Estimating bias
Alex, an engineering graduate student, is conducting assessment as part of his graduate research. He collected and analyzed some samples and his adviser asked him to calculate the bias in his results. Using (11.3), Alex calculated the bias for the set of samples in Table E11.1, compared to a known value of 1.6 mg/L:

Table E11.1 Example of experiment data

Measurement	Value
1	1.74
2	1.65
3	1.90
4	1.70
C_m	1.75

$$b = \frac{(C_{\mathrm{m}} - C_{\mathrm{T}}) \times 100}{C_{\mathrm{T}}} = \frac{(1.75 - 1.6) \times 100}{1.6} = 9.2\%$$

A common cause of bias is sample contamination. Bias as a result of contamination can often be determined by analysis of field blanks. Field blanks are commonly samples of high-purity distilled water, generally provided by the analytical lab, that are filtered, stored, labeled, and analyzed according to the sampling procedure documentation. If analysis of field blanks indicates measured values greater than detection limits, contamination is likely occurring in the sampling, handling, or analysis process. Field blanks can be incorporated during each step of the collection, handling, and analysis process to identify the step in which samples are becoming contaminated.

11.4.4 Precision

Precision is quantified by repeating analysis on the same sample or on samples with the same concentration. Precision is often calculated as a relative standard deviation (RSD), as given in (11.4). For sampling programs, a common practice is to collect a replicate sample for every ten samples and estimate the "pooled" standard deviation from (11.5). The pooled standard deviation is suitable to field sampling because the mean concentration cannot typically be assumed to be constant. An example calculation of precision is given in Example 11.2:

$$\mathrm{RSD} = \frac{s}{C_{\mathrm{m}}}(100\%) \tag{11.4}$$

where

RSD = relative standard deviation
s = standard deviation of the measurements
C_m = average measured value of QC standard

When there the duplicates, for example, the standard deviation is given as:

$$s = \left(\frac{\sum d^2}{2(k-1)} \right)^{\frac{1}{2}} \tag{11.5}$$

where

d = difference of duplicate measurements
k = number of sets of measurements

Example 11.2: Estimating relative standard deviation
Alex, an engineering graduate student, is conducting assessment as part of his graduate research. He collected and analyzed some samples and his adviser asked him to calculate the precision in his results. Using (11.4) and (11.5), Alex calculated the precision for the set of replicated samples in Table E11.2:

Table E11.2 Example experiment data

First result	Second result	d	d^2
3.4	3.6	−0.2	0.04
4.5	4.4	0.1	0.01
10.3	9.7	0.6	0.36
12.8	13.4	−0.6	0.36
5.0	4.5	0.5	0.25
6.1	5.5	0.6	0.36
$C_m = 7.017$			$\sum$=1.38

$$s = \left(\frac{\sum d^2}{2(k-1)} \right)^{\frac{1}{2}} = \left(\frac{1.38}{2 \times 6} \right)^{\frac{1}{2}} = 0.371$$

$$\text{RSD} = \frac{0.371}{7.017}(100\%) = 5.29\%$$

11.4.5 Detection Limit

The limit of detection (LOD) is the lowest concentration that can be reliably distinguished from a blank sample (high-purity distilled water). Operationally, the LOD is determined by repeated analysis of a low-level standard to determine the standard deviation (s), and the LOD is calculated as three standard deviations (LOD = 3 s). The limit of quantification (LOQ) is the smallest concentration that can be reliably quoted as a measured concentration. The LOQ is typically given as ten standard deviations (LOQ = 10 s). Most standard analytical methods include information on detection limits and quantification limits. Most constituents can be analyzed by more than one analytical method, but methods with smaller LOD are commonly more expensive than methods with a larger LOD. Therefore, selecting an analytical method is often a balance between LOD and cost.

11.5 Implementation and Verification

A typical QA program includes analysis of field blanks to determine contamination (bias), replicates to determine precision, and check samples to determine bias. If bias or imprecision is detected during sample collection, handling, or analysis, the sample procedure documentation may need to be revised, personnel may need additional training, or compliance may need stricter enforcement. Example 11.3 is an example of how QA program results can be reviewed to determine bias, precision, and the quality of the gathered data.

Example 11.3: Reviewing and interpreting quality assurance program results

Alex, an engineering graduate student, is conducting assessment as part of his graduate research. He is reviewing results from his QA program for precision and bias. Alex has decided on the following thresholds for quality data: bias within ±10%, precision within ±10%, and contamination bias (field blanks) < 0.10 mg/L for the contaminant he is measuring.

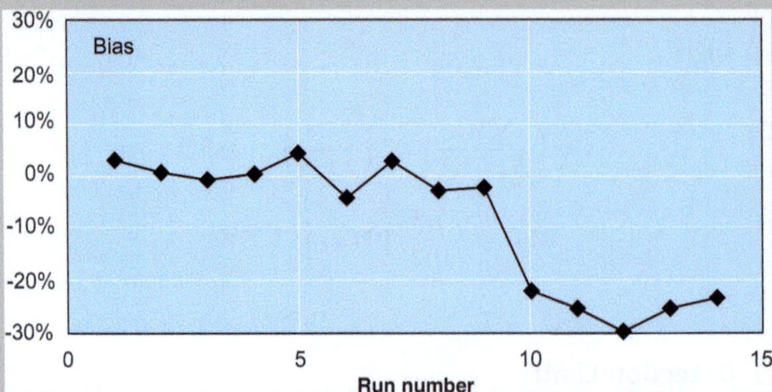

Fig. E11.1 Bias

(continued)

Example 11.3: (continued)

Alex examines the QA program data and finds that bias values were consistently between ±5% through run 9, but bias was greater than the threshold (±10%) in runs 10–14 (<−20%).

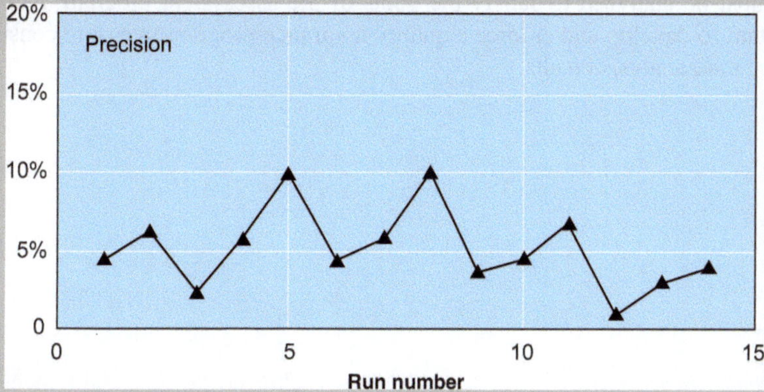

Fig. E11.2 Precision

There is no apparent trend in precision (upward or downward) and the average precision is around 5%, which is within the threshold (±10%).

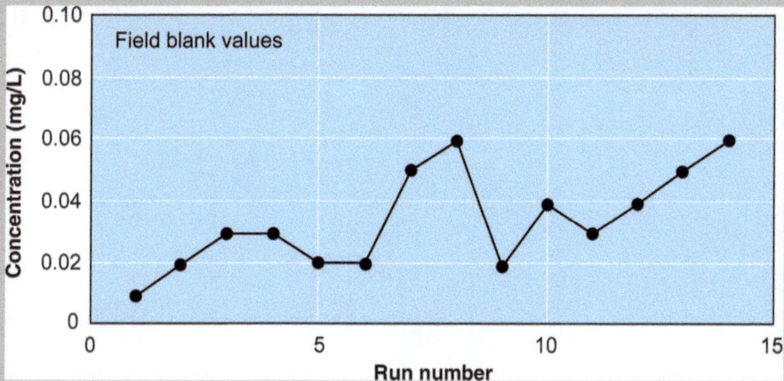

Fig. E11.3 Field blanks

Field blank values are consistently less than 0.06 mg/L, which is within the threshold (< 0.10 mg/L).

If Alex had been reviewing the QA program data continuously, he would have noticed the bias in run 10 immediately. The cause of the bias could have been identified and corrected before conducting runs 11–14. In this example, however, Alex neglected to review the QA program data while conducting the assessment; therefore, all data collected in runs 10–14 are likely unusable.

11.6 Recommendations for Analysis of Water and Soils

Careful planning of the analytical program will increase the success of a stormwater assessment program. It is important to consider which constituents in water and soils must be analyzed to satisfy the goals of the assessment program. It is also important to develop and enforce a quality assurance program to ensure consistent and accurate analysis results.

Data Analysis

12

Abstract

To assess the performance of a stormwater treatment practice, data and/or samples must be collected, samples must be analyzed, and all data, including the results of the sample analysis, must be analyzed. Previous chapters covered how to measure water budget components, collect samples, and analyze the samples to determine relevant water and soil parameters. This chapter discusses and provides examples regarding how to analyze all data to determine the performance level of the practice, including parameters such as sediment capacity and removal, metal and nutrient removal, overall effective saturated hydraulic conductivity, time required for a practice to drain or infiltrate a desired volume of runoff, and others. Incorporating uncertainty of the results is also included.

A critical step in developing an assessment program is choosing and implementing the most appropriate data analysis methods. The following sections include discussion and recommendations of standardized methods of data analysis for each assessment level.

12.1 Data Analysis for Visual Inspection

Visual inspection (level 1) is a qualitative assessment technique that may determine if a stormwater treatment practice is malfunctioning. Data analysis for visual inspection involves interpreting visual observations and recognizing what maintenance action(s) or further assessment efforts are warranted. Data analysis techniques for visual inspection are provided in Chap. 5.

A.J. Erickson et al., *Optimizing Stormwater Treatment Practices: A Handbook of Assessment and Maintenance*, DOI 10.1007/978-1-4614-4624-8_12,

12.2 Data Analysis for Capacity Testing

The primary function of most stormwater treatment practices is either to infiltrate stormwater (a function of saturated hydraulic conductivity) or to capture solids through sediment accumulation. Capacity testing (level 2a) is specifically designed to assess the ability of stormwater treatment practices to filter runoff, reduce runoff volume, and determine the amount of sediment storage capacity remaining.

12.2.1 Assessment of Saturated Hydraulic Conductivity and Retention Time for Filtration Practices

Point measurements can be used to estimate saturated hydraulic conductivity (K_s) values at specific locations within a stormwater treatment practice, as described in Chap. 6. To determine performance of a treatment practice from K_s measurements, equations describing flow through porous media must be used. One such equation is Darcy's Law, which describes one-dimensional flow through porous media, given by (12.1):

$$Q = \frac{\mathrm{d}V}{\mathrm{d}t} = -K_s A \frac{\mathrm{d}H}{\mathrm{d}z} \tag{12.1}$$

where

Q = flow rate through the media
$\mathrm{d}V/\mathrm{d}t$ = change in volume of water with respect to time
K_s = saturated hydraulic conductivity of the media
A = surface area through which flow occurs
H = piezometric head
z = distance in the direction of flow
$\mathrm{d}H/\mathrm{d}z$ = the piezometric gradient across the media

Darcy's Law can be used with results from capacity testing (and synthetic runoff testing) to evaluate the performance of filtration practices after making some assumptions. The first assumption is that both the water surface of the stormwater stored above the filtration practice and the subsurface pipe collection system below the filter media, including the gravel subbase, are at atmospheric pressure because they are exposed to the atmosphere and there is no suction within the media. This assumption is valid in most filtration practices. Therefore, with an arbitrary datum of $z = 0$ at the bottom of the filter media, the piezometric head (H) at the filter surface at any point in time is equal to the water level above the filter surface (z_{w}) plus the length (i.e., vertical depth) of the porous media (L_{m}). As a result, the hydraulic gradient across the filter media varies with time as the water surface drops and, at any given time, is given by (12.2):

$$\frac{\mathrm{d}H}{\mathrm{d}z} = \left(\frac{z_{\mathrm{w}} + L_{\mathrm{m}}}{L_{\mathrm{m}}}\right) \tag{12.2}$$

where

z_w = water level above the media surface at any time
L_m = vertical thickness of sand filter media

The second assumption is that the filter surface area is equally filtering stormwater in one and only one dimension (i.e., vertically downward). For most filtration practices there will be some lateral flow around the perimeter of the practice, but stormwater can generally be assumed to flow in one dimension (i.e., vertically downward) unless a layer within a portion of the media is restricting flow. If the second assumption is valid, the total discharge, $Q = \mathrm{d}V/\mathrm{d}t$, through the filter is equal to the product of the surface area through which flow occurs (A) and the change in water level with respect to time ($\mathrm{d}z_w/\mathrm{d}t$), as given by (12.3):

$$Q = \frac{\mathrm{d}V}{\mathrm{d}t} = A\frac{\mathrm{d}z_w}{\mathrm{d}t} \tag{12.3}$$

For capacity testing, the area (A) corresponds to the area of a single point measurement. Substituting (12.2) and (12.3) into (12.1) and canceling the area (A) terms result in (12.4):

$$\frac{\mathrm{d}z_w}{\mathrm{d}t} = -K_s\frac{z_w + L_m}{L_m} \tag{12.4}$$

where

z_w = water level above the media surface at any time
t = time
K_s = saturated hydraulic conductivity
L_m = vertical thickness of sand filter media

Rearranging (12.4) in preparation for integration results in (12.5):

$$\int_{z_{max}}^{z} \frac{\mathrm{d}z_w}{z_w + L_m} = \int_0^t -\frac{K_s}{L_m}\mathrm{d}t \tag{12.5}$$

where

z_{max} = initial water depth
z = water level above the media surface at time t
t = any time such that $0 < t \le t_f$

The third assumption is that the length of the porous media (L_m) and saturated hydraulic conductivity (K_s) are constant with respect to both time (t) and water level elevation (z_w). The depth of the filter bed media (L_m) is a physical property of the filtration practice that is based on the design and construction of the practice and should therefore be constant. The K_s is a property of the porous media and the fluid that is passing through the porous media. For homogenous synthetic runoff and nondynamic porous media, K_s can be assumed to be constant. Therefore, integration of (12.5) results in (12.6):

$$z + L_m = (z_{max} + L_m)e^{\left(\frac{-K_s}{L_m}\right)t} \quad (12.6)$$

where

z = water level above the media surface at time t
t = any time such that $0 < t \leq t_f$
z_{max} = initial water level above the media surface
K_s = saturated hydraulic conductivity
L_m = vertical thickness of the filter media

A filtration practice will require a certain amount of time to filter stormwater runoff captured within the practice (t_f). This time required can be used as a criterion for evaluating the performance of, or scheduling maintenance for, the filtration practice. To do so, however, requires that (12.6) be rearranged to solve for the time (t), that the drain time (t_f) be substituted for t, and that zero be substituted for z because all of the water has been filtered at time = t_f. The result is (12.7):

$$t_f = -\frac{L_m}{K_s}\left[\ln\left(\frac{L_m}{z_{max} + L_m}\right)\right] \quad (12.7)$$

where

t_f = time required to filter the entire depth of stormwater runoff
L_m = vertical thickness of the filter media
K_s = saturated hydraulic conductivity
z_{max} = initial water level above the media surface

It is important to note that, for capacity testing, the derivation of (12.6) and (12.7) is based on the area through which flow occurs (A), and this area corresponds to the area of a single point measurement. Therefore, the saturated hydraulic conductivity (K_s) corresponds to a single point measurement. To evaluate filtration practices, (12.6) and (12.7) must be applied to individual point measurements to determine where and when maintenance is needed. When calculating drain time (t_f) from (12.7) in this way, the resulting drain time is a "conceptual" drain time that would occur if the max water depth (z_{max}) was isolated over the point where K_s was measured and allowed to filter into the soil in only one (i.e., vertical) direction. This "conceptual" drain time does not relate to the actual drain time of the entire practice that would occur during a runoff event, but can be used to determine if maintenance is necessary at that location. An example of this process is given in Example 12.1. Synthetic runoff testing or monitoring can be used to estimate the actual drain time of the entire practice that would occur during a runoff event.

Example 12.1: Selecting maintenance for a filtration basin using capacity testing of hydraulic conductivity
Lana, a watershed district engineer, assesses a filtration basin with a media thickness of 20 in. (50.8 cm) and a design maximum storage depth above the filter surface of 36 in. (91.4 cm). She conducts capacity testing, collects

(continued)

Example 12.1: (continued)
measurements at 23 locations in the practice, and then calculates saturated hydraulic conductivity (K_s) at each location (see Chap. 6 for capacity testing methods and Chap. 9 for methods for estimating K_s), as listed in Table E12.1. Lana would like to know if maintenance is required anywhere in the filtration basin.

Table E12.1 Saturated hydraulic conductivity data

Measurement location	Saturated hydraulic conductivity, K_s (cm/h)	Conceptual drain time (t_f, hours)
1	0.80	65.1
2	0.98	53.1
3	1.61	32.4
4	0.09	600.3
5	0.61	85.3
6	0.81	64.3
7	22.19	2.4
8	0.54	97.6
9	5.27	9.9
10	0.12	429.5
11	11.59	4.5
12	2.31	22.6
13	5.90	8.9
14	5.76	9.1
15	1.47	35.5
16	0.89	58.7
17	2.42	21.6
18	0.11	489.5
19	0.66	79.4
20	29.38	1.8
21	0.36	146.6
22	0.93	56.2
23	8.75	6.0

Lana uses (12.7) to determine a "conceptual" drain time for each location where saturated hydraulic conductivity was measured (Table E12.1). For example, at the first location Lana measured K_s to be 0.80 cm/h. Using the filter media depth (L_m = 50.8 cm) and the max water depth (z_{max} = 91.4 cm) in (12.7) yields

$$t_f = -\frac{L_m}{K_s}\left[\ln\left(\frac{L_m}{z_{max}+L_m}\right)\right] = -\frac{50.8\,\text{cm}}{0.80\,\text{cm/h}}\left[\ln\left(\frac{50.8\,\text{cm}}{91.4\,\text{cm}+50.8\,\text{cm}}\right)\right]$$

$$t_f = -63.3[\ln(0.357)] = 65.1\,\text{h}$$

(continued)

Example 12.1: (continued)

The watershed district where Lana works requires that stormwater treatment practices drain completely within 48 h after a runoff event. From Lana's analysis, it appears that locations 1, 2, 4, 5, 6, 8, 10, 16, 18, 19, 21, and 22 all have conceptual drain times greater than 48 h, indicating that maintenance would be most effective in these locations. From this analysis, however, Lana cannot precisely determine the overall drain time for the entire practice, and she does not know if the practice as a whole meets or exceeds the watershed district regulations.

12.2.2 Assessment of Saturated Hydraulic Conductivity and Retention Time for Infiltration Practices

Data analysis for stormwater infiltrating into an infiltration practice must consider that the length of the saturated media varies as water infiltrates, and the piezometric head gradient (i.e., $\mathrm{d}H/\mathrm{d}z$) cannot be easily represented with a single equation. An infiltration model can be used if some simplifying assumptions are valid. One such as model is the Green-Ampt equation, which assumes that there is a wetting front moving through the soil such that the soil is either fully saturated or at the initial moisture content before infiltration began. To develop the appropriate form of the Green-Ampt equation, Darcy's Law (12.1) can be used to model flow through the fully saturated media:

$$Q = \frac{\mathrm{d}V}{\mathrm{d}t} = -K_s A \frac{\mathrm{d}H}{\mathrm{d}z} \tag{12.1}$$

where

Q = flow rate through the filter media
$\mathrm{d}V/\mathrm{d}t$ = change in volume of water with respect to time
K_s = saturated hydraulic conductivity of the media
A = surface area
H = piezometric head
z = distance in the vertical direction
$\mathrm{d}H/\mathrm{d}z$ = the piezometric gradient in the media

The following can be substituted into (12.1): the change in piezometric head ($\mathrm{d}H$ = the piezometric head at the water surface minus the piezometric head at the wetting front) and the gradient length ($\mathrm{d}z$ = the length of the saturated media). The piezometric head at the wetting front is as follows:

$$H(\text{wetting front}) = \Psi + L + h(t) \tag{12.8}$$

Table 12.1 Variable definitions for Green-Ampt derivation

Time, t	Water depth, $h(t)$	Cumulative infiltrated stormwater, $F(t)$	Description
$t = 0$	$h = 0$	$F = 0$	Water supply is turned on and begins to fill the basin
$t = t_i$	$h = h_i$	$F = F_i$	Basin is full and the water supply is turned off
$t = t_f$	$h = h_f = 0$	$F = F_f$	Basin has drained completely

where

H = piezometric head
ψ = soil suction at the wetting front (a positive value)
L = length of saturated media
$h(t)$ = water depth above the media surface at time t

If the water surface is open to the atmosphere, then the piezometric head at the water surface is zero and the piezometric gradient in the media (dH/dz) is

$$\frac{dH}{dz} = \frac{-\Psi - L - h(t)}{L} \tag{12.9}$$

The depth of infiltrated water ($F(t)$) at any time, t, is equal to the product of the saturated media length (L) and the change in moisture content, as follows:

$$F(t) = L(\theta_f - \theta_i) = L\Delta\theta \tag{12.10}$$

where

$F(t)$ = cumulative depth of surface water infiltrated at any time t
θ_f = final moisture content
θ_i = initial moisture content
$\Delta\theta$ = change in moisture content = $\theta_f - \theta_i$

Rearranging (12.10) to solve for the saturated media length (L) gives

$$L = \frac{F(t)}{\Delta\theta} \tag{12.11}$$

If the water supply is turned on at time equals zero ($t = 0$), the corresponding water depth (h) and cumulative depth of infiltrated water (F) are also zero at this moment in time. When the infiltration practice has been filled to the desired level the water supply is turned off, which corresponds to the initial time (t_i), initial water depth (h_i), and initial cumulative infiltrated water (F_i). The point in time when all of the stormwater has infiltrated into the media corresponds to the final time (t_f), final water depth (h_f), and final cumulative infiltrated water (F_f). These conditions are summarized in Table 12.1.

The water depth above the media surface at any time ($h(t)$) is equal to the initial water depth (h_i) minus the depth of water infiltrated since time t_i ($F(t)-F_i$). Thus, the water depth at any time ($h(t)$) is given by (12.12)

$$h(t) = h_i - F(t) + F_i \tag{12.12}$$

Substituting (12.9), (12.11), and (12.12) into Darcy's Law (12.1), noting that Q/A is equal to the infiltration rate (f) which is also the change in cumulative infiltrated depth with respect to time (dF/dt), and simplifying yield the infiltration rate, f:

$$f = \frac{dF}{dt} = K_s\left(\frac{\Delta\theta(\Psi + h_i + F_i) + (1 - \Delta\theta)F(t)}{F(t)}\right) \tag{12.13}$$

After separation of variables (12.13), can be integrated from the initial time (t_i) until any time (t) such that $t_i < t \leq t_f$, and from the initial cumulative infiltrated depth (F_i) until the cumulative infiltrated depth at time t ($F(t)$), as follows:

$$\int_{t_i}^{t} K_s dt = \int_{F_i}^{F(t)} \left(\frac{F(t)}{\Delta\theta(\Psi + h_i + F_i) + (1 - \Delta\theta)F(t)}\right) dF \tag{12.14}$$

Integration and simplification of (12.14) yield a form of the Green-Ampt equation that can be used to assess infiltration practices:

$$K_s(t - t_i) = \left(\frac{F(t) - F_i}{1 - \Delta\theta}\right) - \left(\frac{\Delta\theta(\Psi + h_i + F_i)}{(1 - \Delta\theta)^2}\right) \times \ln\left(\frac{\Delta\theta(\Psi + h_i + F_i) + (1 - \Delta\theta)F(t)}{\Delta\theta(\Psi + h_i + F_i) + (1 - \Delta\theta)F_i}\right) \tag{12.15}$$

where

K_s = overall effective saturated hydraulic conductivity of the media in the practice
t = any time such that $t_i < t \leq t_f$
t_i = time when the water supply is turned off
$F(t)$ = cumulative depth of infiltrated water at any time t
F_i = cumulative depth of infiltrated water when the basin is full and the water supply is turned off
$\Delta\theta$ = change in moisture content
ψ = soil suction at the wetting front (a positive value)
h_i = initial water depth above the media surface when the basin is full and the water supply is turned off

The time required for an infiltration practice to infiltrate the stormwater runoff it captures (t_f) can be used as a criterion for evaluating the performance of, or scheduling maintenance for, the infiltration practice. To do so, however, requires that (12.15) be rearranged to solve for the time (t), that the drain time (t_f) be substituted for t, and that all of the assumptions inherent to the Green-Ampt solution are valid (i.e., vertical, one-dimensional infiltration). For this calculation, the cumulative infiltrated water depth after all the stormwater has infiltrated (F_f) is equal to the sum of the initial water depth (h_i) and the initial cumulative infiltrated water depth (F_i), according to (12.12), and thus $F_f - F_i = h_i$. The result is (12.16):

$$t_f = \frac{1}{K_s}\left(\frac{h_i}{1-\Delta\theta}\right) - \left(\frac{\Delta\theta(\Psi + h_i + F_i)}{K_s(1-\Delta\theta)^2}\right) \times \ln\left(\frac{\Delta\theta(\Psi + h_i + F_i) + (1-\Delta\theta)(h_i + F_i)}{\Delta\theta(\Psi + h_i + F_i) + (1-\Delta\theta)F_i}\right) \quad (12.16)$$

where

K_s = saturated hydraulic conductivity of the media
t_f = time when the basin has drained completely
F_i = cumulative depth of infiltrated water when the basin is full and the water supply is turned off
$\Delta\theta$ = change in moisture content
ψ = soil suction at the wetting front (a positive value)
h_i = initial water depth above the media surface when the basin is full and the water supply is turned off

It is important to note that, for capacity testing, the derivation of (12.15) and (12.16) is based on the area through which flow occurs (A), and this area corresponds to the area of a single point measurement. Therefore, the saturated hydraulic conductivity (K_s) corresponds to a single point measurement. To evaluate infiltration practices, (12.15) and (12.16) must be applied to individual point measurements to determine where and when maintenance is needed. When calculating drain time (t_f) from (12.16) in this manner, the resulting drain time is a "conceptual" drain time that would occur if the max water depth (h_i) was isolated over the point where K_s was measured and allowed to infiltrate into the soil in only one (i.e., vertical) direction. This "conceptual" drain time does not relate to the actual drain time of the entire practice that would occur during a runoff event, but can be used to determine if maintenance is necessary at that location. An example of this process is given in Example 12.2. The actual drain time of the entire practice that would occur during a runoff event can be estimated with synthetic runoff testing (Chap. 7) or monitoring (Chap. 8).

Example 12.2: Selecting maintenance for an infiltration basin using capacity testing of hydraulic conductivity

Lana, the watershed district engineer, is interested to know how her filtration practice assessment in (12.1) would be different if the practice was an infiltration practice. She assumes the same design maximum storage depth

(continued)

Example 12.2: (continued)
above the media surface of 36 in. (91.4 cm) and uses the same 23 estimates of K_s as listed in Table E12.2. Lana checks Tables 6.1 and 6.2 and assumes the following infiltration parameters: $\psi = 2$ cm, $\theta_i = 0.04$, and a porosity of 0.45 and thus $\theta_f = 0.45$ ($\Delta\theta = 0.45 - 0.04 = 0.41$). Lana would like to know if maintenance is required anywhere in the infiltration basin.

Table E12.2 Saturated hydraulic conductivity data

Measurement location	Saturated hydraulic conductivity, K_s (cm/h)	Drain time (t_f, hours)
1	0.80	72.5
2	0.98	59.2
3	1.61	36.1
4	0.09	668.6
5	0.61	95.0
6	0.81	71.6
7	22.19	2.6
8	0.54	108.7
9	5.27	11.1
10	0.12	478.3
11	11.59	5.0
12	2.31	25.2
13	5.90	9.9
14	5.76	10.1
15	1.47	39.6
16	0.89	65.4
17	2.42	24.1
18	0.11	545.2
19	0.66	88.5
20	29.38	2.0
21	0.36	163.3
22	0.93	62.6
23	8.75	6.7

Lana uses (12.16) to determine a "conceptual" drain time for each location where saturated hydraulic conductivity was measured (Table E12.2). For example, at the first location Lana measured K_s to be 0.80 cm/h. Using the infiltration parameters mentioned in (12.16) yields:

$$t_f = \frac{1}{K_s}\left(\frac{h_i}{1-\Delta\theta}\right) - \left(\frac{\Delta\theta(\Psi + h_i + F_i)}{K_s(1-\Delta\theta)^2}\right)$$
$$\times \ln\left(\frac{\Delta\theta(\Psi + h_i + F_i) + (1-\Delta\theta)(h_i + F_i)}{\Delta\theta(\Psi + h_i + F_i) + (1-\Delta\theta)F_i}\right)$$

(continued)

Example 12.2: (continued)

$$t_f = \frac{1}{0.80\ \text{cm/h}}\left(\frac{91.4\ \text{cm}}{1-0.41}\right) - \left(\frac{0.41(2\ \text{cm}+91.4\ \text{cm})}{0.80\ \text{cm/h}(1-0.41)^2}\right)$$
$$\times \ln\left(\frac{0.41(2\ \text{cm}+91.4\ \text{cm})+(1-0.041)(91.4\ \text{cm})}{0.41(2\ \text{cm}+91.4\ \text{cm})}\right)$$

$$t_f = 1.24(155) - \left(\frac{38.3}{0.28}\right)\ln\left(\frac{92.3}{38.3}\right) = 72.5\ \text{h}$$

The watershed district where Lana works requires that stormwater treatment practices drain completely within 48 h after a runoff event. From Lana's analysis, it appears locations 1, 2, 4, 5, 6, 8, 10, 16, 18, 19, 21, and 22 all have conceptual drain times greater than 48 h, indicating that maintenance would be most effective in these locations. From this analysis, however, Lana cannot determine the overall drain time for the entire practice and she does not know if the practice as a whole meets or exceeds the watershed district regulations.

12.2.3 Sediment Accumulation Testing

Point measurements of sediment accumulation depth can be averaged arithmetically to determine the overall sediment accumulation depth. Using computer-aided drafting (CAD) software provides a more accurate estimation of sediment accumulation because the software can directly compare the current sediment depth to historical sediment depths and can estimate the accumulated sediment volume and the remaining water storage capacity. This method of testing can be used to track the change in sediment accumulation over time.

12.3 Data Analysis for Synthetic Runoff Testing

Synthetic runoff testing (level 2b) can be used to measure stormwater treatment practice effectiveness for runoff volume reduction, retention time, and pollutant removal. When performing synthetic runoff testing to assess volume reduction or retention time of stormwater treatment practices, the most important criterion is often whether the stormwater treatment practice can drain or infiltrate the design storm volume in the required time, which is typically 48 h. The next several sections discuss methods that may be used to estimate the time required for a filtration practice to drain a specified volume of runoff (i.e., retention time), estimating the drain time of infiltration practices (i.e., volume reduction), and assessment of pollutant removal using synthetic runoff testing.

12.3.1 Assessment of Saturated Hydraulic Conductivity and Retention Time for Filtration Practices

The assessment of a filtration practice for retention time determines if the practice can drain the design storm volume within the design time (e.g., 48 h). Water flow through a filter can be modeled with Darcy's Law (12.1) as described above. In the case of a sand filter that is assessed using synthetic runoff testing, the same assumptions and derivation used to develop (12.6) and (12.7) are applicable.

Equation (12.6) can be used to determine the overall effective saturated hydraulic conductivity (K_s) of the porous media in a filtration practice, and (12.7) can be used to determine the overall drain time using synthetic runoff testing data. An example of this process is given in Example 12.3.

Example 12.3: Analyzing data from synthetic runoff testing of a filtration practice for retention time

Lana, the watershed district engineer, used synthetic runoff testing to evaluate the retention time of the design runoff volume in a filtration practice. The filter media is 20 in. (50.8 cm) thick, and the data from five synthetic runoff tests, all of which had significantly less water depth and volume than the design runoff event, are shown in Fig. E12.1. Note that the data overlap when plotted so that it is difficult to distinguish between the tests (for graphical representations, 1 in. = 2.54 cm).

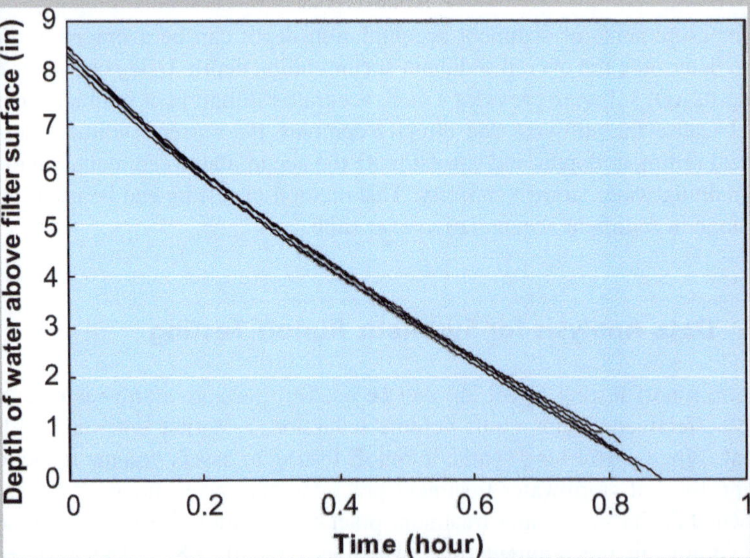

Fig. E12.1 Synthetic runoff testing data

Lana must determine the overall effective saturated hydraulic conductivity (K_s) of the filter media before she determines the retention time of the design

(continued)

Example 12.3: (continued)
runoff volume. To determine the effective K_s, Lana will fit her synthetic runoff testing data to (12.6). To do this, she separates (12.6) into a general exponential equation:

$$y = ae^{-bx}$$

where

$y = z + L_m$.
$a = z_{max} + L_m$
$b = K_s/L_m$
$x = t$

With her testing data, Lana adds the porous media length (L_m) to each water level value (z) recorded during the synthetic runoff tests, and plots all values of this sum versus the corresponding time (t) for each data point in Microsoft Excel™. She then uses the "add trendline" function, chooses an "exponential" function, and changes the options to "Display equation on chart" and "Display R-squared" value on chart. The result is the best-fit exponential function displayed in Fig. E12.2.

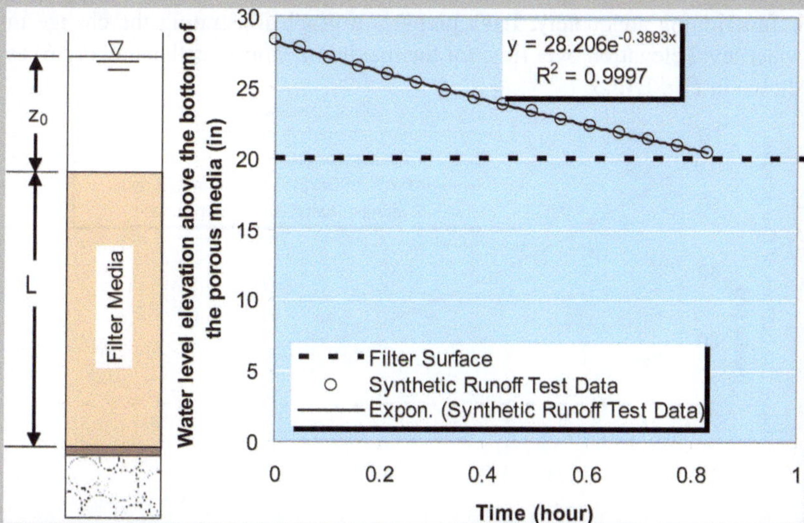

Fig. E12.2 Testing data and equation fit

Lana determines the effective saturated hydraulic conductivity (K_s) by setting the coefficients of the displayed equation equal to the corresponding parts of (12.6) (i.e., $28.206 = z_{max} + L_m$; $-0.3893 = -K_s/L_m$). With a porous media length (L_m) of 20 in. (50.8 cm), Lana calculates the initial

(continued)

Example 12.3: (continued)
water level elevation (z_{max}) to be 8.2 in. (20.8 cm), which corresponds with the testing data (see Fig. E12.1). Lana calculates the overall effective K_s to be 7.79 in./h (19.8 cm/h).

Lana can use the effective saturated hydraulic conductivity ($K_s = 19.8$ cm/h), the porous media length, and the initial water level elevations given in (12.7) to determine the drain time of other stormwater depths in the filtration practice. Lana knows the design maximum storage depth above the filter surface is 36 in. (91.4 cm). Therefore, using (12.7), Lana determines the drain time to be 2.64 h, as follows:

$$t_d = -\frac{L_m}{K_s}\left[\ln\left(\frac{L_m}{z_{max} + L_m}\right)\right] = -\frac{50.8\ \text{cm}}{19.8\ \text{cm/h}}\left[\ln\left(\frac{50.8\ \text{cm}}{91.4\ \text{cm} + 50.8\ \text{cm}}\right)\right]$$

$$t_d = -2.57\left[\ln(0.357)\right] = 2.64\ \text{h}$$

The calculated drain time of 2.64 h is less than the recommended design value of 48 h. Lana also considers the possibility that macropores may be present, as recommended in Chaps. 6 and 7, and finds that the overall effective K_s she calculated (7.8 in./h) is significantly less than that of gravel (85 m/day = 140 in./h). Therefore, Lana concludes that the filtration practice is functioning adequately. Lana prepares a graph illustrating the change in water level elevation over time for the maximum storm depth using (12.6) as shown in Fig. E12.3.

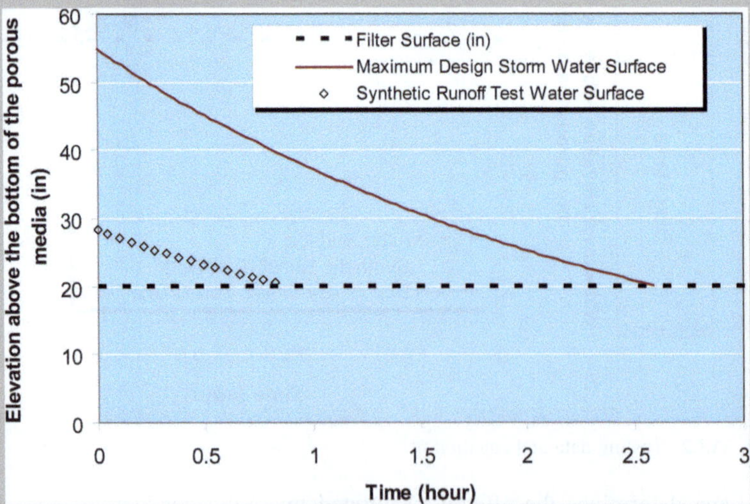

Fig. E12.3 Synthetic runoff test data compared to maximum design storm

12.3.2 Assessment of Saturated Hydraulic Conductivity and Retention Time for Infiltration Practices

The assessment of an infiltration practice for retention time determines if the practice can drain the design storm volume within the design time (e.g., 48 h). The derivation of the Green-Ampt equation (12.15) and (12.16) is applicable to assessment of infiltration basins with synthetic runoff testing. Assuming all infiltration occurs downward and there is no lateral infiltration, (12.15) can be used with data from synthetic runoff testing to understand infiltration in a stormwater treatment practice. Note that (12.15) is valid for any limits of cumulative infiltration depth (F_i, $F(t)$) or time (t_i, t) under the condition set by (12.12). For example, if the water surface in the practice has dropped 2 cm at 1.0 h after being filled, then it can be assumed that the depth of infiltrated water ($F(t)-F_i$) is 2 cm and that the water depth ($h(t)$) is the initial water depth (h_i) minus the infiltrated water ($F(t)-F_i$). Thus, for every recorded water elevation in a synthetic runoff test, the right-hand side (RHS) of (12.15) can be plotted as a function of time. The slope of the best-fit line through the data is the saturated hydraulic conductivity (K_s). This process is illustrated in Example 12.4.

For synthetic runoff testing, K_s for an infiltration practice can be used to estimate the time required to infiltrate the design water quality volume of the practice. Assuming that the basin has no outlet other than an emergency or overflow spillway that will not be active when runoff volumes do not exceed the water quality volume (WQV), the walls of the practice are vertical (i.e., $S = 0$), and that all of the assumptions inherent to the Green-Ampt solution apply (i.e., vertical, one-dimensional infiltration), (12.15) can be solved for the drain time ($t = t_f$) as given by (12.16). Calculating the drain time is also illustrated in Example 12.4.

Example 12.4: Determining K_s

Lana, the watershed district engineer, used synthetic runoff testing to determine the overall effective saturated hydraulic conductivity (K_s) and drain time (t_f) of a bioinfiltration practice. The basin was filled to a depth of 31.5 cm. The water depth as a function of time was recorded every 3 min for almost 1 h. For this basin and test, Lana uses (12.15) and assumes that $\psi = 2$ cm, $\theta_i = 0.04$, and the porosity or $\theta_f = 0.45$ ($\Delta\theta = 0.45-0.04 = 0.41$). Lana is unsure how much water infiltrated before she started measuring water depth; therefore, she assumes (conservatively) that $F_i = 0$.

From (12.12), Lana can calculate the depth of water that has infiltrated ($F(t)$) at any time after the water supply is shut off ($t-t_i$) as $F(t) = h_i-h(t) + F_i$ for every measured water depth ($h(t)$, see Table E12.3). For example, at time = 0.25 h:

$$F(t) = h_i - h(t) + F_i = 31.5\text{ cm} - 28.7\text{ cm} + 0 = 2.8\text{ cm}$$

(continued)

Example 12.4: (continued)

$$K_s(t) = \left(\frac{F(t)}{1-\Delta\theta}\right) - \left(\frac{\Delta\theta(\Psi + h_i)}{(1-\Delta\theta)^2}\right)\ln\left(\frac{\Delta\theta(\Psi + h_i) + (1-\Delta\theta)F(t)}{\Delta\theta(\Psi + h_i)}\right)$$

$$K_s(t) = \left(\frac{28.7\text{ cm}}{1-0.41}\right) - \left(\frac{0.41(2\text{ cm} + 31.5\text{ cm})}{(1-0.41)^2}\right)$$
$$\times \ln\left(\frac{0.41(2\text{ cm} + 31.5\text{ cm}) + (1-0.41)(28.7\text{ cm})}{0.41(2\text{ cm} + 31.5\text{ cm})}\right)$$

$$K_s(t) = (4.75) - (39.4)\ln\left(\frac{15.4}{13.7}\right) = 0.26\text{ cm}$$

This set of calculations is performed for each data point as shown in Table E12.3.

Table E12.3 Synthetic runoff testing data and calculations (RHS = right hand side)

Time (h)	Water depth (cm)	$F(t)$ (cm)	RHS of (12.15)
0	31.5	0.0	0
0.05	30.8	0.7	0.02
0.1	30.2	1.2	0.05
0.15	29.6	1.8	0.12
0.2	29.2	2.3	0.18
0.25	28.7	2.8	0.26
0.3	28.1	3.4	0.39
0.35	28.0	3.4	0.39
0.4	27.5	4.0	0.53
0.45	26.9	4.6	0.69
0.5	26.5	4.9	0.78
0.55	26.0	5.5	0.96
0.6	25.5	6.0	1.13
0.65	25.2	6.3	1.21
0.7	24.8	6.7	1.38
0.75	24.2	7.3	1.61
0.8	23.9	7.6	1.74
0.85	23.2	8.3	2.02
0.9	22.8	8.6	2.19
0.95	22.6	8.9	2.31

Lana recognizes that (12.15) can be approximated as a linear function of the form:

$$y = mx + b$$

(continued)

Example 12.4: (continued)

where

y = independent variable [right hand side (RHS) of (12.15)]
m = slope (overall effective saturated hydraulic conductivity, K_s)
x = dependent variable (time, t)
b = intercept (zero for this application)

The right-hand side (RHS) of (12.15) is graphed versus time (hours) and the slope of a linear regression (Fig. E12.4) with the intercept equal to zero is found to be 2.06 cm/h. Thus, the best estimate of the effective saturated hydraulic conductivity (K_s) for this basin is 2.06 cm/h.

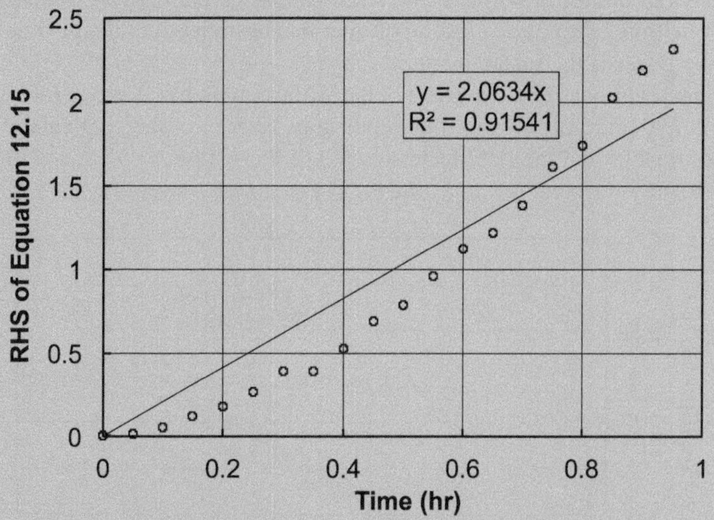

Fig. E12.4 Determination of effective saturated hydraulic conductivity from synthetic runoff test data (*RHS* = right-hand side)

Lana also wants to estimate the time required to completely drain the water quality volume (WQV). She knows the basin has a maximum depth of 180 cm when filled with the WQV, and she assumes that the overall effective saturated hydraulic conductivity (K_s) is 2.06 cm/h, as determined above. Lana assumes that $F_i = 0$, which is conservative (i.e., the estimated drain time is shorter than actual drain time). Lana uses $h_i = 180$ cm, $K_s = 2.06$ cm/hr, $\psi_f = 2$ cm, and $\Delta\theta = 0.41$ and inputs these values into Equation (12.16):

$$t_f = \frac{1}{K_s}\left(\frac{h_i}{1-\Delta\theta}\right) - \left(\frac{\Delta\theta(\Psi + h_i + F_i)}{K_s(1-\Delta\theta)^2}\right)$$

$$\times \ln\left(\frac{\Delta\theta(\Psi + h_i + F_i) + (1-\Delta\theta)(h_i + F_i)}{\Delta\theta(\Psi + h_i + F_i) + (1-\Delta\theta)F_i}\right)$$

(continued)

Example 12.4: (continued)

$$t_{\rm f} = \frac{1}{2.06\ \text{cm/h}}\left(\frac{180\ \text{cm}}{1-0.41}\right) - \left(\frac{0.41(2\ \text{cm}+180\ \text{cm})}{2.06\ \text{cm/h}(1-0.41)^2}\right)$$
$$\times \ln\left(\frac{0.41(2\ \text{cm}+180\ \text{cm})+(1-0.041)(180\ \text{cm})}{0.41(2\ \text{cm}+180\ \text{cm})}\right)$$

$$t_{\rm f} = 0.485(305) - \left(\frac{74.6}{0.717}\right)\ln\left(\frac{180.8}{74.6}\right) = 55.9\ \text{h}$$

Lana ascertains that the soil of this basin may be clogged because it takes 2.3 days to drain the WQV. If the requirement for the basin is to drain the WQV within 2 days, Lana can report that this basin has a drainage time that slightly exceeds the requirement.

Lana is curious about the effect of the initial cumulative depth of infiltrated water ($F_{\rm i}$), and guesses that the depth may be 5 cm. She recalculates the values for $F(t)$ and the right-hand side of (12.15) with $F_{\rm i} = 5$ cm and finds a significantly different result, as shown in Fig. E12.5.

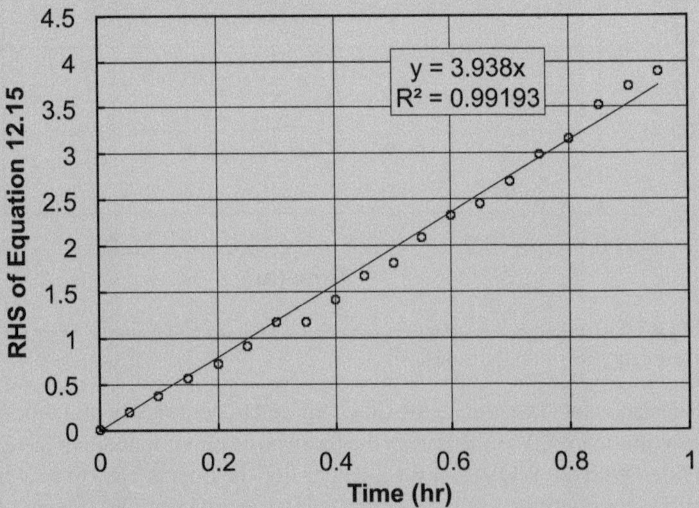

Fig. E12.5 Determination of effective saturated hydraulic conductivity from synthetic runoff test data, assuming a value for $F_{\rm i}$ (*RHS* = right-hand side)

Lana's recalculation results in an effective saturated hydraulic conductivity (K_s) of 3.9 cm/h which is nearly twice as large as her previous estimate, and the fit of the estimate as measured by the R^2 value is significantly better (0.99 vs. 0.92). Lana also remembers that her assumption about the initial infiltrated water depth ($F_{\rm i}$) results in a conservative estimate of drain time.

(continued)

Example 12.4: (continued)

Thus, the actual drain time is likely shorter than 2.3 days. She uses a value of $F_i = 5$ cm and recalculates the drain time ($K_s = 3.9$ cm/h) to be approximately 30 h. This is considerably shorter than her previous estimate of 2.3 days and well within the two-day requirement. This convinces Lana that she must measure or estimate the initial infiltrated water depth (F_i) in this and future assessments.

Lana knows, however, that this assessment does not guarantee there are no localized areas of clogging, nor does it guarantee that there are no preferential flow paths with excessive infiltration rates. To determine if either of these scenarios exists, Lana would perform capacity testing (level 2a).

12.3.3 Assessment of Volume Reduction for Infiltration Practices

To estimate the long-term volume reduction for an infiltration practice, the depth of rain that will generate enough runoff to fill the practice to its design capacity (i.e., WQV) must be compared to local precipitation exceedance frequencies. For example, analyzing historical rainfall data in Minneapolis-St. Paul, MN, USA, and summing rainfall depths over each consecutive 2-day period throughout the recorded data, the percentage of 2-day rainfall depth totals (i.e., events) that exceed a given total depth of rain are plotted in Fig. 12.1. In this example, each non-zero 2-day precipitation depth total is considered a rainfall event because stormwater treatment practices in this region are designed to drain within 2 days. Figure 12.2 is based on the same data and shows the percent exceedance as a

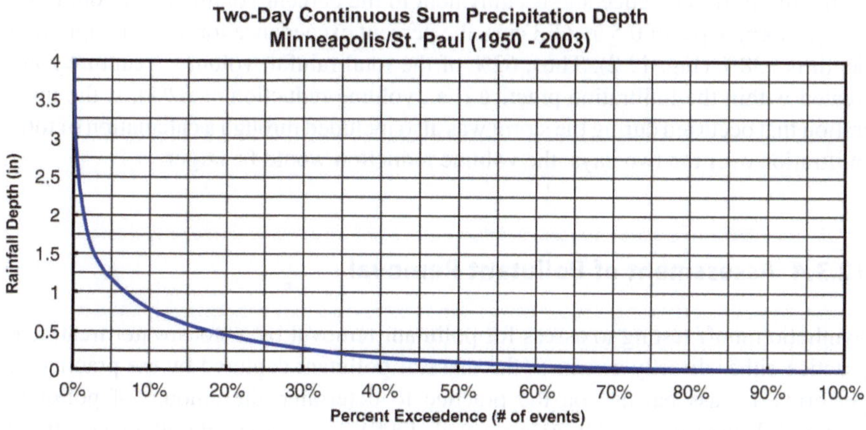

Fig. 12.1 Percent exceedance of 2-day continuous sum precipitation events in Minneapolis-St. Paul, Minnesota, USA (Weiss et al. 2005)

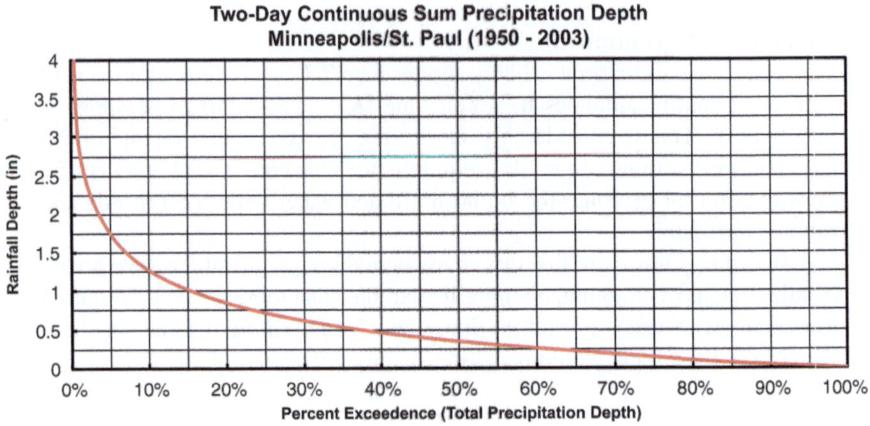

Fig. 12.2 Percent exceedance for total 2-day continuous sum of total precipitation depth in Minneapolis-St. Paul, Minnesota, USA (Weiss et al. 2005)

function of total 2-day continuous sum rainfall depth instead of number of events. If the water quality volume (WQV) of an infiltration practice is generated by a rainfall depth of 0.5 in. (1.3 cm) the percent of storm events that exceed the capacity of the practice is ~18% (Fig. 12.1) and thus ~82% (100−18%) of storm events are completely captured by the infiltration practice.

The total volume reduction includes both the total volume of storms completely captured by the practice and the portion of larger storms that is captured before overflow. For example, if the design storm depth for an infiltration practice is 0.5 in. (1.3 cm), the volume reduction is the summation of all storm events less than or equal to 0.5 in. (1.3 cm) in depth plus the first 0.5 in. (1.3 cm) of depth from all storms larger than 0.5 in. (1.3 cm). This is the percent of total precipitation depth captured by the infiltration practice and is equivalent to the percent volume reduction. For a design storm depth of 0.5 in. (1.3 cm), the percent exceedance for total precipitation depth is ~38% (Fig. 12.2). Thus, 62% of the total rainfall volume is captured and treated within the infiltration practice (i.e., volume reduction = 62%). If the infiltration that occurred during the storm was also included through a calculation of total infiltration over the two days, the volume reduction would be larger.

12.3.4 Assessment of Pollutant Removal

Synthetic runoff testing to assess for pollutant removal by a stormwater treatment practice either directly measures the mass of pollutant captured by the practice or performs a mass balance on the practice to determine the amount of pollutant captured (Wilson et al. 2009; Asleson et al. 2009). The amount of pollutant captured by the practice must be compared to the total amount of pollutant that entered the practice using (12.17) to determine the percent of pollutant captured.

$$\text{Removal efficiency(summation of load)} = \frac{M_{\mathrm{C}}}{M_{\mathrm{I}}} \times 100\% \tag{12.17}$$

where

M_{C} = total pollutant mass captured by the practice
M_{I} = total pollutant mass that entered the practice

Alternatively, the influent and effluent runoff volume and pollutant concentrations can be measured and the pollutant removal effectiveness estimated using (12.18) and (12.19):

$$M = \sum_{i=1}^{n} V_i C_i \tag{12.18}$$

where

M = total mass of pollutant
V_i = discharge amount corresponding to sample i
C_i = pollutant concentration in sample i
i = sample number
n = total number of samples collected

$$\text{Removal efficiency(summation of load)} = \left(1 - \frac{M_{\mathrm{E}}}{M_{\mathrm{I}}}\right) \times 100\% \tag{12.19}$$

where

M_{E} = effluent pollutant mass load as calculated by (12.18)
M_{I} = influent pollutant mass load as calculated by (12.18)

12.4 Data Analysis for Monitoring

Monitoring (level 3) is used to assess stormwater treatment practice performance within a watershed for natural storm events in which the influent and effluent discharge and pollutant concentrations are not controlled and therefore vary with time. "Urban Stormwater BMP Performance Monitoring" (US EPA 2002) discusses ten methods for assessing performance from monitoring assessment data and recommends the effluent probability method. In addition to effluent probability, two other methods from US EPA (2002) are described here: (1) summation of load and (2) event mean concentration (EMC) efficiency. A fourth method described in this chapter is the exceedance method. Most monitoring studies (e.g., Anderson et al. 1985; Kovacic et al. 2000; Winer 2000; Lin and Terry 2003; Silvan et al. 2004; Bell et al. 1995) report pollutant removal or retention efficiencies based on EMC, but current (e.g., TMDLs) or future regulations may require retention calculations to be based on pollutant load reductions. Reporting data using the effluent probability and

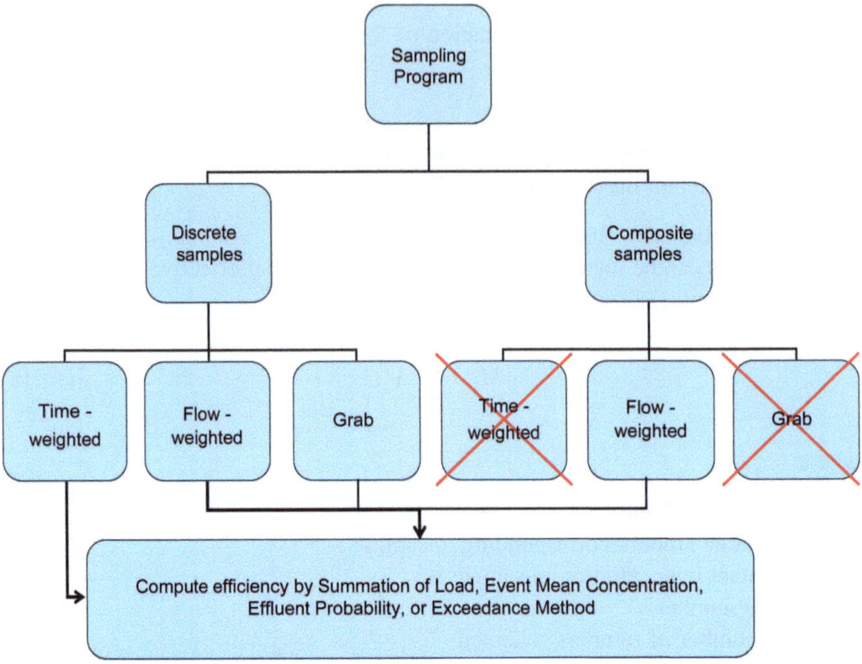

Fig. 12.3 Data analysis flow chart for sampling programs

exceedance methods will become more common as practitioners become more familiar with these methods.

As will be discussed in more detail later in this chapter, both summation of load and event mean concentration efficiency methods can be used to analyze data for a single storm event or for multiple storm events if flow rates are monitored and samples collected at all influent and effluent locations. While the effluent probability and exceedance methods can also be applied to single storm event data, the strength of these methods is apparent when applied to data from several storm events when assessing long-term performance.

As discussed in Chap. 10 there are several methods for collecting and storing stormwater samples. The organizational flow for the rest of this section is based on these methods and shown in Fig. 12.3.

12.4.1 Assessment of Stormwater Volume Reduction

One assessment goal for monitoring a stormwater treatment practice may be to determine the reduction in runoff volume achieved by the stormwater treatment practice. This can be determined by performing a water budget on the stormwater treatment practice as described in Chap. 9. This section provides water budget recommendations and an example of a water budget analysis as it pertains to a monitored stormwater treatment practice.

In order to perform a water budget, the discharge of all open channel and conduit flow entering and exiting the stormwater treatment practice must be recorded as a function of time. Also, if direct rainfall, evapotranspiration, or infiltration are significant (> 5% of water budget), the volume of the significant components must be measured or estimated. A water balance for a typical stormwater treatment practice is given by (9.1) (repeated here):

$$\Delta S = \sum V_{\text{in}} - \sum V_{\text{out}} = S_2 - S_1 = \left(\sum_{i=1}^{N} Q_i \Delta t_i\right) + P(A_{\text{W}}) - \left(\sum_{k=1}^{Z} Q_k \Delta t_k\right) - V_{\text{ET}} - V_{\text{infiltration}} \tag{9.1}$$

where

ΔS = change in water volume stored in the stormwater treatment practice
ΣV_{in} = sum of all water volumes that enter the stormwater treatment practice
ΣV_{out} = sum of all water volumes that exit the stormwater treatment practice
S_1 = volume of water stored in the stormwater treatment practice prior to runoff event
S_2 = volume of water stored in the stormwater treatment practice after runoff event
Q_i = influent flow rate data point
i = influent data point number
Δt_i = time duration between data point i and $i + 1$
P = depth of precipitation falling directly into the stormwater practice
A_{W} = surface area of the stormwater treatment practice
Q_k = effluent flow rate data point
k = effluent data point number
Δt_k = time duration between data point k and $k + 1$
V_{ET} = volume of water exported by evapotranspiration
$V_{\text{infiltration}}$ = volume of water exported by infiltration
N = number of influent data points
Z = number of effluent data points

The volume of water contained within a stormwater treatment practice can be determined from the water depth (see Chap. 9) if the stormwater treatment practice surface geometry (or bathymetry) is known. Water depth measurements can usually be stored on the same data logger as flow rate, precipitation, and other data. Otherwise, water depth must be recorded manually during site visits just prior to and immediately after a runoff event.

The most direct assessment of volume reduction is achieved by measuring all the components in (9.1) while ensuring that the mass balance of water is satisfied (i.e., $\Delta S = \Sigma V_{\text{in}} - \Sigma V_{\text{out}}$). In theory, the volume exported by evapotranspiration (V_{ET}) and infiltration ($V_{\text{infiltration}}$) can be summed to determine the volume reduction. It is often difficult and costly, however, to measure all water budget components, especially the volume lost to evapotranspiration and infiltration. Therefore, a common practice is to measure the initial and final storage values and the other terms in (9.1) and then solve (9.1) for the net sum of the volume exported by evapotranspiration and infiltration ($V_{ET} + V_{infiltration} = V_{loss}$), as given in (12.20):

$$V_{loss} = S_1 - S_2 + \left(\sum_{i=1}^{N} Q_i \Delta t_i\right) + P(A_W) - \left(\sum_{k=1}^{Z} Q_k \Delta t_k\right) \tag{12.20}$$

where

V_{loss} = volume of stormwater lost through infiltration and evapotranspiration
S_1 = volume of water stored in the stormwater treatment practice prior to runoff event
S_2 = volume of water stored in the stormwater treatment practice after runoff event
Q_i = influent flow rate data point
i = influent data point number
Δt_i = time duration between data point i and $i + 1$
P = depth of precipitation falling directly into the stormwater practice
A_W = surface area of the stormwater treatment practice
Q_k = effluent flow rate data point
k = effluent data point number
Δt_k = time duration between data point k and $k + 1$
N = number of influent data points
Z = number of effluent data points

Although the V_{loss} term in (12.20) contains the volume of runoff lost through infiltration and evapotranspiration, the losses due to evapotranspiration are small and can be assumed to be zero if the duration of the runoff event is small (i.e., a few days or less). Thus, the V_{loss} term in (12.20) is an estimate of the volume of stormwater runoff that has been infiltrated by the stormwater treatment practice and can be used to estimate the runoff volume reduction performance as demonstrated in Example 12.5.

Example 12.5: Volume reduction effectiveness of a dry pond with underdrains

Lana, the watershed district engineer, has data from a monitoring program for a 3-acre (1.21 ha) dry pond with underdrains. She wants to calculate the volume reduction performance for a 1.39-in. (3.5 cm) storm event for which the inflow, outflow, and hourly precipitation values are plotted in Fig. E12.6. To determine the stormwater runoff volume reduction, the water balance given in (12.20) must be solved (for graphical representations, 1 cfs = 0.028 m^3/s; 1 in. = 2.54 cm).

(continued)

Example 12.5: (continued)

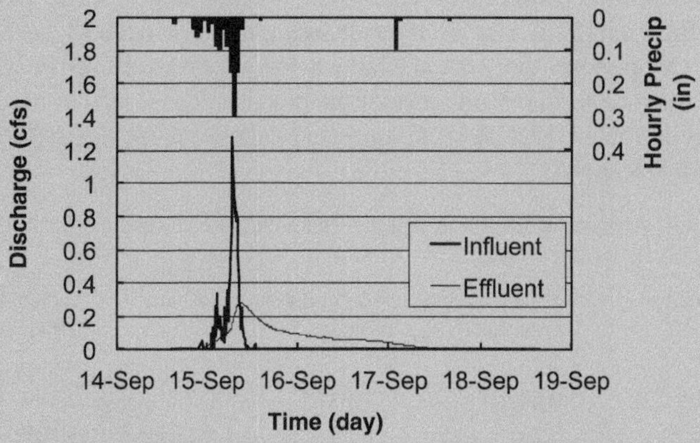

Fig. E12.6 Example storm event data

Lana knows that the area encompassed by the inflow and outflow hydrographs corresponds to the influent and effluent volumes, respectively. She also knows that the incremental volume is the product of the discharge and the time step for each data point on the hydrograph and that the total volume is the summation of the incremental volumes for the entire hydrograph. For this dry pond and storm event, Lana calculates the influent and effluent volumes to be 11,312 ft^3 and 18,967 ft^3 (320,321 L and 537,086 L), respectively. She also calculates the rainfall contribution from the total rainfall depth of 1.39 in. over the entire 3-acre dry pond, which is 0.35 ac−ft = 15,137 ft^3 (428,632 L) of water. Because the stormwater treatment practice is a dry pond, Lana assumes that the storage within the pond before and after a storm is zero, which in fact is the case for this dry pond, assuming the storage is zero when the outflow is zero (see Fig. E12.6). Therefore, Lana can simplify (12.20) and estimate V_{loss} as:

$$V_{loss} = S_1 - S_2 + \left(\sum_{i=1}^{N} Q_i \Delta t_i\right) + P(A_W) - \left(\sum_{k=1}^{Z} Q_k \Delta t_k\right)$$

$$V_{loss} = 0 - 0 + 11,312\ ft^3 + 15,137\ ft^3 - 18,967\ ft^3$$

$$V_{loss} = 7,482\ ft^3\ (211,867\ L)$$

Lana assumes this volume lost is exported by evapotranspiration and/or infiltration. Lana also wants a measure of performance and therefore calculates the effectiveness as the volume reduction percentage, which is

(continued)

Example 12.5: (continued)
simply the infiltrated volume divided by the total influent volume $= 7{,}482\ \text{ft}^3/(11{,}312\ \text{ft}^3 + 15{,}137\ \text{ft}^3) = 28.3\%$ volume reduction efficiency.

Lana can perform similar calculations for all storm events in a year to estimate the long-term performance of the practice. The results of her analysis are given in Table E12.4 (note: storm event data shown in Fig. E12.6 are are from storm event #6).

Table E12.4 Analysis of long-term volume reduction performance

Storm event #	Influent volume (10^6 L)	Direct rainfall volume (10^6 L)	Total influent volume (10^6 L)	Effluent volume (10^6 L)	Volume reduction efficiency (%)
1	2.16	1.26	3.42	1.98	42.0%
2	0.44	0.69	1.13	0.70	37.9%
3	0.34	0.22	0.56	0.11	80.6%
4	1.13	0.69	1.82	0.91	49.8%
5	0.88	0.49	1.37	0.25	81.8%
6	0.32	0.43	0.75	0.54	28.3%
7	1.11	0.51	1.62	0.86	47.0%
8	0.26	0.13	0.39	0.15	62.3%
9	0.72	0.36	1.08	0.92	15.0%
10	0.14	0.12	0.26	0.06	76.9%
11	0.24	0.16	0.40	0.20	51.2%
12	0.04	0.06	0.10	0.01	93.1%
Total	7.79	5.11	12.90	6.68	

Lana calculates the total influent and effluent stormwater volume by summing the volume from each storm event and estimates the long-term volume reduction efficiency using a variation of (12.19) in which the mass of pollutants is replaced with volume of stormwater:

$$\text{Removal efficiency(volume reduction)} = \left(1 - \frac{V_\text{E}}{V_\text{I}}\right) \times 100\%$$

$$\text{Removal efficiency} = \left(1 - \frac{6.68 \times 10^6\ L}{12.90 \times 10^6\ L}\right) \times 100\% = 48.2\%$$

It is important to note that calculating the overall removal efficiency from the total stormwater volumes is different than averaging the volume reduction efficiencies from each storm event. If the volume reduction efficiencies in Table E12.4 are averaged for all 12 storm events, the long-term volume reduction efficiency would be 55.5% instead of 48.2%. The discrepancy is caused by storm events with large stormwater volumes that have a significant impact on overall results.

12.4.2 Assessment of Pollutant Removal

Many stormwater monitoring programs are implemented with the goal of assessing the amount of pollutants retained by the stormwater treatment practice. In addition to measuring discharge, assessing the capture of pollutants also requires sampling of all stormwater treatment practice influent and effluent locations. Data for several storms, often for two or more rainy seasons, are typically required to accurately assess pollutant removal performance with an acceptable range of uncertainty (see Chap. 10, for more information).

The process for analyzing monitoring data starts with a single storm event. The pollutant concentrations from the samples collected during the storm event are used in conjunction with influent and effluent runoff volumes to determine the pollutant removal efficiency for that storm event. It cannot, however, be assumed that the calculated efficiency from a single storm event is applicable to all other storm events. Therefore, several storm events representing a range of conditions (e.g., discharge and pollutant concentration) must be monitored to accurately assess the long-term performance of stormwater treatment practices.

12.4.2.1 Analysis of Individual Storm Events

Summation of Load Efficiency

The summation of load method is used to determine the average reduction of pollutant mass (i.e., load). The sum of the mass load (kilograms or pounds) for both influent and effluent samples can be calculated using (12.18) which is applicable to any number of samples (n) that correspond to discharge volume (V) and concentration measurements (C).

After the influent and effluent loads have been summed, the stormwater treatment practice performance can be calculated based on the summation of load method according to (12.19). An example of how to apply the summation of load method to assess stormwater treatment practice removal efficiency using discrete-sampled monitoring data is given in Example 12.6. The results can be compared to other storm events for the same stormwater treatment practice, storm events for a different treatment practice, results obtained from other methods of analysis (e.g., event mean concentration efficiency), or they can be combined with other storm event data for the same stormwater treatment practice in an analysis of long-term performance.

Example 12.6: Storm event analysis by summation of load method
Lana, the watershed district engineer, is analyzing monitoring data to estimate performance of a dry pond with underdrains. She wants to estimate phosphorus removal by the summation of load method, using (12.18) and (12.19) for a 1.39-in. (3.5 cm) storm event shown in Fig. E12.7 (for graphical representation, 1 cfs = 0.028 m^3/s; 1 in. = 2.54 cm).

(continued)

Example 12.6: (continued)

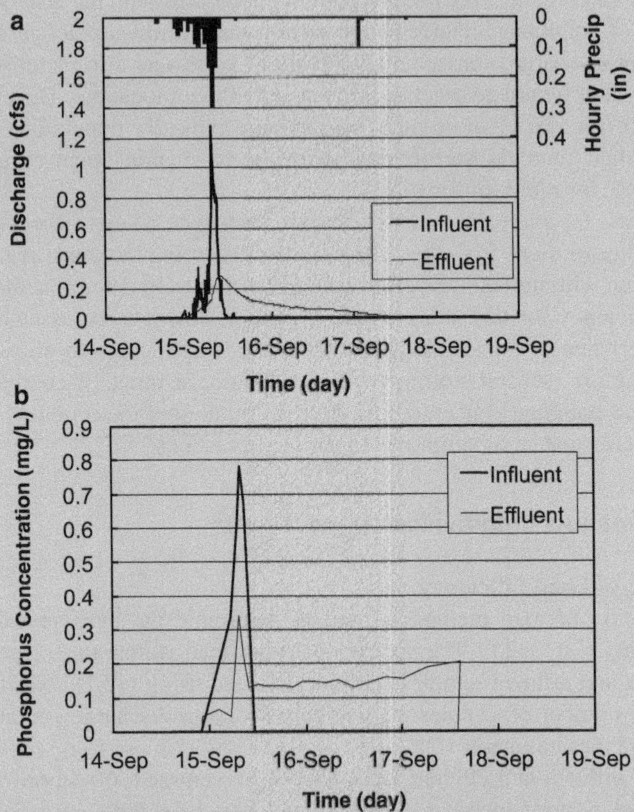

Fig. E12.7 Example storm hydrograph (a) and pollutograph (b)

The summation of load method requires discharge volume and pollutant concentration data throughout the entire storm event. Therefore, Lana must first list the pollutant concentration of each sample with the corresponding discharge volume according to (12.18) (Tables E12.5 and E12.6). To determine the pollutant load (mass) corresponding to each sample, Lana multiplies the sample pollutant concentration (mass/volume) by the corresponding discharge volume. For example, the calculation for the first influent data point is

$$\text{Mass} = \text{Concentration} \times \text{Volume}$$

$$\text{Mass}_1 = 0.238\ \text{mg/L} \times 62,396\ \text{L} = 14,850\ \text{mg} = 14.9\ \text{g}\ (0.033\ \text{lb})$$

Lana then sums the pollutant load from each sample to determine the total pollutant load for both the influent and effluent load. For example, the calculation for the influent pollutant load is

$$\text{Total mass} = \text{Mass}_1 + \text{Mass}_2 + \ldots + \text{Mass}_n$$

(continued)

Example 12.6: (continued)

Table E12.5 Influent load calculations

Influent data			
Collection time (mm/dd hh:mm)	Volume represented by sample (liters)	Pollutant concentration (mg/L)	Incremental mass load (g)
9/15 5:14 AM	62,396	0.238	14.9
9/15 6:55 AM	102,808	0.434	44.6
9/15 7:45 AM	75,893	0.551	41.8
9/15 8:43 AM	56,093	0.475	26.6
9/15 11:00 AM	22,014	0	0.0
9/15 3:30 PM	1,117	0	0.0
Totals	320,321		127.9

Table E12.6 Effluent load calculations

Effluent data			
Collection time (mm/dd hh:mm)	Volume represented by sample (liters)	Pollutant concentration (mg/L)	Incremental mass load (g)
9/14 10:00 PM	0	0.048	0.0
9/15 2:22 AM	10,421	0.067	0.7
9/15 5:23 AM	23,888	0.045	1.1
9/15 7:16 AM	26,941	0.343	9.2
9/15 8:30 AM	31,875	0.233	7.4
9/15 9:39 AM	32,406	0.133	4.3
9/15 10:49 AM	31,689	0.137	4.3
9/15 12:04 PM	30,341	0.142	4.3
9/15 1:25 PM	28,560	0.138	3.9
9/15 2:54 PM	26,843	0.137	3.7
9/15 4:33 PM	25,318	0.139	3.5
9/15 6:28 PM	25,168	0.139	3.5
9/15 8:36 PM	24,546	0.132	3.2
9/15 11:03 PM	25,839	0.149	3.9
9/16 1:43 AM	25,167	0.149	3.7
9/16 4:46 AM	25,422	0.147	3.7
9/16 8:02 AM	23,980	0.152	3.6
9/16 11:49 AM	24,614	0.132	3.2
9/16 3:26 PM	23,132	0.144	3.3
9/16 7:40 PM	24,449	0.163	4.0
9/17 12:22 AM	22,199	0.157	3.5
9/17 6:23 AM	17,554	0.191	3.4
9/17 2:21 PM	6,734	0.205	1.4
Totals	537,086		83.0

(continued)

Example 12.6: (continued)

With the total pollutant loads of the influent and effluent known, Lana can then estimate the pollutant removal efficiency by (12.19). Using the results from this monitoring study, Lana estimates the pollutant removal efficiency according to the summation of load method (12.19) as

$$\text{Removal} = \left(1 - \frac{\text{Load}_{\text{out}}}{\text{Load}_{\text{in}}}\right) \times 100\% = \left(1 - \frac{83.0\text{g}}{127.9\text{g}}\right) \times 100\%$$

$$\text{Removal} = 35.1\%$$

Therefore, Lana has determined the phosphorus removal efficiency of this dry pond with underdrains for this specific storm event to be 35.1% by using the summation of load method.

The application of (12.18) depends on the type of samples collected. Therefore, the summation of load method is described for each type of sampling method (e.g., flow-weighted discrete, flow-weighted composite) in the following sections.

Flow-Weighted Discrete Samples

When samples are collected based on a user-specified constant incremental volume of discharge that passes the sampler (e.g., every 5,000 gallons) that passes the sampler, the samples are defined as flow weighted. Each flow-weighted sample is assumed to represent the average pollutant concentration for the entire incremental volume of water to which it corresponds. Each discrete sample is stored in an individual container, and the contents of each container are analyzed separately. Equation (12.18) can be simplified if flow-weighted discrete samples are collected because the volume increment ($V=V_i$) is the same for each sample. Therefore, the summation of load method for flow-weighted discrete samples can be calculated using (12.21):

$$M = V_{\text{T}} \frac{\sum_{i=1}^{n} C_i}{n} \tag{12.21}$$

where

M = total mass of pollutant
V_{T} = total discharge volume ($V_{\text{T}} = nV$)
C_i = pollutant concentration in sample i
i = sample number
n = total number of samples

Flow-Weighted Composite Samples

Flow-weighted composite samples are collected every time a user-specified constant volume of flow passes the sampler, and all samples are stored in a single container. To determine pollutant concentration, an aliquot is collected from the composite sample, and the concentration is assumed to represent the entire composite sample. Equation (12.18) can be simplified if flow-weighted composite samples are collected because the volume increment ($V=V_i$) is the same for each sample, and the composite sample concentration is a volume-weighted average of all the individual samples that were collected. Therefore, the summation of load method for flow-weighted composite samples can be calculated using (12.22):

$$M = V_{\mathrm{T}} C_{\mathrm{c}} \tag{12.22}$$

where

M = total mass of pollutant
V_{T} = total discharge volume ($V_{\mathrm{T}} = nV$)
C_{c} = composite sample pollutant concentration.

Note that in this situation, C_{c} equals the event mean concentration (EMC).

Time-Weighted Discrete Samples

Time-weighted discrete samples are collected at a user-specified, constant time interval (e.g., 30 min), and each sample is stored in a separate container and analyzed separately. Because the magnitude of the discharge during a natural storm event varies over time, each time-weighted sample does not represent a constant volume of discharge. Equation (12.18) cannot be simplified if time-weighted discrete samples are collected because the volume increment (V_i) and concentrations of the discrete samples (C_i) will vary.

Time-Weighted Composite Samples

Time-weighted composite samples are collected at equal time increments, and all samples are stored in a single container. For reasons discussed in Chap. 10 time-weighted composite sampling is not recommended.

Discrete Grab Samples

A grab sample is a single sample collected at one location over a relatively short time period, typically sampling the entire cross section of water. Discharge must be accurately and continuously measured and the time of each grab sample must be recorded to assess pollutant removal performance. Discrete samples are stored in individual containers, and the contents of each container are analyzed separately. Equation (12.18) must be used if time-weighted discrete samples are collected because the volume increment (V_i) and concentrations of the discrete samples (C_i) will vary.

Composite Grab Samples

Composite grab samples are typically collected at variable time and volume increments and stored in a single sample storage container. For reasons discussed in Chap. 10 composite grab sampling is not recommended.

Event Mean Concentration Efficiency

The event mean concentration (EMC) efficiency method is used to determine the average reduction in pollutant concentration for a given stormwater treatment practice. EMC in units of mass per volume (e.g., mg/L) can be calculated using (12.23), which is applicable to any number of samples within a storm event. When comparing (12.18) and (12.23), it is apparent that EMC can be calculated simply by dividing the total mass of pollutant (12.18) by the total volume of stormwater:

$$\mathrm{EMC} = \frac{\sum_{i=1}^{n} V_i C_i}{\sum_{i=1}^{n} V_i} \tag{12.23}$$

where

EMC = event mean concentration
V_i = discharge amount corresponding to sample i
C_i = pollutant concentration in sample i
i = sample number
n = total number of samples collected

The total volume of stormwater entering and exiting the treatment practice must be accounted for in the calculation of EMC. The influent EMC is determined by summing the influent mass load and dividing by the total influent stormwater volume, as described by (12.23). The influent EMC must then be adjusted to account for rain that falls directly onto the stormwater treatment practice by multiplying the influent EMC by the ratio of measured influent volume to the sum of measured influent and the volume of rainfall, as described by (12.24). This procedure assumes that the rainfall contains no pollutant and is well-mixed with the stormwater stored in the treatment practice:

$$\text{Adjusted EMC}_\mathrm{I} = \mathrm{EMC}_\mathrm{I} \frac{V_\mathrm{I}}{V_\mathrm{I} + PA_\mathrm{W}} \tag{12.24}$$

where

EMC_I = influent event mean concentration
V_I = total volume of measured influent (excluding rainfall)
P = rainfall depth
A_W = surface area of stormwater treatment practice

After the influent and effluent EMCs have been calculated, the EMC efficiency of the stormwater treatment practice can be calculated using (12.25). This process is demonstrated for discrete-sampled monitoring data in Example 12.7. After storm event data have been analyzed, they can be compared to other storm events for the same stormwater treatment practice, storm events for a different treatment practice, results obtained from other methods of analysis (e.g., summation of load), or they can be combined with other storm event data for the same stormwater treatment practice to analyze long-term performance, which is discussed later in this chapter:

$$\text{EMC efficiency} = \frac{\text{EMC}_\text{I} - \text{EMC}_\text{E}}{\text{EMC}_\text{I}} \times 100\% \qquad (12.25)$$

where

EMC_I = influent event mean concentration as calculated by (12.23)
EMC_E = effluent event mean concentration as calculated by (12.23)

Example 12.7: Storm event analysis by the event mean concentration (EMC) efficiency method

Lana, the watershed district engineer, is analyzing monitoring data to estimate performance of a dry pond with underdrains. She wants to estimate phosphorus removal by the event mean concentration efficiency method, using (12.23), (12.24), and (12.25) for a 1.39-in. (3.5 cm) storm event (see data in Example 12.6). First, Lana uses (12.23) to calculate the event mean concentration of the influent (EMC_I) as follows:

$$\text{EMC}_\text{I} = \frac{\sum_{i=1}^{n} V_i C_i}{\sum_{i=1}^{n} V_i} = \frac{\sum_{i=1}^{n} (62{,}396\,\text{L} \times 0.238\,\text{mg/L} + 102{,}808\,\text{L} \times 0.434\,\text{mg/L} + \ldots)}{\sum_{i=1}^{n} (62{,}396\,\text{L} + 102{,}808\,\text{L} + \ldots)}$$

$$\text{EMC}_\text{I} = \frac{\sum_{i=1}^{n} (14.9\text{ g} + 44.6\text{ g} + 41.8\text{ g} + 26.6\text{ g})}{320{,}321\text{ L}} = \frac{127.9\text{ g}}{320{,}321\text{ L}}$$

$$= 0.399\text{ mg/L}$$

Fortunately, Lana has already tabulated the influent and effluent data in the summation of load calculation (see Example 12.6), which simplifies the event mean concentration calculation. Lana must, however, account for the rainfall that fell directly into the dry pond from this 1.39-in. (3.5 cm) event. To do so, she uses (12.24) knowing that the measured influent runoff volume is 320,321 L and the rainfall volume (from Example 12.5) is 428,632 L:

$$\text{Adjusted EMC}_\text{I} = \text{EMC}_\text{I} \frac{V_\text{I}}{V_\text{I} + PA_\text{W}}$$

$$\text{Adjusted EMC}_\text{I} = 0.399\text{ mg/L} \times \frac{320{,}321\text{ L}}{320{,}321\text{ L} + 428{,}632\text{ L}}$$

$$\text{Adjusted EMC}_\text{I} = 0.171 \frac{\text{mg}}{\text{L}}$$

(continued)

Example 12.7: (continued)

Note that (12.24) assumes that the pollutant concentration in the rainfall is zero and that the rainfall is instantaneously well mixed throughout the pond. Lana calculates the effluent event mean concentration (EMC_E) using (12.23), which results in a value of 0.154 mg/L. Lana then estimates EMC efficiency using (12.25) as follows:

$$\text{EMC efficiency} = \frac{\text{Adjusted } EMC_I - EMC_E}{\text{Adjusted } EMC_I} \times 100\%$$

$$\text{EMC efficiency} = \frac{(0.171\text{mg/L}) - (0.154\text{mg/L})}{(0.171\text{mg/L})} \times 100\% = 9.9\%$$

Note that when the EMC efficiency method (Example 12.7) is used to analyze the same data as the summation of load method (Example 12.6), the results are different. The summation of load method is based on the total mass of pollutant that enters and exits the stormwater treatment practice, whereas the event mean concentration efficiency method is based on the pollutant concentration entering and exiting the stormwater treatment practice. The difference between the two methods arises because samples are collected at monitored locations and the volume of runoff entering through monitored inlets is usually different than the volume leaving through monitored outlets due to direct rainfall, infiltration, and evapotranspiration. For example, if an infiltration practice infiltrates half of the influent runoff volume with all of the pollutant mass load associated with the infiltrated water (i.e., 50% of the total mass load) retained by the practice (e.g., solids filtered at the soil surface), the efficiency would be 50% based on the summation of load method (assuming no settling of particles). For the same case and assumption of no settling, the EMC of the effluent would be equal to the EMC of the influent (assuming no direct rainfall onto the practice), and the EMC efficiency would be zero.

The application of Equation (12.23) depends on the type of samples collected. Therefore, the event mean concentration method is described for each type of sampling method (e.g., flow-weighted discrete, flow-weighted composite, etc.) in the following sections.

Flow-Weighted Discrete Samples

Flow-weighted discrete samples are collected every time a user-specified constant volume of flow passes the sampler and are stored in individual containers that are analyzed separately. Equation (12.23) can be simplified if flow-weighted discrete samples are collected because the volume increment (V_i) is the same for each sample and the summation of the volumes is equal to the total volume. Therefore, the EMC for flow-weighted discrete samples can be calculated using (12.26):

$$\text{EMC} = \frac{\sum_{i=1}^{n} C_i}{n} \tag{12.26}$$

where

EMC = event mean concentration
C_i = pollutant concentration in sample i
i = sample number
n = total number of samples collected

Flow-Weighted Composite Samples

Flow-weighted composite samples are collected at equal volumes of stormwater runoff and stored in a single container. Equation (12.23) can be simplified for flow-weighted composite samples because the volume increment (V_i) is the same for each sample and the composite sample concentration is a volume-weighted average of all the individual samples that were collected. Therefore, the EMC for flow-weighted composite samples is simply the concentration of the composite sample (C_c).

Time-Weighted Discrete Samples

Time-weighted discrete samples are collected at equal time increments, stored in individual containers, and analyzed separately. Equation (12.23) cannot be simplified if time-weighted discrete samples are collected because the volume increment (V_i) and concentrations of the discrete samples (C_i) will vary.

Time-Weighted Composite Samples

Time-weighted composite samples are collected at equal time increments, and all samples are stored in a single container. For reasons discussed in Chap. 10 time-weighted composite sampling is not recommended.

Discrete Grab Samples

Discrete grab samples are typically collected at variable time and volume increments, stored in individual containers, and analyzed separately. A stop watch may be used to minimize the variability in time intervals between sample collection. Equation (12.23) cannot be simplified if time-weighted discrete samples are collected because the volume increment (V_i) and concentrations of the discrete samples (C_i) will vary.

Composite Grab Samples

Composite grab samples are typically collected at variable time and volume increments and stored in a single sample storage container. For reasons discussed in Chap. 10 composite grab sampling is not recommended.

12.4.2.2 Analysis of Long-Term Performance

After assessment data from multiple storm events have been analyzed, the long-term performance of a stormwater treatment practice can be calculated. Long-term performance can be expressed as a single value for performance with associated uncertainty (e.g., average phosphorus capture = 72% ± 17% confidence interval for $\alpha = 0.05$) or expressed graphically for the entire range of data (e.g., exceedance method). Applying both single value and graphical approaches to monitoring data yields a detailed description of performance. Results from analysis of long-term performance represent only the period of time encompassed by the storm events (e.g., 3 months, 1 year, 2 years) and that specific treatment practice. Analysis of monitoring data from many storms can be used to investigate relationships between stormwater treatment practice performance and runoff intensity, pollutant load or concentration, or other variables.

Summation of Load

Analysis of long-term performance by summation of load is similar to analysis of a single storm event, except that the data comprise pollutant load from multiple storm events instead of loads from individual samples. To do this, influent and effluent load is calculated separately for each storm event using (12.18), as illustrated in Example 12.6. Long-term performance can then be calculated using (12.19), with the total mass of influent load and total mass of effluent load being the sum of the influent and effluent mass load for all storm events.

In the summation of load method, the pollutant mass entering and exiting the stormwater treatment practice for each runoff event is summed, and the removal efficiency is computed from the total influent and effluent load. Thus, a storm with a relatively small pollutant load will contribute less to the total load than a storm with a relatively large pollutant load, as shown in Example 12.8. Therefore, assessment data from a stormwater treatment practice that is analyzed using the summation of load method may be biased by storms with large pollutant load.

Example 12.8: Analysis of long-term performance by summation of loads
Lana, the watershed district engineer, is analyzing monitoring data to estimate performance of a 3-acre (1.21 ha) dry pond with underdrains. She wants to estimate the long-term performance of the dry pond for phosphorus capture by the summation of load method, using (12.19). Lana calculates the influent and effluent load for each of 12 storm events, similar to Example 12.6, as shown in Table E12.7.

(continued)

Example 12.8: (continued)

Table E12.7 Analysis of long-term data by summation of load

Storm event #	Influent TP load (kg)	Effluent TP load (kg)	Summation of load efficiency (%)
1	0.554	0.175	68.4%
2	0.279	0.106	62.0%
3	0.059	0.009	84.7%
4	0.422	0.190	55.0%
5	0.363	0.046	87.3%
6	0.128	0.083	35.2%
7	0.277	0.194	30.0%
8	0.078	0.018	76.9%
9	0.285	0.218	23.5%
10	0.041	0.014	65.9%
11	0.065	0.028	56.9%
12	0.007	0.001	85.7%
Total	2.558	1.082	

Lana calculates the total influent and effluent phosphorus load by summing the load from each storm event and estimates the long-term capture efficiency using (12.19).

$$\text{Removal efficiency(summation of load)} = \left(1 - \frac{M_E}{M_I}\right) \times 100\%$$

$$\text{Removal efficiency} = \left(1 - \frac{1.082\ \text{kg}}{2.558\ \text{kg}}\right) \times 100\% = 57.7\%$$

It is important to note that calculating the overall removal efficiency from the total mass loads is different than averaging the pollutant removal efficiencies from each storm event. If the pollutant removal efficiencies in Table E12.7 are averaged for all 12 storm events, the long-term efficiency by summation of load would be 61.0% instead of 57.7%. The discrepancy is caused by storm events with large pollutant loads that have a significant impact on overall results.

The average pollutant removal from the assessment of a given stormwater treatment practice's long-term performance can be used to compare different time periods or watershed conditions for the same stormwater treatment practice, to other stormwater treatment practices of the same type (e.g., dry pond vs. dry pond), or to other stormwater treatment practices (e.g., dry pond vs. rain garden). Removal efficiencies obtained from the assessment of long-term performance can also be compared with efficiencies obtained from other analysis methods (e.g., EMC efficiency), as described in the next section and shown in Example 12.9.

Event Mean Concentration Efficiency

Analysis of long-term performance by the event mean concentration (EMC) efficiency method involves calculating the influent and effluent EMC for each storm event (12.24), Example 12.7), determining the average influent and effluent EMC of all storms, and calculating the long-term performance as the percent reduction in concentration based on the average influent and effluent EMC (US EPA 2002). Long-term removal efficiency by the EMC method can be calculated using (12.26) for all storm events, as shown in Example 12.9:

$$\text{Long} - \text{term efficiency}_{\text{EMC}} = 1 - \left[\frac{\text{average EMC}_{\text{E}}}{\text{average EMC}_{\text{I}}}\right] \times 100\% \tag{12.27}$$

where

EMC = event mean concentration for all storms
EMC_I = adjusted influent event mean concentration (12.24)
EMC_E = effluent event mean concentration

Example 12.9: Analysis of long-term performance by event mean concentration (EMC) efficiency

Lana, the watershed district engineer, is analyzing monitoring data to estimate performance of a 3-acre (1.21 ha) dry pond with underdrains. She wants to estimate the long-term performance of the dry pond for phosphorus capture by the event mean concentration (EMC) efficiency method, using (12.27). Lana calculates the influent and effluent EMC and adjusts the influent EMC for each of 12 storm events for rainfall that falls directly on the practice, similar to Example 12.7, as shown in Table E12.8.

Table E12.8 Analysis of long-term data by event mean concentration (EMC)

Storm event #	Sampled influent TP EMC (mg/L)	Adjusted influent TP EMC (mg/L)	Effluent TP EMC (mg/L)	EMC efficiency (%)
1	0.257	0.162	0.088	45.7%
2	0.632	0.247	0.151	38.9%
3	0.171	0.105	0.082	21.9%
4	0.375	0.232	0.208	10.3%
5	0.412	0.265	0.183	31.0%
6	0.400	0.171	0.155	9.4%
7	0.250	0.171	0.225	−31.8%
8	0.298	0.201	0.125	37.9%
9	0.393	0.263	0.237	9.9%
10	0.294	0.157	0.236	−50.5%
11	0.266	0.162	0.142	12.3%
12	0.198	0.077	0.086	−11.7%
Average	0.329	0.184	0.160	

(continued)

Example 12.9: (continued)

Lana calculates the average influent and effluent EMC by summing the EMC from each storm event (e.g., Influent: 0.162 + 0.247 + 0.105 + ... + 0.077 mg/L) and dividing by the number of storm events (e.g., $n = 12$). She can then estimate long-term performance using (12.27):

$$\text{Long} - \text{term efficiency}_{\text{EMC}} = 1 - \left[\frac{\text{average EMC}_\text{E}}{\text{average EMC}_\text{I}}\right] \times 100\%$$

$$\text{Long} - \text{term efficiency}_{\text{EMC}} = 1 - \left[\frac{0.160\ \text{mg/L}}{0.184\ \text{mg/L}}\right] \times 100\% = 13.0\%$$

It is important to note that calculating the overall removal efficiency from the average influent and effluent EMC is different than averaging the percent EMC reductions from each storm event. If the percent EMC reductions in Table E12.8 are averaged for all 12 storm events, the long-term efficiency by EMC would be 10.3% instead of 13.0%. The discrepancy is caused by storm events with large influent and effluent EMC that outweigh storm events with negative removal efficiencies.

The results of long-term efficiency from Examples 12.8 and 12.9 differ significantly. During the 12 storm events that were monitored, 57.5% of the pollutant load was removed by the stormwater treatment practice, but the EMC was only reduced on average by 20.5%. Discrepancies between summation of load and EMC efficiency is caused by significant water budget components (e.g., infiltration, evapotranspiration) that would result in larger retention efficiency for load than EMC. For the data shown in Examples 12.5, 12.6, and 12.7, it can be concluded that a water budget export component (e.g., infiltration) is significant due to the estimated 28.3% runoff volume reduction. Infiltration of stormwater within a treatment practice will reduce the mass of dissolved pollutants (e.g., due to adsorption in the soils) but is not likely to reduce the EMC in the effluent because both the volume of runoff and the dissolved pollutants are infiltrating. This conclusion is only possible because the data were analyzed using three methods; volume reduction, summation of load, and the EMC efficiency methods. Typically, the long-term performance as estimated by the summation of load method is numerically larger than the EMC efficiency (e.g., 57.5% vs. 20.5%). If there is no addition or loss of water (e.g., due to direct rainfall, evapotranspiration, and infiltration, etc.), the summation of load and event mean concentration methods will yield identical performance estimations.

Estimating Uncertainty

The uncertainty of long-term performance analysis by summation of load and event mean concentration (EMC) is related to the total number and variation of storms

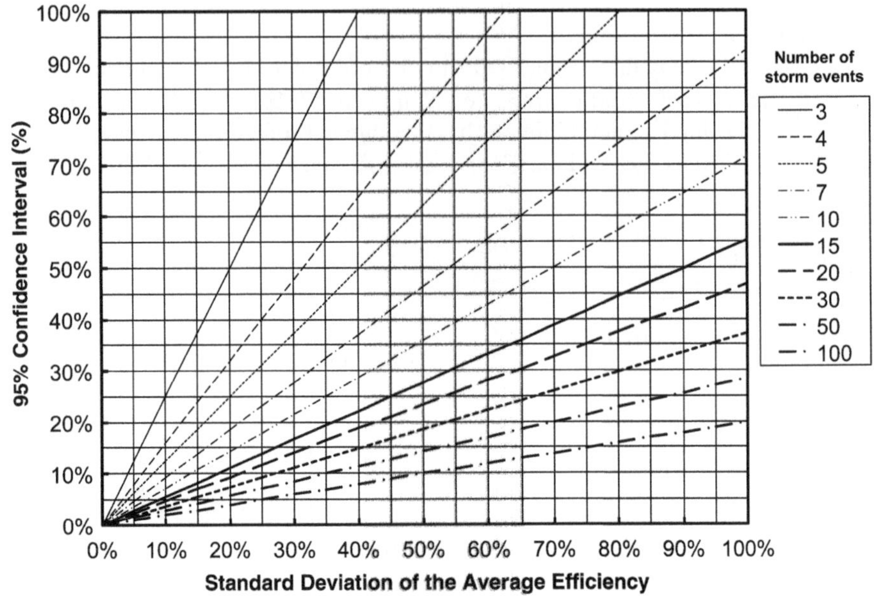

Fig. 10.1 Relationship between number of storm events, standard deviation, and 95 % confidence interval

assessed. With all other variables held constant, the uncertainty in the average percent removal decreases as the number of analyzed storm events increases. One requirement for calculating uncertainty is that a percent removal and standard deviation for all incorporated storm events can be calculated.

The 95% confidence interval is recommended to adequately represent uncertainty in average pollutant removal efficiency because it indicates that there is a 95% probability that the actual average performance will be within the confidence interval. For example, a stormwater treatment practice with an average pollutant capture rate of 72% ± 17% confidence interval ($\alpha = 0.05$) will have a 95% (19 out of 20) probability that the actual average pollutant capture rate is between 55 and 89%. The range of the confidence interval (in this case, ±17% for $\alpha = 0.05$) is dependent on the standard deviation and the number of monitored storm events. The relationship between standard deviation, number of storm events, and 95% confidence interval is shown in Fig. 10.1 (repeated here).

A simple method (12.28) for calculating an estimate of uncertainty is based on the Student (Gosset 1908) t-distribution. The Student (Gosset 1908) t-distribution, given in Table 12.2, is a probability distribution used to estimate the average of a normally distributed population from a sample of the population and is more accurate than the similar z-distribution for small ($n < 30$) sample sizes. Thus, the Student (Gosset 1908) t-distribution is used because the number of storms assessed will likely be fewer than 30. For more information on distributions,

Table 12.2 Student (Gosset 1908) *t* values

	Probability of failure, α							
Degrees of freedom	**0.5**	**0.333**	**0.25**	**0.1**	**0.05**	**0.01**	**0.005**	**0.001**
1	1.00	1.73	2.41	6.31	12.7	63.7	127	637
2	0.82	1.26	1.60	2.92	4.30	9.92	14.1	31.6
3	0.76	1.15	1.42	2.35	3.18	5.84	7.45	12.9
4	0.74	1.10	1.34	2.13	2.78	4.60	5.60	8.61
5	0.73	1.07	1.30	2.02	2.57	4.03	4.77	6.87
6	0.72	1.05	1.27	1.94	2.45	3.71	4.32	5.96
7	0.71	1.04	1.25	1.89	2.36	3.50	4.03	5.41
8	0.71	1.03	1.24	1.86	2.31	3.36	3.83	5.04
9	0.70	1.02	1.23	1.83	2.26	3.25	3.69	4.78
10	0.70	1.02	1.22	1.81	2.23	3.17	3.58	4.59
11	0.70	1.01	1.21	1.80	2.20	3.11	3.50	4.44
12	0.70	1.01	1.21	1.78	2.18	3.05	3.43	4.32
13	0.69	1.00	1.20	1.77	2.16	3.01	3.37	4.22
14	0.69	1.00	1.20	1.76	2.14	2.98	3.33	4.14
15	0.69	1.00	1.20	1.75	2.13	2.95	3.29	4.07
16	0.69	1.00	1.19	1.75	2.12	2.92	3.25	4.01
17	0.69	1.00	1.19	1.74	2.11	2.90	3.22	3.97
18	0.69	0.99	1.19	1.73	2.10	2.88	3.20	3.92
19	0.69	0.99	1.19	1.73	2.09	2.86	3.17	3.88
20	0.69	0.99	1.18	1.72	2.09	2.85	3.15	3.85
21	0.69	0.99	1.18	1.72	2.08	2.83	3.14	3.82
22	0.69	0.99	1.18	1.72	2.07	2.82	3.12	3.79
23	0.69	0.99	1.18	1.71	2.07	2.81	3.10	3.77
24	0.68	0.99	1.18	1.71	2.06	2.80	3.09	3.75
25	0.68	0.99	1.18	1.71	2.06	2.79	3.08	3.73
26	0.68	0.99	1.18	1.71	2.06	2.78	3.07	3.71
27	0.68	0.99	1.18	1.70	2.05	2.77	3.06	3.69
28	0.68	0.98	1.17	1.70	2.05	2.76	3.05	3.67
29	0.68	0.98	1.17	1.70	2.05	2.76	3.04	3.66
30	0.68	0.98	1.17	1.70	2.04	2.75	3.03	3.65

consult a statistics text (e.g., MacBerthouex and Brown 1996, Moore and McCabe 2003):

$$U = \frac{t\sigma}{\sqrt{n}} \tag{12.28}$$

where

U = uncertainty
t = Student t value from Table 12.2
σ = standard deviation
n = number of storm events

Uncertainty can be estimated with (12.28), using the number of storm events, the standard deviation of the performance data, and the Student t value (from Table 12.2). The standard deviation can be calculated in Microsoft Excel™ (stdev function) or as described in many statistical textbooks. The Student t value can also be obtained in Microsoft Excel™ (tinv function) or from Table 12.2 using the degrees of freedom (d.f. $= n-1$) and the probability of failure (α). For the 95% confidence interval, $\alpha = 0.05$. For example, if $n = 15$, the Student t value for the 67% confidence interval would be 1.00 (d.f. $= 14$, $\alpha = 0.33$). Alternatively, uncertainty can be estimated directly using Fig. 10.1 for a known standard deviation, number of storm events, and an assumed 95% confidence interval, as shown in Example 12.10.

Example 12.10: Determining the 95% confidence interval

Lana, the watershed district engineer, is analyzing monitoring data to estimate performance of a 3-acre (1.21 ha) dry pond with underdrains. She wants to estimate the uncertainty in her analysis of long-term performance of the dry pond for phosphorus capture using Fig. 10.1. Lana has calculated the average volume reduction and phosphorus capture performance from 12 storm events to be 48.2% (volume reduction), 57.7% (summation of load), and 13.0% (event mean concentration), as described in Examples 12.5, 12.8, and 12.9.

To estimate uncertainty, she calculates the standard deviation of the performance for each of the 12 storms (e.g., 68.4%, 62.0%, 84.7%, 55.0%, etc. from Example 12.8) as 23.8%, 21.9%, and 29.2% for volume reduction, summation of load, and EMC efficiency methods, respectively. Lana uses these values along with the number of storm events ($n = 12$) in Fig. 10.1 to estimate the uncertainty.

Lana estimates the uncertainty to be approximately 15% for the volume reduction, 14% for the summation of load, and 18% for the EMC efficiency. Lana wants to make sure she is reading Fig. 10.1 correctly by using (12.28) and Table 12.2 to verify her results using the values for long-term volume reduction. From Table 12.2, with a probability of failure equal to 0.05 ($\alpha = 0.05$) and $n = 12$ (degrees of freedom $= n-1 = 11$), Lana determines that $t = 2.20$. From above, the standard deviation is 23.8% ($\sigma = 0.238$) and the uncertainty, U, is calculated using (12.28).

$$U = \frac{t\sigma}{\sqrt{n}} = \frac{2.20 \times 0.238}{\sqrt{12}} = 0.151 = 15.1\%$$

This result corresponds well with the value obtained graphically in Fig. E12.8 (15%). Therefore, using the results from Examples 12.5, 12.8, and 12.9, Lana can report the long-term performance of the dry pond with underdrains as

(continued)

Example 12.10: (continued)

- 48.2% ± 15% ($\alpha = 0.05$) runoff volume reduction.
- 57.5% ± 14% ($\alpha = 0.05$) phosphorus load reduction.
- 13.0% ± 18% ($\alpha = 0.05$) phosphorus EMC reduction.

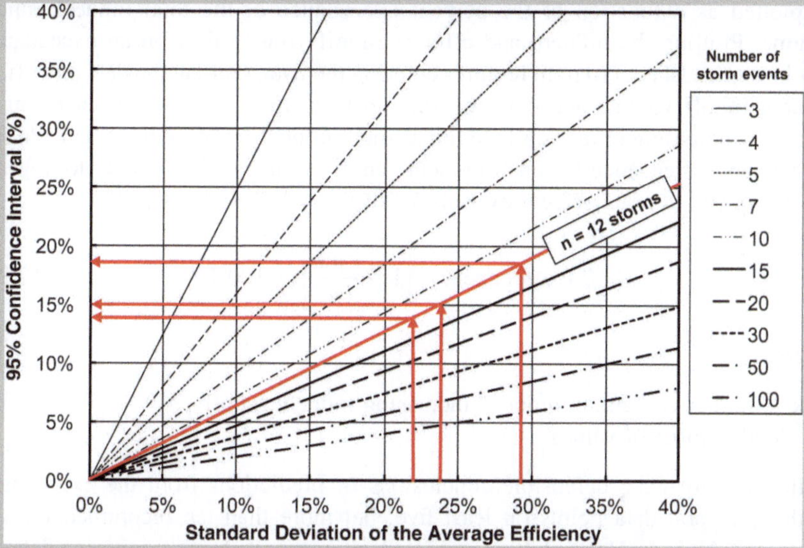

Fig. E12.8 Estimating uncertainty for long-term performance

Influent Exceedance Method

The influent exceedance method is a graphical representation of the long-term performance of a stormwater treatment practice and is especially useful in estimating expected effluent characteristics (e.g., pollutant mass load, runoff volume) and visually comparing and illustrating trends in performance over a range of data values (e.g., runoff volume, pollutant concentration, and pollutant load). This method does not, however, result in a single value of performance for rapid comparison to other monitoring data (other sites, time periods, etc.). The results of this method can be graphically compared to other monitoring data and often provide more insight and useful information than a single numerical measure of performance (e.g., retention efficiency).

As discussed below, the exceedance method involves plotting multiple storm event data for influent and effluent data pairs of one variable (i.e., influent–effluent runoff volumes, influent–effluent pollutant mass loads, influent–effluent pollutant event mean concentrations, etc.) for a practice as a function of percent influent volume exceedance on a single graph. This allows the user to observe trends in the performance of the practice. Or, if similar graphs can be generated for two or more practices, the graphs can be used to compare the performance of multiple practices.

The exceedance method is applied to monitoring data by plotting the influent and effluent runoff volume data for each runoff event versus the percent of time that the influent volume is exceeded. This will indicate what happens to the runoff in the practice during different size storms. The pollutant load and concentration can then be plotted as a function of the percent exceedance of the total influent runoff volume. Plotting the influent and effluent runoff volume data on an exceedance plot in a spreadsheet program requires entering the data, ranking the data pairs (i.e., influent runoff volume and corresponding effluent runoff volume of each runoff event) in increasing order as a function of total influent runoff volume, calculating the percent exceedance for each influent runoff volume (12.29), and plotting the runoff volumes versus percent exceedance of the influent volume.

$$\%\ \text{Exceedance} = \left[1 - \frac{\text{rank}}{n}\right] \times 100\% \tag{12.29}$$

where

rank = numerical rank in order of increasing influent runoff
n = total number of values

In order to make defensible conclusions or predictions from the exceedance method, several data points (at least five, but more than ten recommended) are necessary. An example of plotting and interpreting results from monitoring data for runoff volume using the exceedance method is shown in Example 12.11.

Example 12.11: Long-term analysis of volume reduction using the exceedance method

Lana, the watershed district engineer, is analyzing monitoring data to estimate performance of a 3-acre (1.21 ha) dry pond with underdrains that can infiltrate runoff. She wants to evaluate the long-term performance of the dry pond for reduction of runoff volume using the influent exceedance method. Lana has entered the influent and effluent runoff volume data into a spreadsheet program. She then ranks the paired data in increasing order of influent runoff volume, calculates the percent exceedance using (12.29) (Table E12.9), and plots the data on an exceedance plot as shown in Fig. E12.9. Note that the data pairs are never separated; the effluent volume is always paired with the influent volume to which it originally corresponded.

(continued)

Example 12.11: (continued)

Table E12.9 Unranked and ranked runoff volume data

Unranked data			Ranked (by influent volume) data				
Storm event #	Influent volume (10^6 L)	Effluent volume (10^6 L)	Storm event #	Influent volume (10^6 L)	Effluent volume (10^6 L)	Rank	Percent exceedance (%)
1	3.42	1.98	12	0.09	0.01	1	92%
2	1.13	0.70	10	0.26	0.06	2	83%
3	0.56	0.11	8	0.39	0.15	3	75%
4	1.82	0.91	11	0.40	0.20	4	67%
5	1.37	0.25	3	0.56	0.11	5	58%
6	0.75	0.54	6	0.75	0.54	6	50%
7	1.62	0.86	9	1.08	0.92	7	42%
8	0.39	0.15	2	1.13	0.70	8	33%
9	1.08	0.92	5	1.37	0.25	9	25%
10	0.26	0.06	7	1.62	0.86	10	17%
11	0.40	0.20	4	1.82	0.91	11	8%
12	0.09	0.01	1	3.42	1.98	12	0%

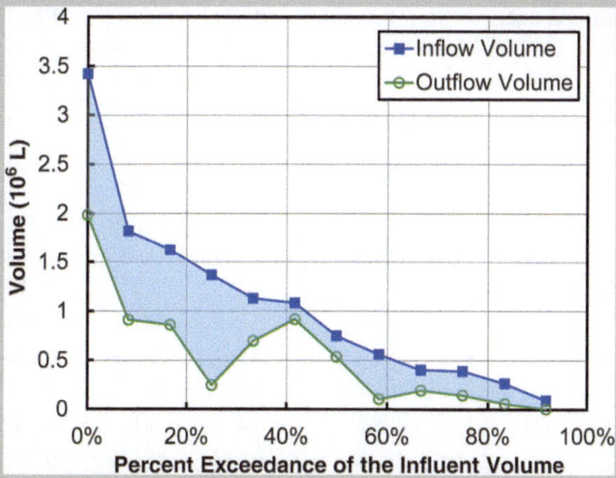

Fig. E12.9 Exceedance method for runoff volume reduction

From the exceedance plot of influent/effluent runoff volume, Lana notes that the volume reduction, which is the difference between the influent and effluent runoff volume (shaded area), is largest for the largest runoff events (smallest percent exceedance) and is minimal for runoff events less than about 40,000 ft^3 (1,132.7 m^3). She also notes that ~60% of the storms monitored were less than 40,000 ft^3 (1,132.7 m^3).

When using the exceedance method to plot variables such as pollutant load and event mean concentrations, the data are again ranked in order of increasing influent runoff volume. The data are then plotted as a function of percent exceedance of total influent runoff volume. Example 12.12 demonstrates this process.

Example 12.12: Long-term analysis of pollutant mass load and EMC reduction using the exceedance method

Lana, the watershed district engineer, wishes to use the influent exceedance method to evaluate the long-term performance of a dry pond with underdrains with respect to total phosphorus mass load reduction and total phosphorus event mean concentration (EMC) reduction. In Example 12.11, Lana sorted the storm events in increasing order of influent runoff volume. In her spreadsheet, she made sure to also select the pollutant mass load and EMC data when she sorted the data, resulting in all of her data ranked in order of increasing influent volume, as shown in Table E12.10. Lana then plots the influent and effluent total phosphorus load data pairs as a function of percent exceedance of total influent volume, as shown in Fig. E12.10.

Table E12.10 Ranked (by influent runoff volume) runoff volume, pollutant mass load, and EMC data

Storm event #	Influent volume (10^6 L)	Effluent volume (10^6 L)	Rank	Percent exceedance	Influent TP load (kg/event)	Effluent TP load (kg/event)	Influent TP EMC (mg/L)	Effluent TP EMC (mg/L)
12	0.09	0.01	1	92%	0.007	0.001	0.077	0.086
10	0.26	0.06	2	83%	0.041	0.014	0.157	0.236
8	0.39	0.15	3	75%	0.078	0.018	0.201	0.125
11	0.40	0.20	4	67%	0.065	0.028	0.162	0.142
3	0.56	0.11	5	58%	0.059	0.009	0.105	0.082
6	0.75	0.54	6	50%	0.128	0.083	0.171	0.155
9	1.08	0.92	7	42%	0.285	0.218	0.263	0.237
2	1.13	0.70	8	33%	0.279	0.106	0.247	0.151
5	1.37	0.25	9	25%	0.363	0.046	0.265	0.183
7	1.62	0.86	10	17%	0.277	0.194	0.171	0.225
4	1.82	0.91	11	8%	0.422	0.19	0.232	0.208
1	3.42	1.98	12	0%	0.554	0.175	0.162	0.088

From the exceedance plot of influent/effluent runoff volume, Lana notes all storm events removed TP load (influent > effluent, i.e., positive removal). Also, Lana notes that the largest load reduction (shaded area) occurs for the largest runoff events, although load reduction (as a percent) is still significant for most of the smaller storms.

(continued)

Example 12.12: (continued)

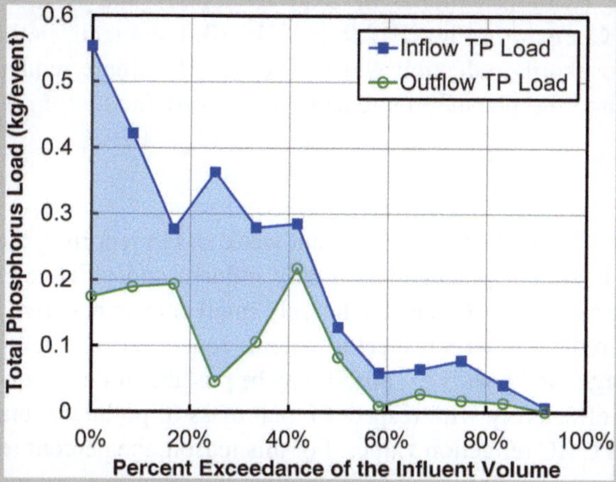

Fig. E12.10 Influent Exceedance method for TP Load

Lana then plots the EMC data pairs as a function of percent exceedance of total influent runoff volume, as shown in Fig. E12.11.

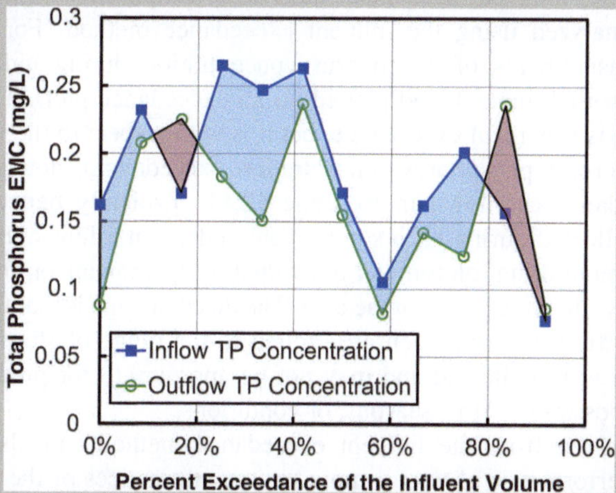

Fig. E12.11 Influent Exceedance method for TP EMC reduction

Lana is surprised to find that two runoff events (~18% exceedance, ~85% exceedance) had negative removal rates (influent < effluent, shaded red) and that EMC reduction (shaded blue) was relatively small for most runoff events. Lana quickly realizes that if she relied on EMC as the only measure of

(continued)

Example 12.12: (continued)
performance (without volume or load reduction), the results would be misleading. Comparing Figs. E12.9 and E12.10, Lana realizes that much of the total phosphorus reduction was due to runoff volume reduction due to infiltration, and the phosphorus reduction due to reduction in concentration was limited.

Example 12.12 demonstrates an inherent weakness in reporting the performance of stormwater treatment practices with only influent and effluent EMCs. Percent removal with respect to EMC can be relatively small or even negative in some cases, yet the percent reduction with respect to volume reduction and total pollutant mass load is much larger and positive. Thus, it may be possible for a treatment practice to be functioning effectively with respect to total mass of pollutant retained but have relatively small EMC reduction values. For this reason, the percent reduction in the total pollutant mass load retained by a practice is typically more indicative of the impact the practice has on surface water quality within the watershed.

The influent exceedance method can be applied to any data, including runoff volume, flow rate, pollutant concentration, pollutant load, residence time, and drain time, among others. In addition to using storm event data such as event mean concentration and total pollutant load, individual sample and flow measurement data can be analyzed using the influent exceedance method. For example, all individual measurements of phosphorus concentration during individual storm events can be sorted, ranked, and plotted on an exceedance plot. In this case, the horizontal X-axis is percent exceedance, but it is with respect to the variable being plotted (i.e., percent phosphorus concentration exceedance), not influent runoff volume exceedance as shown in Example 12.11. Typically between 5 and 50 samples are collected during each storm event and when a few storms have been assessed with monitoring, plotting all individual measurements on the exceedance plot can quickly illustrate trends in the data. The disadvantage is that the predictions or conclusions from this approach are limited to the range of storm events from which the data were collected and may not be appropriate for predicting performance for all possible storms, seasons, or conditions.

Graphs obtained from the influent exceedance method can also be used to compare the performance of two or more treatment practices or the same practice over different time spans. The size (i.e., design water quality volume) of each compared practice must be considered. For example, if practice A has a design WQV corresponding to approximately 95% influent volume exceedance and is to be compared to practice B with a design WQV corresponding to approximately 90% influent volume exceedance, then comparisons should not be made at specific values of percent exceedance for influent runoff volume (i.e., comparing both practices at 50% exceedance). Rather, a more useful approach would be to compare the performance of the practices relative to their respective WQVs. In this case, it would be more appropriate to compare the performance of practice A at 95% exceedance to practice B at 90% exceedance.

Effluent Probability Method

The US EPA (2002) recommends using the effluent probability method for analysis of monitoring data. The effluent probability method is a graphical representation of long-term performance that is useful in visually illustrating trends in performance over the range of data values (e.g., runoff volume, flow rate, pollutant concentration, or pollutant load). The effluent probability method does not, however, result in a single value of performance that is easily comparable to other monitoring data (other sites, time periods, etc.). Therefore, comparison of monitoring data using the effluent probability method is typically limited to graphical comparison (e.g., plotting inflow and outflow for multiple practices). Pollutant removal can be estimated by integrating the curve or best-fit distribution function for both the inflow and outflow data and then calculating the difference between in the inflow and outflow. Due to the complexity of this method, it is more common to report performance for several probabilities (i.e., inflow and outflow data for 10%, 50%, and 90% probabilities).

One advantage of the effluent probability method is that monitoring data are plotted on a standard parallel probability plot, which results in a straight line if the measurements are normally (normal-probability plot) or log-normally (log-probability plot) distributed. Also, the inflow and outflow data will be parallel if the percent difference between the inflow and outflow is constant for all ranked data pairs. One disadvantage is that special software may be required to generate probability plots. In order to make defensible conclusions or predictions from the effluent probability method, several data points (at least five, but more than ten recommended) are necessary.

The primary difference between the effluent probability method and the exceedance method is that plotting the data using the exceedance method can be done using most commercially available spreadsheet software (e.g., Microsoft Excel™). One disadvantage of the exceedance method is that the normality of the data cannot be assessed visually.

Maintenance of Stormwater Treatment Practices

13

Abstract

Without maintenance, the performance of any stormwater treatment practice will decline over time until it reaches an unacceptable level. Thus, every stormwater management plan should include an estimated schedule for maintenance activities, and funds should be budgeted to support this schedule. This chapter provides recommendations for maintenance activities based on the treatment practice and assessment results and also presents typical suggested corresponding frequencies of such activities. In addition, this chapter provides the results of a maintenance activity survey, again grouped by the type of practice, that offers insights on typical issues that trigger the need for maintenance, maintenance complexity, and maintenance frequency.

As described in Chap. 1, stormwater treatment practice maintenance is purposeful management intended to ensure proper function and extend useable life by maintaining a treatment practice at the desired level of performance. Maintenance consists of routine (regular and relatively frequent), nonroutine (irregular and less frequent), and major (irregular and rare) activities (Fig. 13.1). The purpose of routine and nonroutine maintenance activities is to prevent or limit the need for major maintenance; therefore, the combination of these activities is called preventative maintenance.

The usable life of stormwater treatment practices from their creation (design and construction) through operative stages depends on maintenance actions. Maintenance requires significant resources (personnel, equipment, materials, sediment disposal expense, etc.). The more that is learned about stormwater treatment practice performance, the easier it will be to make appropriate and cost-effective maintenance decisions to optimize performance and extend the useable life of the practice. Therefore, assessment using a combination of the three levels to determine performance of a stormwater treatment practice is a necessary step in developing a cost-effective maintenance program.

Assessment program results are used to adaptively develop a maintenance program for stormwater treatment practices. For example, visual inspection

A.J. Erickson et al., *Optimizing Stormwater Treatment Practices: A Handbook of Assessment and Maintenance*, DOI 10.1007/978-1-4614-4624-8_13,

Fig. 13.1 Stormwater treatment practice operation & maintenance pyramid

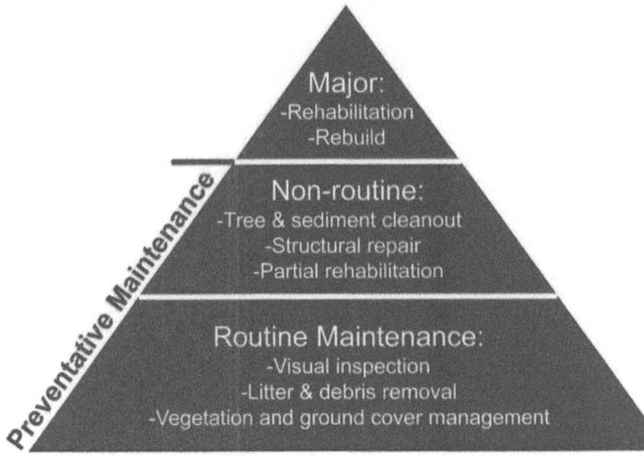

(level 1) of a bioretention practice (rain garden) could indicate that a significant amount of sediment has accumulated near the inlet of the practice. Capacity testing (level 2a) could indicate that approximately 20 cubic feet of sediment has accumulated (sediment retention), and that the saturated hydraulic conductivity (K_s) in the area of sediment accumulation is minimal compared to the design value. Therefore, maintenance should be scheduled to remove the accumulated sediment and restore the water and sediment storage capacity, as well as the infiltration capacity in the area near the inlet of the practice. Conducting visual inspection and needed capacity testing regularly could determine that sediment removal maintenance is required every 2–3 years for this practice. Conducting maintenance on this schedule could prevent costly rehabilitation or rebuilding of the entire practice and extend the useable life of the practice by maintaining infiltration. Note that additional guidance for scheduling maintenance has been provided in Chaps. 5–8. Recommendations for scheduling maintenance and common maintenance frequencies, efforts, and costs are provided in the rest of this chapter.

A general rule of thumb discerned from the overall construction-maintenance costs in this book is that the cumulative present value of maintenance costs for a stormwater treatment practice will equal design and construction costs (excluding land costs) after a certain number of years, depending on the size of the practice. For example, maintenance costs will equal construction costs after only 5 years for a $1,000 (2005 dollars) installation. For a $10,000 installation (2005 dollars), maintenance costs will equal construction costs after only 10 years. For a $100,000 installation (2005 dollars), maintenance costs will equal construction costs after 25 years. Thus, there is an economy of scale for the maintenance of stormwater treatment practices because the maintenance cost-relative to construction cost will generally decrease for a larger stormwater treatment practice.

13.1 Maintenance of Sedimentation Practices

Maintenance of sedimentation devices often involves sediment and trash removal, repairing misaligned pipes, or addressing invasive vegetation. The performance of a sedimentation practice often depends upon the continuous maintenance activities undertaken. The performance level of a practice will eventually drop below the level it had when new unless it is maintained on a regular basis.

13.1.1 Actions

13.1.1.1 Dry Ponds

Dry ponds can be effective at retaining suspended solids and pollutants associated with these solids. After continued operation, retained solids will eventually need to be removed. The dry pond must be regularly inspected to determine its condition and the amount of solids retained in the pond. The required frequency of inspection and maintenance is dependent on the watershed land use (e.g., urban, rural, and farm, among others), construction activities present, and rainfall amounts and intensity. It is recommended, however, that visual inspection and any associated maintenance be performed at least once per year.

If any level of assessment reveals that a dry pond is not draining a runoff event within the specified design time, the following measures should be taken:

1. Inspect all outlet structures for clogging and/or structural damage. Remove debris and repair/replace outlet structure, if necessary.
2. Inspect the outflow location(s) to ensure an elevated downstream water level is not impeding discharge from the pond. If this is the case, the downstream water elevation should be reduced or the outlet modified such that drainage occurs within the desired time.
3. If the pond still does not drain within the specified design time, the hydraulics of the pond should be reevaluated and the geometry and outlet structure redesigned. Hunt and Lord (2006b) discuss the maintenance requirements of wetlands and wet ponds. Even though Hunt and Lord (2006b) do not specifically discuss dry ponds, their recommendations that apply to both wet ponds and dry ponds are revised in Table 13.1.

Dry ponds are most effective in retaining suspended solids and pollutants that are associated with solids and are usually not implemented to reduce temperature impacts or achieve retention of dissolved pollutants. Thus, the following discussion only considers solids and pollutants which typically adsorb to or are incorporated in solids. If a dry pond is not retaining suspended solids (or other adsorbed pollutants) at expected levels, the following steps should be taken:

1. Check that the desired levels of pollutant retention are realistic. For example, if the target pollutant is total phosphorus and the runoff entering the pond contains a large fraction of dissolved phosphorus, large retention rates of total phosphorus

Table 13.1 Maintenance requirements and frequencies for dry ponds (modified from Hunt and Lord 2006b)

Task	Frequency	Notes
Inspection	Annually or after every 2-year storm	
Remove retained sediment	Variable (once every 5–10 years is typical in stable watersheds)	In unstable watersheds (i.e., those with active construction), the frequency is typically once a year
Monitor sediment depth	Once per year	Can be performed with capacity testing
Inspect outlet structures	Annually or after every 2-year storm	Follow visual inspection guidelines
Remove trash and debris	Annually	Increase frequency, if needed
Remove vegetation from dam top and faces, if applicable	Once per year	Increase frequency, if needed

may not be possible for a device that retains pollutant mostly by sedimentation. Or, if the sediment size distribution contains an uncharacteristically large fraction of fines, the hydraulic retention time may not be adequate to achieve the desired pollutant retention rate. If retention of the desired pollutant is not realistic, consider implementing another stormwater treatment practice to achieve desired results.

2. Perform a sediment capacity test to determine the remaining sediment storage capacity of the pond. If there is no remaining capacity or if the capacity is nearly exhausted, retained sediment should be removed to restore sediment storage capacity.
3. If there is adequate storage capacity remaining in the pond and pollutant removal is still less than expected values, a tracer study should be performed (see synthetic runoff testing for wet ponds in Chap. 7) to determine if substantial short-circuiting is occurring. If short-circuiting is occurring, consider adding one or more baffles or retrofitting the pond to redirect the flow of runoff in a way that eliminates or minimizes short-circuiting.

13.1.1.2 Wet Ponds

Wet ponds can be effective at retaining suspended solids and pollutants that typically adsorb to or are incorporated in solids but are not as effective at retaining dissolved pollutants. After continued operation, retained solids will need to be removed. Thus the wet pond must be inspected regularly to determine its condition and the amount of solids retained in the pond. The required frequency of inspection and maintenance is dependent on the watershed land use (e.g., urban, rural, and farm, among others), construction activities in the watershed, and rainfall amounts and intensity. It is recommended, however, that visual inspection and any associated maintenance be performed at least once per year.

Hunt and Lord (2006b) discuss the maintenance requirements of wetlands and wet ponds. The revised recommendations that apply to wet ponds are reproduced in Table 13.2.

Table 13.2 Maintenance requirements and frequencies for wet ponds (modified from Hunt and Lord 2006b)

Task	Frequency	Notes
Inspection	Annually or after every 2-year storm	
Remove all sediment from forebay and deep pool (dredging)	Variable (once every 5–10 years is typical in stable watersheds)	In unstable watersheds (i.e., those with active construction), the frequency is typically once a year
Monitor sediment depth in forebay and deep pools	Once per year	Can be performed with capacity testing
Inspect outlet structures	Annually or after every 2-year storm	Follow visual inspection guidelines
Remove trash and debris	Annually	Increase frequency, if needed
Remove vegetation from dam top and faces, if applicable	Once per year	Increase frequency, if needed
Mow wet pond perimeter	As needed	
Remove muskrats and beavers, if present	As needed	Destroy muskrat holes whenever present. Contact a professional trapper to remove beavers

Capacity testing or monitoring are required to determine if a wet pond is retaining pollutants as expected. If a wet pond is not retaining suspended solids or other adsorbed pollutants at expected levels, the following steps should be taken:

1. Check that the desired levels of pollutant retention are realistic. For example, if the target pollutant is total phosphorus and the runoff entering the pond contains a large fraction of dissolved phosphorus, large retention rates of total phosphorus may not be possible for a device that retains pollutant mostly by sedimentation. Or, if the sediment size distribution contains an uncharacteristically large fraction of fines, the hydraulic retention time may not be adequate to achieve the desired pollutant retention rate. If retention of the desired pollutant is not realistic, consider implementing another stormwater treatment practice to achieve desired results.
2. Perform a sediment capacity test to determine the remaining sediment storage capacity of the pond. If there is little remaining capacity, the retained sediment must be removed to restore sediment storage capacity.
3. If there is adequate storage capacity remaining in the pond and pollutant removal is still less than expected values, a tracer study should be performed to determine if short-circuiting is occurring (see synthetic runoff testing for wet ponds in Chap. 7). If short-circuiting is occurring, consider adding one or more baffles or retrofitting the pond to redirect the flow of water in a way that eliminates or minimizes short-circuiting.

13.1.1.3 Underground Sedimentation Devices

Underground sedimentation devices generally target the same pollutants as dry and wet ponds, but because they are typically located underground, they are used more often in urban areas where land space for a treatment practice is less available. Compared to dry and wet ponds, underground sedimentation devices are typically smaller and have smaller design flows and total solid capture capacity.

If an underground sedimentation device is underperforming, the following steps should be taken:

1. Inspect all outlet structures for clogging and/or structural damage. Remove debris and repair/replace outlet structure, if necessary.
2. Inspect outflow location to ensure an elevated downstream water level is not impeding discharge from the device. If this is the case, the downstream water elevation must be reduced or the outlet modified such that the device performs as expected.
3. If the device is perforated, check that the perforations are not clogged with sediment or debris. Remove sediment and debris, if present.
4. If the device is proprietary, contact the manufacturer.

If assessment reveals that an underground sedimentation device is not draining as expected, the following steps should be taken:

1. Inspect all outlet structures for clogging and/or structural damage. Remove debris and repair/replace outlet structure, if necessary.
2. Inspect outflow location to ensure an elevated downstream water level is not impeding discharge from the device. If this is the case, the downstream water elevation must be reduced or the outlet modified such that drainage occurs within the desired time.
3. If the device is proprietary, contact the manufacturer.

13.1.2 Frequency, Effort, and Cost

Adequate inspection frequency is dependent on many variables, including rainfall amounts and intensity, watershed use, size of the practice, size of the sediment and sediment loads, and other variables. Survey results (Erickson et al. 2010) indicate that the frequency of inspection for a majority of sedimentation practices of the responding municipalities is once a year or less. Details of the survey results (Erickson et al. 2010) related to the frequency of inspection of sedimentation practices are given in Table 13.3. The median annual staff hours devoted to the inspection and maintenance of sedimentation practices are also shown in Table 13.3 to be between 1 and 2 h per site. It must be noted, however, that these results are estimates and do not indicate that this level of maintenance was sufficient to keep the stormwater treatment practice performing properly.

Municipalities were also asked to gauge the level of complexity of their maintenance efforts as minimal, simple, moderate, or complicated (Erickson et al. 2010). The results given in Table 13.3 indicate that maintenance of sedimentation devices is the least complex of the practices considered in this manual. This may be because they are the most common, with a well-developed knowledge base.

Table 13.3 Inspection frequency and efforts related to sedimentation practices

	Dry ponds	Wet ponds	Underground sedimentation devices
Inspected less than once a year	52% ($n = 27$)	52% ($n = 32$)	13% ($n = 17$)
Inspected once a year	48% ($n = 27$)	44% ($n = 32$)	56% ($n = 17$)
Inspected twice a year	0% ($n = 27$)	0% ($n = 32$)	31% ($n = 17$)
Inspected more than twice a year	0% ($n = 27$)	4% ($n = 32$)	0% ($n = 17$)
Median annual staff hours spent on inspection and maintenance	1 ($n = 19$)	2 ($n = 24$)	1.25 ($n = 14$)
Minimal complexity[a]	67% ($n = 25$)	57% ($n = 27$)	50% ($n = 16$)
Simple complexity[a]	28% ($n = 25$)	30% ($n = 27$)	20% ($n = 16$)
Moderate complexity[a]	0% ($n = 25$)	4% ($n = 27$)	10% ($n = 16$)
Complicated complexity[a]	5% ($n = 25$)	9% ($n = 27$)	20% ($n = 16$)

[a]Levels of complexity: minimal = stormwater professional or consultant is seldom needed; simple = stormwater professional or consultant is occasionally needed; moderate = stormwater professional or consultant is needed about half the time; and complicated = stormwater professional or consultant is always needed

The US EPA (1999a) estimated typical annual operating and maintenance (O & M) costs for dry ponds, based on expected maintenance schedules, to be less than 1% of the total construction cost. Weiss et al. (2007) reviewed the literature and found that, based on expected maintenance schedules and estimated prices, the expected annual O & M costs for dry ponds ranged from 1.8 to 2.7% of the total construction cost. For wet ponds, the US EPA (1999a) did not report a value, but the literature review of Weiss et al. (2007) found the range to be 1.9–10.2% of the total construction cost. It must be noted that these cost estimates are not based on actual O & M cost data but rather on typical maintenance schedules and activities combined with the estimated costs to perform these activities. Based on the data in Table 13.3, the inspection and maintenance costs of underground devices are expected to fall between the inspection and maintenance costs of dry ponds and wet ponds.

13.1.2.1 Factors Affecting Performance

Maintenance efforts for sedimentation practices are typically focused on sediment and trash removal, fixing clogged pipes, and addressing invasive vegetation. Table 13.4 lists the percent of responding municipalities (Erickson et al. 2010) that indicated the listed factors most frequently caused deterioration of stormwater treatment practice performance. The most frequent factors causing deterioration of performance were sediment buildup, litter and debris, pipe clogging, and invasive vegetation. Bank erosion, groundwater level, and structural problems, however, can cause serious and rapid deterioration of performance when present.

Table 13.4 Percent of respondents who indicated the listed factor frequently caused deterioration of stormwater treatment practice performance

	Dry ponds (n = 49) (%)	Wet ponds (n = 90) (%)	Underground sedimentation devices (n = 19) (%)
Sediment buildup	24%	26%	58%
Litter/debris	31%	19%	21%
Pipe clogging	18%	21%	11%
Invasive vegetation	16%	10%	0%
Bank erosion	8%	11%	0%
Groundwater level	2%	7%	5%
Structure problems	0%	7%	5%

13.1.3 Recommendations

Sedimentation practices are designed to be effective at retaining suspended solids and pollutants that typically adsorb to or are incorporated in solids but not to effectively retain dissolved pollutants. After continued operation, the retained solids need to be removed. The sedimentation practice must therefore be regularly inspected to determine its condition. The required maintenance frequency is dependent on the watershed land use (e.g., urban, rural, and farm, among others), construction activities in the watershed, and rainfall amounts and intensity.

13.2 Maintenance of Filtration Practices

Maintenance of filtration practices can range from the relatively easy task of trash removal to the much more expensive and time consuming complete removal and replacement of the filter media and underlying system. This section is intended to provide information on the frequency, complexity, and time required for maintenance efforts related to filtration practices as well as typical factors that commonly inhibit the performance of filtration treatment practices.

The two primary failure mechanisms for filtration practices are clogging and the presence of macropores. Clogging can result in long periods of standing water, flooding of surrounding areas, bypassing of the filter by untreated stormwater, lack of measurable effluent, or any combination thereof. Macropores, such as wormholes, can cause short-circuiting of the filtration practice and subsequently result in reduced solids retention efficiency and less peak flow reduction.

13.2.1 Actions

Maintenance actions for all sand and soil filters, whether under or above ground or the filtration portion of hybrid systems, are generally similar. Thus, maintenance actions for all sand and soil filters are discussed together in this section.

Surface and underground sand filters and hybrid systems are typically implemented to reduce suspended solids in stormwater runoff. A separate pretreatment practice just upstream of the filtration practice will improve performance significantly; it must be maintained, as well. Like all stormwater treatment practices, filters need regular inspection and maintenance to ensure proper operation.

If any level of assessment indicates that the filter is not draining or will not drain the design runoff volume within the required time, the following steps can be taken to address the problem:

1. Inspection of outlet structures and underdrain system with removal of any objects obstructing flow and/or replacement of structural components, if necessary.
2. Performing capacity or synthetic runoff tests to determine filtration rates and saturated hydraulic conductivity (K_s) values of the filter media. If filtration rates are small, the following steps may correct the problem
   (a) Rototilling of the top 6 in. of filter media
   (b) Removal of the sediment layer if it exists and removal and replacement of the top 6–8 in. of filter media
   (c) Removal and replacement of the entire media bed

If filtration rates are too large or the total suspended solids retention rate is too small, it is likely that there is short-circuiting in the filter media. The following steps can be taken to confirm and address the problem.

1. Visual inspection of the filter media to ensure there are no macropores or large holes, ruts, or other openings in the media that would allow runoff to pass without being sufficiently filtered. If any such areas are present the media in the suspect area should be replaced only after any underlying causes (e.g., filter media layer insufficiently deep, macropores in filter media, and insufficient gravel subbase, among others) are found and corrected.
2. Capacity testing to determine filtration rates at various locations on the filter surface. Any locations where the saturated hydraulic conductivity (K_s) is determined to be larger than 280 ft/day should be corrected by removing and replacing the filter media in that area.
3. Synthetic runoff tests to determine an overall effective saturated hydraulic conductivity (K_s) value for the filter, which could be completed during or following precipitation. A disadvantage of this test is that it will not identify specific locations in the media where short-circuiting is likely occurring. As a result, if synthetic runoff tests identify short-circuiting and no additional testing is performed (i.e., capacity tests), the entire media bed must be replaced.

While frequency of filter media replacement will vary depending on watershed and filter size, land use, rainfall amounts and intensities, etc., Wossink and Hunt (2003) report that removal of the top layer of filter media typically is required from once per year to once every 3 years. Landphair et al. (2000) report that surface sand filter media typically needs to be replaced every 3–5 years.

Table 13.5 Inspection frequency and efforts related to filtration practices

	Surface sand or soil filter	Underground filtration practice
Inspected less than once a year	67% ($n = 9$)	44% ($n = 9$)
Inspected once a year	33% ($n = 9$)	56% ($n = 9$)
Inspected twice a year	0% ($n = 9$)	0% ($n = 9$)
Inspected more than twice a year	0% ($n = 9$)	0% ($n = 9$)
Median annual staff hours spent on inspection and maintenance	1 ($n = 7$)	1 ($n = 7$)
Minimal complexity[a]	63% ($n = 8$)	50% ($n = 10$)
Simple complexity[a]	0% ($n = 8$)	20% ($n = 10$)
Moderate complexity[a]	25% ($n = 8$)	10% ($n = 10$)
Complicated complexity[a]	13% ($n = 8$)	20% ($n = 10$)

[a]Levels of complexity: minimal = stormwater professional or consultant is seldom needed; simple = stormwater professional or consultant is occasionally needed; moderate = stormwater professional or consultant is needed about half the time; and complicated = stormwater professional or consultant is always needed

13.2.2 Frequency, Effort, and Cost

The inspection frequency required to maintain a filtration practice depends on variables that include rainfall amounts and intensity, watershed use, other stormwater treatment practices within the watershed, and other factors. Survey results (Erickson et al. 2010) indicate that responding municipalities inspect filtration practices once a year or less (see Table 13.5). In addition to the frequency of inspection, the median annual staff hours devoted to the inspection and maintenance of filtration devices is also given in Table 13.5. It must be noted, however, that these results are estimates of the hours spent and do not indicate that this level of maintenance was sufficient to keep the stormwater treatment practice performing properly. The authors experience is that more frequent inspections and greater annual staff hours are often required to keep filtration practices operating properly.

In the survey (Erickson et al. 2010), municipalities were also asked to gauge the level of complexity of their maintenance efforts as minimal, simple, moderate, or complicated. These results are also given in Table 13.5 as the percentage of respondents selecting each complexity level.

The US EPA (1999a) estimated that typical annual operating and maintenance (O & M) costs for sand filters, based on expected maintenance schedules, range from 11 to 13% of the total construction cost. Weiss et al. (2007) reviewed the literature and found that, based on expected maintenance schedules and estimated prices, the expected annual O & M costs ranged from 0.9 to 9.5% of the total construction cost. It must be noted that these cost estimates are not based on actual O & M cost data but rather on typical maintenance schedules and activities, combined with the estimated costs to perform these activities.

Table 13.6 Percent of respondents who indicated the listed factor frequently caused deterioration of stormwater treatment practice performance

	Surface sand or soil filter ($n = 10$)	Underground infiltration practice ($n = 8$)
Sediment buildup	50%	50%
Litter/debris	30%	25%
Pipe clogging	10%	13%
Groundwater level	0%	13%
Oil spill	10%	0%

13.2.2.1 Factors Affecting Performance

Maintenance efforts for filtration practices are typically focused on the removal of filtered sediment buildup and litter/debris. Survey (Erickson et al. 2010) results indicate that, at times, maintenance efforts were also needed to address the issues of groundwater and oil spills. Table 13.6 lists the percentage of respondents that indicated the corresponding factor most frequently caused deterioration of the stormwater treatment practice performance. Other factors not listed in Table 13.6, such as invasive vegetation and bank erosion, were listed as possible survey choices but received zero responses.

13.2.3 Recommendations

The required frequency of maintenance is dependent on the watershed land use (e.g., urban, rural, and farm, among others), construction activities in the watershed, and rainfall amounts and intensity. A pretreatment system such as a sediment forebay, however, can significantly reduce the frequency and extent of maintenance by removing settleable solids upstream of the filtration practice. Maintenance of a sediment forebay is typically less time intensive and less costly than maintenance of a filtration practice.

Even with a pretreatment system, retained solids will eventually need to be removed from the filter. If the retained solids are at or near the surface of the filter media, the practice can often be repaired by removing the top 2–5 in. (5.08–12.7 cm) of filter media, rototilling the surface, and replacing the removed media with similar or approved alternative media. If this procedure does not resolve the problem, the entire filter bed may need to be replaced to restore functionality.

13.3 Maintenance of Infiltration Practices

In the past, infiltration practices have been shown to have a relatively short life span. In one study, over 50% of infiltration systems either partially or completely failed within the first 5 years of operation (US EPA. 1999a). In a Maryland study on infiltration trenches (Lindsey et al. 1991), 53% were not operating as designed, 36% were clogged, and 22% showed reduced infiltration. In a study of 12 infiltration basins (Galli 1992), none of which had built-in pretreatment systems, all had failed within the first 2 years of operation. With proper assessment and maintenance, however, the life span of infiltration practices can be significantly extended.

13.3.1 Actions

13.3.1.1 Infiltration Basins and Trenches

The most common reason for failure of infiltration structures is clogging due to sediment and organic debris. Due to susceptibility of clogging, pretreatment of stormwater upstream of an infiltration basin or trench is recommended to remove sediments and debris (US EPA 1999b). Pretreatment structures, such as a plunge pool, sump pit, filter strip, sedimentation basin, grass channel, or a combination of these practices, should be installed upstream of the infiltration practice.

In order to maintain proper function and optimal pollutant removal, infiltration practices require regular maintenance and inspection. Table 13.7 provides guidance on typical maintenance practices and time frames.

If inspection/assessment indicates unacceptable levels of infiltration in basins or trenches, any sediment layer or crust on the surface should be removed. If this does not sufficiently restore the practice, the filter fabric, if present, should be inspected and replaced, if warranted. If infiltration rates still do not increase to acceptable levels, the entire basin or trench should be renovated and/or replaced.

Table 13.7 Typical maintenance activities for infiltration basins and trenches (Modified from WMI 1997)

Activity	Schedule
• Remove sediment and oil/grease from pretreatment devices and overflow structures • Mow and remove litter and debris • Stabilize eroded banks, repair undercut, and eroded areas at inflow and outflow structure	Standard maintenance (as needed)
• Inspect pretreatment devices and diversion structures for signs of sediment buildup and structural damage • If dead or dying grass is evident at the bottom or the basin/trench, check to ensure water percolates within 2 days following significant rain events	Semiannual inspection
• Disc or otherwise aerate bottom • De-thatch basin bottom	Annual maintenance
• Provide an extended dry period, if bypass capability is available, to regain or increase the infiltration rate in the short term	5-year maintenance
• Perform total rehabilitation of the trench to maintain storage capacity within 2/3 of the design treatment volume and 48-h exfiltration rate • Excavate trench walls to expose clean soil	Upon failure

13.3.1.2 Permeable Pavements

In order to remove accumulated sediment, permeable pavements should be inspected, vacuumed, and pressure washed at least three to four times per year (Landphair et al. 2000; NJDEP 2004). More frequent maintenance will be required in particularly dirty areas, such as a watershed with a construction site. Solids removed in this process must be disposed of as required by all governing regulations. If any level of assessment reveals inadequate infiltration capacity and vacuuming and washing does not increase infiltration rates to acceptable levels, the entire surface may have to be removed and replaced.

13.3.2 Frequency, Effort, and Cost

Survey results (Erickson et al. 2010) indicate that a majority of responding municipalities inspect their infiltration basins/trenches and permeable pavements once a year or more. However, regular and frequent inspection of infiltration practices is recommended. Detailed survey results related to inspection frequency are given in Table 13.8.

The median annual staff hours devoted to the inspection and maintenance of infiltration practices are also shown in Table 13.8. Municipalities (Erickson et al. 2010) were also asked to gauge the level of complexity of their maintenance efforts as either minimal, simple, moderate, or complicated. The percentage of respondents that selected the four categories is also shown in Table 13.8. It must be noted, however, that these results are estimates and do not indicate that this level of maintenance was sufficient to keep the stormwater treatment practice performing properly.

Table 13.8 Inspection frequency and efforts related to infiltration practices

	Infiltration basin or trench	Permeable pavement
Inspected less than once a year	18% ($n = 19$)	25% ($n = 14$)
Inspected once a year	76% ($n = 19$)	50% ($n = 14$)
Inspected twice a year	0% ($n = 19$)	8% ($n = 14$)
Inspected more than twice a year	6% ($n = 19$)	17% ($n = 14$)
Median annual staff hours spent on inspection and maintenance	1 ($n = 17$)	2 ($n = 9$)
Minimal complexity[a]	34% ($n = 18$)	61% ($n = 16$)
Simple complexity[a]	40% ($n = 18$)	31% ($n = 16$)
Moderate complexity[a]	13% ($n = 18$)	0% ($n = 16$)
Complicated complexity[a]	13% ($n = 18$)	8% ($n = 16$)

[a]Levels of complexity: minimal = stormwater professional or consultant is seldom needed; simple = stormwater professional or consultant is occasionally needed; moderate = stormwater professional or consultant is needed about half the time; and complicated = stormwater professional or consultant is always needed

Table 13.9 Percent of respondents who indicated the listed factor frequently caused deterioration of stormwater treatment practice performance

	Infiltration basin or trench ($n = 39$) (%)	Permeable pavement ($n = 9$) (%)
Sediment buildup	36%	67%
Litter/debris	21%	11%
Pipe clogging	10%	11%
Invasive vegetation	5%	0%
Bank erosion	5%	0%
Groundwater level	13%	0%
Structure problems	5%	0%
Oil spill	3%	11%
Mechanical problems	3%	0%

The US EPA (1999a) estimated that typical annual operating and maintenance (O & M) costs for infiltration basins, based on expected maintenance schedules, range from 1 to 10% of the total construction cost. Weiss et al. (2007) reviewed the literature and found that, based on expected maintenance schedules and estimated prices, the expected annual O & M costs ranged from 2.8 to 4.9% of the total construction cost. For infiltration trenches the corresponding values were 5–20% (US EPA 1999a) and 5.1–126% (Weiss et al. 2007). No such values were reported for permeable (or porous) pavements. It must be noted that these cost estimates are not based on actual O & M cost data but rather on typical maintenance schedules and activities combined with the estimated costs to perform these activities.

13.3.2.1 Factors Affecting Performance

Maintenance efforts for infiltration practices are typically focused on removal of accumulated sediment and litter/debris. For a list of all factors that were deemed to frequently cause deterioration of stormwater treatment practice performance and the percent of responding municipalities (Erickson et al. 2010) that indicated the factor most frequently affected their stormwater treatment practices performance, see Table 13.9. Landphair et al. (2000) note that the performance of an infiltration trench is expected to decrease with time as the void spaces in the surrounding native soil fill with fines from the runoff which has infiltrated.

13.3.3 Recommendations

All infiltration practices should be designed to minimize required regular maintenance in order to remain effective. The required frequency of inspection and maintenance is dependent on the watershed land use (e.g., urban, rural, and farm, among others), construction practices in the watershed, and rainfall amounts and intensity.

13.4 Maintenance of Biologically Enhanced Practices

Maintenance of biologically enhanced practices can range from the relatively simple task of trash removal to more complicated tasks such as controlling invasive vegetation and repairing and stabilizing eroded banks. This section is intended to provide information on the maintenance of biologically enhanced practices and factors that typically inhibit performance of the practices.

13.4.1 Actions

13.4.1.1 Bioretention Practices (Rain Gardens)

As with all stormwater treatment practices, bioretention practices require regular maintenance if they are to remain effective. The required frequency of inspection and maintenance is dependent on the watershed land use (e.g., urban, rural, and farm, among others), construction in the watershed, and rainfall amounts and intensity. Visual inspection and any associated maintenance should be performed at least once per year. Additional recommended maintenance includes annual inspection for sediment accumulation and removal, if necessary. Hunt and Lord (2006a) discuss the maintenance requirements of bioretention practices (Table 13.10).

Table 13.10 Maintenance requirements and frequencies for bioretention practices (modified from Hunt and Lord 2006a)

Task	Frequency	Notes
Inspection	Annually or after every 2-year storm	
Inspection and maintenance of pretreatment unit(s)	Variable	Frequency and tasks depend on the pretreatment unit(s)
Pruning	As needed	Nutrients in runoff often cause bioretention vegetation to flourish
Mowing	Variable	Frequency is dependent on location and desired aesthetics
Mulching	As needed	
Mulch and top layer of soil removal	As needed	Mulch accumulation reduces available water storage and decreases infiltration rates. The top layer usually is the cause of clogging; entire rain gardens rarely need to be replaced
Watering	One time every 2–3 days for first 1–2 months. As needed afterward	
Replacement and removal of dead plants	Annually	Within first year, 10% of plants may die. Survival rates increase with time
Weeding, trash collection, clearing overflow structures, etc.	Annually	

Table 13.11 Maintenance requirements and frequencies for constructed wetlands (from Hunt and Lord 2006b)

Task	Frequency	Notes
Inspection	Annually or after every 2-year storm	
Remove all sediment from forebay and deep pool (dredging)	Variable (once every 5–10 years is typical in stable watersheds)	In unstable watersheds (i.e., those with active construction), the frequency is typically once a year
Monitor sediment depth in forebay and deep pools	Once per year	Can be performed with capacity testing
Inspect outlet structures	Annually or after every 2-year storm	Follow visual inspection guidelines
Remove trash and debris	Annually	Increase frequency, if needed
Remove vegetation from dam top and faces, if applicable	Annually	Increase frequency, if needed
Remove invasive species (particularly cattails)	Annually or as needed	Large cattail colonies should be removed with a backhoe. Chemical application may be used for small or new cattail growth
Remove muskrats and beavers, if present	As needed	Destroy muskrat holes whenever present. Contact a professional trapper to remove beavers

If any level of assessment reveals that a bioretention practice is not adequately infiltrating runoff, the following steps should be taken:

1. Replacement of mulch, if present, and the top layer of material.
2. If the previous step does not correct the situation, the entire practice may need to be replaced.

13.4.1.2 Constructed Wetlands

Constructed wetlands require regular maintenance to remain effective. For example, constructed wetlands can lose their capacity to remove phosphorus over time (Oberts 1999), which may be attributable to vegetation reaching a maximum density (Faulkner and Richardson 1991) or to the soils reaching a maximum adsorption capacity. Furthermore, overabundant and decaying vegetation can become a source of soluble and particulate phosphorus that may be released with the effluent. While regularly harvesting wetland vegetation to 'remove' phosphorus from the wetland system appears to be the logical solution, research has shown that only minimal amounts of phosphorus are removed when wetland vegetation is harvested (Kadlec and Knight 1996). Eventually, reconstruction may be required for the constructed wetland to remain effective at retaining pollutants. If assessment reveals that existing sediment storage volumes are too large, the stored sediment must be removed from the wetland and disposed of properly. Hunt and Lord (2006b) discuss maintenance requirements for wet ponds and wetlands (Table 13.11).

If a constructed wetland is not retaining pollutants at expected levels, the following steps should be taken:

1. Check that the desired levels of pollutant capture are realistic. For example, if the sediment size distribution contains an uncharacteristically large fraction of fines, the hydraulic retention time may not be adequate to achieve the desired pollutant retention rate. If retention of the desired pollutant is not realistic, consider implementing another stormwater treatment practice to achieve desired results. Or, if the pollutant is primarily in dissolved form and the vegetation in the wetland is known not to uptake the pollutant at significant levels, it is unrealistic to expect significant levels of retention.
2. Perform a sediment capacity test to determine the remaining sediment storage capacity of the wetland. If the storage capacity is exhausted or nearly exhausted, the retained sediment should be removed.
3. If there is adequate storage capacity remaining in the pond and pollutant removal is still below expected values, a tracer study should be performed to determine if short-circuiting is occurring (see synthetic runoff testing for wet ponds in Chap. 7). If short-circuiting is occurring, consider adding baffles or retrofitting the wetland to redirect the flow of water in a way that minimizes short-circuiting.

13.4.1.3 Filter Strips and Swales

Filter strips and swales can retain suspended solids and reduce stormwater runoff volumes through infiltration. The required frequency of inspection and maintenance is dependent on the watershed land use (e.g. urban, rural, and farm, among others), construction practices in the watershed, and rainfall amounts and intensity. Visual inspection and any associated maintenance should be performed at least once per year.

If infiltration rates or sediment retention rates of a filter strip are unacceptable, the top-soil may have to be broken up and the surface reconstructed. Also, increasing the density of vegetation can provide benefits by three mechanisms:

1. Decreasing water velocities, which will allow more time for infiltration
2. Creating more flow paths for the water to infiltrate into the soil
3. Providing increased surface area (e.g., stem and leaf) in the flow to enhance sedimentation, which occurs as suspended solids hit the surfaces and drop out of the flow or adhere to the surface.

Finally, decreasing the slope of the filter strip will slow velocities and allow for more sedimentation and more time for infiltration.

Landphair et al. (2000) states that maintenance requirements for roadside swales are minimal, other than typical mowing and trash pickup. Immediate replacement of any dead, dying, or missing vegetation, however, is imperative.

13.4.2 Frequency, Effort, and Cost

Adequate inspection frequency depends on many variables, including rainfall amount and intensity, watershed use, sediment loads, and many other factors. Survey

Table 13.12 Inspection frequency and efforts related to sedimentation practices

	Rain gardens	Constructed wetlands	Filter strips and swales
Inspected less than once a year	21% ($n = 22$)	38% ($n = 16$)	58% ($n = 13$)
Inspected once a year	42% ($n = 22$)	56% ($n = 16$)	33% ($n = 13$)
Inspected twice a year	16% ($n = 22$)	6% ($n = 16$)	0% ($n = 13$)
Inspected more than twice a year	21% ($n = 22$)	0% ($n = 16$)	8% ($n = 13$)
Median annual staff hours spent on inspection and maintenance	1 ($n = 13$)	1.5 ($n = 14$)	($n = 11$)
Minimal complexity[a]	41% ($n = 22$)	40% ($n = 15$)	46% ($n = 14$)
Simple complexity[a]	29% ($n = 22$)	13% ($n = 15$)	8% ($n = 14$)
Moderate complexity[a]	12% ($n = 22$)	40% ($n = 15$)	38% ($n = 14$)
Complicated complexity[a]	18% ($n = 22$)	7% ($n = 15$)	8% ($n = 14$)

[a]Levels of complexity: minimal = stormwater professional or consultant is seldom needed; simple = stormwater professional or consultant is occasionally needed; moderate = stormwater professional or consultant is needed about half the time; and complicated = stormwater professional or consultant is always needed

results (Erickson et al. 2010) indicate that rain gardens and constructed wetlands are most commonly inspected once or more a year, while filter strips and swales are most commonly inspected less than once a year. Table 13.12 shows details related to frequency of inspection of biologically enhanced practices. The median annual staff hours devoted to the inspection and maintenance of biologically enhanced practices is between 1 and 2 h per year per site, as given in Table 13.12. It must be noted, however, that these results are estimates and do not indicate that this level of maintenance was sufficient to keep the stormwater treatment practice performing properly.

Municipalities were also asked to gauge the level of complexity of their maintenance efforts as minimal, simple, moderate, or complicated (Erickson et al. 2010). These results are also given in Table 13.12 as the percentage of respondents that selected the complexity level. Between 30 and 47% of respondents stated that their maintenance efforts were moderate to complicated, indicating that stormwater professionals were often needed. This is the highest percentage of all groups of stormwater treatment practices.

The US EPA (1999a) estimated that typical annual operating and maintenance (O & M) costs for bioretention systems (i.e., rain gardens), based on expected maintenance schedules, range from 5 to 7% of the total construction cost. Weiss et al. (2007) reviewed the literature and found that, based on expected maintenance schedules and estimated prices, the expected annual O & M costs of rain gardens ranged from 0.7 to 10.9% of the total construction cost.

For constructed wetlands, the corresponding values were 2% (US EPA 1999a) and 4–14.1% (Weiss et al. 2007). For filter strips, expected annual O & M costs were $320 per maintained acre in 1999 (US EPA 1999a), with no value being reported by Weiss et al. (2007). Finally, for swales, the US EPA (1999a) reported expected

Table 13.13 Percent of respondents who indicated the listed factor frequently caused deterioration of stormwater treatment practice performance

	Rain gardens (n = 27) (%)	Constructed wetlands (n = 37) (%)	Filter strips and swales (n = 19) (%)
Sediment buildup	33%	24%	21%
Litter/debris	22%	19%	26%
Pipe clogging	7%	14%	5%
Invasive vegetation	26%	22%	26%
Bank erosion	0%	11%	11%
Groundwater level	7%	8%	5%
Structural problems	0%	3%	5%
Oil spill	4%	0%	0%

annual O & M costs to be 5–7% of the total construction cost, whereas Weiss et al. (2007) found the range to be 4–178%. It must be noted that these cost estimates are not based on actual O & M cost data but rather typical maintenance schedules and activities and the estimated costs to perform these activities.

13.4.2.1 Factors Affecting Performance

Survey results (Erickson et al. 2010) indicate that factors that may reduce performance and require maintenance for biologically enhanced practices include the presence of retained sediment, trash, invasive vegetation, and other less common factors such as pipe clogging and bank erosion. The percentage of respondents that indicated these and other issues frequently caused deterioration of performance are given in Table 13.13. Sediment buildup, invasive vegetation, and litter or debris were the primary factors for most of the biologically enhanced practices.

13.4.3 Recommendations

Biologically enhanced practices, which can be effective in reducing stormwater runoff volume as well as retaining suspended solids and dissolved pollutants, require regular maintenance to remain effective. The required frequency of inspection and maintenance depends on the watershed land use (e.g., urban, rural, and farm, among others), construction practices in the watershed, and rainfall amounts and intensity.

For any biologically enhanced stormwater treatment practice it is important to maintain the desired vegetation in a healthy state at appropriate densities. For systems that infiltrate stormwater, it may be periodically necessary to break up the soil surface to allow infiltration to occur. Other practices, such as constructed wetlands, may need to have accumulated sediment removed if suspended solid removal rates are unacceptable due to storage volumes that are at or near capacity.

Appendix A: Visual Inspection Checklists

A.J. Erickson et al., *Optimizing Stormwater Treatment Practices: A Handbook of Assessment and Maintenance*, DOI 10.1007/978-1-4614-4624-8,

Optimizing Stormwater Treatment Practices:

A Handbook of Assessment and Maintenance

Field Data Sheet for Level 1 Assessment: Visual Inspection Filtration Practices

Inspector's Name(s):______________________
Date of Inspection:______________________
Location of the filtration practice:______________________
Address or Intersection:______________________
Latititude, Longitude:______________________
Date the filtration practice began operation: ______________________
Filter Size (ft. x ft.): ______________________
Time since last rainfall (hr): ______________________
Quantity of last rainfall (in): ______________________
Rainfall Measurement Location: ______________________

Site Sketch (include inlets, outlets, etc.)

Based on visual assessment of the site, answer the following questions and make photographic or video-graphic documentation:

1. Has visual inspection been conducted at this location before? □ Yes □ No □ **I don't know**
   1. a) If yes, enter date:______________________
   1. b) Based on previous visual inspections, have any corrective actions been taken?
      □ Yes □ No □ **I don't know** **(If yes, describe actions in comments box)**

2. Has it rained within the last 48 hours at this location? □ Yes □ No □ **I don't know**

3. Does this filtration practice untilize pretreatment practices upstream?
   □ Yes □ No □ **I don't know (If yes, describe pretreatment practices** in comment box)

4. Access
   4. a) Access to the filtration practice is:
      □ Clear □ Partially obstructed □ Mostly obstructed □ Inaccessible
   4. b) If obstructed, the obstruction is (choose and provide comments) :
      □ temporary **and** □ no action needed **or** □ action needed
      □ permanent **and** □ before or during installation **or** □ new since installation
   4. c) Access to the upstream and downstream drainage is:
      □ Clear □ Partially obstructed □ Mostly obstructed □ Inaccessible
   4. d) If obstructed, the obstruction is (choose and provide comments) :
      □ temporary **and** □ no action needed **or** □ action needed
      □ permanent **and** □ before or during installation **or** □ new since installation

Comments

5. Inlet Structures

5. a) How many inlet structures are present? □ 0 □ 1 □ 2 □ 3 □ 4 □ 5 □ > 5

5. b) Are any of the inlet structures clogged? (If yes, mark location on site sketch above and fill in boxes below with items causing clogging (ie. debris, sediment, vegetation, etc.)

	Inlet #:	Inlet #:	Inlet #:	Inlet #:	Inlet #:
Partially					
Completely					
Not Applicable					

5. c) Are any of the inlet structures askew or misaligned from the original design or otherwise in need of maintenance? (if yes, write in reason: frost heave, vandalism, unknown, etc.)

	Inlet #:	Inlet #:	Inlet #:	Inlet #:	Inlet #:
Yes					
No					

6. Is there standing water in the filtration practice? □ Yes □ No

6. a) If yes, does the water have:
- □ Surface sheen (from oils or gasoline)
- □ Murky color (from suspended solids)
- □ Green color (from algae or other biological activity)
- □ Other (describe In comment box)

7. Is there evidence of illicit storm sewer discharges?

□ Yes □ No □ **I don't know (if yes, describe in comment box)**

8. What is the approximate percentage of vegetation coverage in the practice? ________ %

9. Are there indications of any of the following in the filtration practice? (If yes, mark on site sketch)
- □ Sediment deposition
- □ Erosion or channelization
- □ Excessive or undesirable vegetation (that needs mowing or removal)
- □ Bare soil or lack of healthy vegetation significantly different from the original design
- □ Litter or debris
- □ Other
- □ No

9. a) If sediment deposition is evident, what is the source?
- □ Erosion or channelization inside the filtration practice
- □ Erosion or channelization outside the filtration practice
- □ Construction site erosion
- □ Other
- □ Unknown

Comments

10. Are there indications of any of the following on the banks of the filtration practice:
    - □ Erosion or channelization
    - □ Soil slides or bulges
    - □ Excessive animal burrows
    - □ Seeps and wet spots
    - □ Poorly vegetated areas
    - □ Trees on constructed slopes

11. Is the bottom of the filtration practice covered with a layer of silts and/or clays?
    □ Yes □ No

12. Are any outlet structures or the emergency spillway clogged? □ No □ Partially □ Completely □ NA
    12. a) If yes, specify the clogging material (i.e. debris, sediment, vegetation, etc.) in the box below.

	Outlet #:	Outlet #:	Outlet #:
Material			

    12. b) Are any of the outlet structures askew or misaligned from the original design or otherwise in need of maintenance? (if yes, write in reason: frost heave, vandalism, unknown, etc.)

	Outlet #:	Outlet #:	Outlet #:
Reason			

13. Is there any evidence of any of the following downstream of the outlet structure?
    □ Sediment deposition □ Erosion or channelization □ Other □ No
    13. a) If sediment deposition is evident, what is the source?
    - □ Erosion or channelization inside the filtration practice
    - □ Erosion or channelization outside the filtration practice
    - □ Construction site erosion
    - □ Other, Specify______________________________
    - □ Unknown

14. Inspector's Recommendations. When is maintenance needed?
    - □ Before the next rainfall
    - □ Before the next rainy season
    - □ Within a year or two
    - □ No sign that any is required

Comments

15. Summarize the results of this inspection and write any other observations in the box below.

Summary and other observations

Optimizing Stormwater Treatment Practices:

A Handbook of Assessment and Maintenance

Field Data Sheet for Level 1 Assessment: Visual Inspection
Infiltration Basins and Trenches

Inspector's Name(s):______________________
Date of Inspection:______________________
Location of the infiltration practice:______________________
Address or Intersection:______________________
Latititude, Longitude:______________________
Date the infiltration practice began operation: ______________________
Filter Size (ft. x ft.): ______________________
Time since last rainfall (hr): ______________________
Quantity of last rainfall (in): ______________________
Rainfall Measurement Location: ______________________

Site Sketch (include inlets, north arrow, etc.)

Based on visual assessment of the site, answer the following questions and make photographic or video-graphic documentation:

1. Has visual inspection been conducted at this location before? □ Yes □ No □ **I don't know**
   1. a) If yes, enter date:______________________
   1. b) Based on previous visual inspections, have any corrective actions been taken?
      □ Yes □ No □ **I don't know** **(If yes, describe actions in comments box)**

2. Has it rained within the last 48 hours at this location? □ Yes □ No □ **I don't know**

3. Does this infiltration practice untilize pretreatment practices upstream?
   □ Yes □ No □ **I don't know (If yes, describe pretreatment practices in comment box)**

4. Access
   4. a) Access to the infiltration basin or trench is:
      □ Clear □ Partially obstructed □ Mostly obstructed □ Inaccessible
   4. b) If obstructed, the obstruction is (choose and provide comments) :
      □ temporary **and** □ no action needed **or** □ action needed
      □ permanent **and** □ before or during installation **or** □ new since installation
   4. c) Access to the upstream and downstream drainage is:
      □ Clear □ Partially obstructed □ Mostly obstructed □ Inaccessible
   4. d) If obstructed, the obstruction is (choose and provide comments) :
      □ temporary **and** □ no action needed **or** □ action needed
      □ permanent **and** □ before or during installation **or** □ new since installation

Comments

5. Inlet Structures

5. a) How many inlet structures are present? □ 0 □ 1 □ 2 □ 3 □ 4 □ 5 □ > 5

5. b) Are any of the inlet structures clogged? (If yes, mark location on site sketch above and fill in boxes below with items causing clogging (ie. debris, sediment, vegetation, etc.)

	Inlet #:	Inlet #:	Inlet #:	Inlet #:	Inlet #:
Partially					
Completely					
Not Applicable					

5. c) Are any of the inlet structures askew or misaligned from the original design or otherwise in need of maintenance? (if yes, write in reason: frost heave, vandalism, unknown, etc.)

	Inlet #:	Inlet #:	Inlet #:	Inlet #:	Inlet #:
Reason					

6. Is there standing water in the filtration practice? □ Yes □ No

6. a) If yes, does the water have:
- □ Surface sheen (from oils or gasoline)
- □ Murky color (from suspended solids)
- □ Green color (from algae or other biological activity)
- □ Other (describe In comment box)

7. Is there evidence of illicit storm sewer discharges?

□ Yes □ No □ **I don't know (if yes, describe in comment box)**

8. Does the infiltration basin or trench smell like gasoline or oil? □ Yes □ No

9. What is the approximate percentage of vegetation coverage in the practice? ________ %

10. Are there indications of any of the following in the infiltration practice? (If yes, mark on site sketch)
- □ Sediment deposition that will significantly impede infiltration
- □ Erosion or channelization
- □ Bare soil or lack of healthy vegetation significantly different from the original design
- □ Litter or debris
- □ Standing water more than 48 hours after the end of the most recent runoff event
- □ Other
- □ No

10. a) If sediment deposition is evident, what is the source?
- □ Erosion or channelization inside the infiltration practice
- □ Erosion or channelization outside the infiltration practice
- □ Construction site erosion
- □ Other
- □ Unknown

Comments

11. Are there indications of any of the following on the banks of the infiltration basin or trench:
    - □ Erosion or channelization
    - □ Soil slides or bulges
    - □ Excessive animal burrows
    - □ Seeps and wet spots
    - □ Poorly vegetated areas
    - □ Trees on constructed slopes

12. Is the bottom of the infiltration basin or trench covered with a layer of silts and/or clays?
    □ Yes □ No

13. Are any overflow structures clogged? □ No □ Partially □ Completely □ NA

13. a) If yes, specify the clogging material (i.e. debris, sediment, vegetation, etc.) in the box below.

	Outlet #:	Outlet #:	Outlet #:
Material			
Partial or Comp.			

13. b) Are any of the overflow structures askew or misaligned from the original design or otherwise in need of maintenance? (if yes, write in reason: frost heave, vandalism, unknown, etc.)

	Outlet #:	Outlet #:	Outlet #:
Reason			

14. Inspector's Recommendations. When is maintenance needed?
    - □ Before the next rainfall
    - □ Before the next rainy season
    - □ Within a year or two
    - □ No sign that any is required

Comments

15. Summarize the results of this inspection and write any other observations in the box below.

Summary and other observations

Optimizing Stormwater Treatment Practices:

A Handbook of Assessment and Maintenance

Field Data Sheet for Level 1 Assessment: Visual Inspection
Permeable Pavements

Inspector's Name(s):____________________________
Date of Inspection:____________________________
Location of the permeable pavement:____________________
Address or Intersection:____________________
Latititude, Longitude:____________________
Date the permeable pavement began operation: _______________
Pavement area (ft. x ft.): ____________________
Time since last rainfall (hr): ____________________
Quantity of last rainfall (in): ____________________
Rainfall Measurement Location: ____________________

Site Sketch (include curbs, islands, trees, north arrow, etc.)

Based on visual assessment of the site, answer the following questions and make photographic or video-graphic documentation:

1. Has visual inspection been conducted at this location before? □ Yes □ No □ **I don't know**
   1. a) If yes, enter date:____________________
   1. b) Based on previous visual inspections, have any corrective actions been taken?
      □ Yes □ No □ **I don't know** **(If yes, describe actions in comments box)**
2. Has it rained within the last 48 hours at this location? □ Yes □ No □ **I don't know**
3. Is there standing water on top of or water within the permeable pavement?
   □ Yes □ No
4. Are there indications of any of the following on top of or within the permeable pavement?
   (If yes, mark on site sketch)
   □ Sediment deposition
   □ Litter or debris
   □ Other
   □ No

Comments

4. a) If sediment deposition is evident, what is the source?
   - □ Erosion or channelization inside the permeable pavement
   - □ Erosion or channelization outside the permeable pavement
   - □ Construction site erosion
   - □ Other
   - □ Unknown

Comments

5. Inspector's Recommendations. When is maintenance needed?
   - □ Before the next rainfall
   - □ Before the next rainy season
   - □ Within a year or two
   - □ No sign that any is required

6. Summarize the results of this inspection and write any other observations in the box below.

Summary and other observations

Optimizing Stormwater Treatment Practices:

A Handbook of Assessment and Maintenance

Field Data Sheet for Level 1 Assessment: Visual Inspection
Dry Ponds

Inspector's Name(s):______________________________
Date of Inspection:______________________________
Location of the wet pond:______________________
Address or Intersection:______________________
Latititude, Longitude:______________________
Date the wet pond began operation: __________________
Wet pond dimensions. Depth (ft.): __________________
Area (ft. x ft.) ________________________
Time since last rainfall (hr): ____________________
Quantity of last rainfall (in): ____________________
Rainfall Measurement Location: __________________

Site Sketch (include inlets, outlets, north arrow, etc.)

Based on visual assessment of the site, answer the following questions and make photographic or video-graphic documentation:

1. Has visual inspection been conducted at this location before? □ Yes □ No □ **I don't know**
   1. a) If yes, enter date:____________________
   1. b) Based on previous visual inspections, have any corrective actions been taken?
      □ Yes □ No □ **I don't know** **(If yes, describe actions in comments box)**

2. Has it rained within the last 48 hours at this location? □ Yes □ No □ **I don't know**

3. Access
   3. a) Access to the dry pond is:
      □ Clear □ Partially obstructed □ Mostly obstructed □ Inaccessible
   3. b) If obstructed, the obstruction is (choose and provide comments) :
      □ temporary **and** □ no action needed **or** □ action needed
      □ permanent **and** □ before or during installation **or** □ new since installation
   3. c) Access to the upstream and downstream drainage is:
      □ Clear □ Partially obstructed □ Mostly obstructed □ Inaccessible
   3. d) If obstructed, the obstruction is (choose and provide comments) :
      □ temporary **and** □ no action needed **or** □ action needed
      □ permanent **and** □ before or during installation **or** □ new since installation

Comments

4. Inlet Structures
   4. a) How many inlet structures are present? □ 0 □ 1 □ 2 □ 3 □ 4 □ 5 □ > 5
   4. b) Are any of the inlet structures clogged? (If yes, mark location on site sketch above and fill in boxes below with items causing clogging (ie. debris, sediment, vegetation, etc.)

	Inlet #:	Inlet #:	Inlet #:	Inlet #:	Inlet #:
Partially					
Completely					
Not Applicable					

   4. c) Are any of the inlet structures askew or misaligned from the original design or otherwise in need of maintenance? (if yes, write in reason: frost heave, vandalism, unknown, etc.)

	Inlet #:	Inlet #:	Inlet #:	Inlet #:	Inlet #:
Reason					

5. Is there standing water in the dry pond? □ Yes □ No
   5. a) If yes, does the water have:
      □ Surface sheen (from oils or gasoline)
      □ Murky color (from suspended solids)
      □ Green color (from algae or other biological activity)
      □ Other (describe In comment box)

6. Is there evidence of illicit storm sewer discharges?
   □ Yes □ No □ **I don't know (if yes, describe in comment box)**

7. Are there indications of any of the following in the dry pond? (If yes, mark on site sketch)
   □ Sediment deposition
   □ Erosion or channelization
   □ Excessive or undesirable vegetation (that needs mowing or removal)
   □ Bare soil or lack of healthy vegetation significantly different from the original design
   □ Litter or debris
   □ Other
   □ No
   7. a) If sediment deposition is evident, what is the source?
      □ Erosion or channelization inside the dry pond
      □ Erosion or channelization outside the dry pond
      □ Construction site erosion
      □ Other
      □ Unknown

Comments

8. Are there indications of any of the following on the banks of the dry pond:
   - □ Erosion or channelization
   - □ Soil slides or bulges
   - □ Excessive animal burrows
   - □ Seeps and wet spots
   - □ Poorly vegetated areas
   - □ Trees on constructed slopes

9. Are any outlet or overflow structures clogged? □ No □ Partially □ Completely □ NA

9. a) If yes, specify the clogging material (i.e. debris, sediment, vegetation, etc.) in the box below.

	Outlet #:	Outlet #:	Outlet #:
Material			
Partial or Comp.			

9. b) Are any of the outlet or overflow structures askew or misaligned from the original design or otherwise in need of maintenance? (if yes, write in reason: frost heave, vandalism, unknown, etc.)

	Outlet #:	Outlet #:	Outlet #:
Reason			

10. Inspector's Recommendations. When is maintenance needed?
    - □ Before the next rainfall
    - □ Before the next rainy season
    - □ Within a year or two
    - □ No sign that any is required

Comments

11. Summarize the results of this inspection and write any other observations in the box below.

Summary and other observations

Optimizing Stormwater Treatment Practices:

A Handbook of Assessment and Maintenance

Field Data Sheet for Level 1 Assessment: Visual Inspection
Wet Ponds

Inspector's Name(s):______________________
Date of Inspection:______________________
Location of the wet pond:______________________
Address or Intersection:______________________
Latititude, Longitude:______________________
Date the wet pond began operation: ______________________
Wet pond dimensions. Depth (ft.): ______________________
Area (ft. x ft.) ______________________
Time since last rainfall (hr): ______________________
Quantity of last rainfall (in): ______________________
Rainfall Measurement Location: ______________________

Site Sketch (include inlets, outlets, north arrow, etc.)

Based on visual assessment of the site, answer the following questions and make photographic or video-graphic documentation:

1. Has visual inspection been conducted at this location before? □ Yes □ No □ **I don't know**
 1. a) If yes, enter date:______________________
 1. b) Based on previous visual inspections, have any corrective actions been taken?
 □ Yes □ No □ **I don't know** **(If yes, describe actions in comments box)**

2. Has it rained within the last 48 hours at this location? □ Yes □ No □ **I don't know**

3. Access
 3. a) Access to the wet pond is:
 □ Clear □ Partially obstructed □ Mostly obstructed □ Inaccessible
 3. b) If obstructed, the obstruction is (choose and provide comments) :
 □ temporary **and** □ no action needed **or** □ action needed
 □ permanent **and** □ before or during installation **or** □ new since installation
 3. c) Access to the upstream and downstream drainage is:
 □ Clear □ Partially obstructed □ Mostly obstructed □ Inaccessible
 3. d) If obstructed, the obstruction is (choose and provide comments) :
 □ temporary **and** □ no action needed **or** □ action needed
 □ permanent **and** □ before or during installation **or** □ new since installation

Comments

4. Inlet Structures

4. a) How many inlet structures are present? □ 0 □ 1 □ 2 □ 3 □ 4 □ 5 □ > 5

4. b) Are any of the inlet structures clogged? (If yes, mark location on site sketch above and fill in boxes below with items causing clogging (ie. debris, sediment, vegetation, etc.)

	Inlet #:	Inlet #:	Inlet #:	Inlet #:	Inlet #:
Partially					
Completely					
Not Applicable					

4. c) Are any of the inlet structures askew or misaligned from the original design or otherwise in need of maintenance? (if yes, write in reason: frost heave, vandalism, unknown, etc.)

	Inlet #:	Inlet #:	Inlet #:	Inlet #:	Inlet #:
Reason					

5. How many cells are in the wet pond system? ____________________

5. a) Does the water in the pond have:
- □ Surface sheen (from oils or gasoline)
- □ Murky color (from suspended solids)
- □ Green color (from algae or other biological activity)
- □ Other (describe In comment box)

6. Is there evidence of illicit storm sewer discharges?

□ Yes □ No □ **I don't know (if yes, describe in comment box)**

7. Does the wet pond smell like gasoline or oil? □ Yes □ No

8. Are there indications of any of the following in the wet pond? (If yes, mark on site sketch)
- □ Sediment deposition in excess of 50% of the sediment storage capacity
- □ Erosion or channelization
- □ Excessive or undesirable vegetation (that needs mowing or removal)
- □ Bare soil or lack of healthy vegetation significantly different from the original design
- □ Litter or debris
- □ Other
- □ No

8. a) If sediment deposition is evident, what is the source?
- □ Erosion or channelization inside the wet pond
- □ Erosion or channelization outside the wet pond
- □ Construction site erosion
- □ Other
- □ Unknown

Comments

9. Are there indications of any of the following on the banks of the wet pond:
- □ Erosion or channelization
- □ Soil slides or bulges
- □ Excessive animal burrows
- □ Seeps and wet spots
- □ Poorly vegetated areas
- □ Trees on constructed slopes

10. Are any outlet or overflow structures clogged? □ No □ Partially □ Completely □ NA

10. a) If yes, specify the clogging material (i.e. debris, sediment, vegetation, etc.) in the box below.

	Outlet #:	Outlet #:	Outlet #:
Material			
Partial or Comp.			

10. b) Are any of the outlet or overflow structures askew or misaligned from the original design or otherwise in need of maintenance? (if yes, write in reason: frost heave, vandalism, unknown, etc.)

	Outlet #:	Outlet #:	Outlet #:
Reason			

11. Is there any evidence of any of the following downstream of the outlet structure?

□ Sediment deposition □ Erosion or channelization □ Other □ No

11. a) If sediment deposition is evident, what is the source?
- □ Erosion or channelization inside the filtration practice
- □ Erosion or channelization outside the filtration practice
- □ Construction site erosion
- □ Other, Specify________________________
- □ Unknown

12. Inspector's Recommendations. When is maintenance needed?
- □ Before the next rainfall
- □ Before the next rainy season
- □ Within a year or two
- □ No sign that any is required

Comments

12. Summarize the results of this inspection and write any other observations in the box below.

Summary and other observations

Optimizing Stormwater Treatment Practices:

A Handbook of Assessment and Maintenance

Field Data Sheet for Level 1 Assessment: Visual Inspection
Underground Sedimentation Devices

Inspector's Name(s):______________________________
Date of Inspection:________________________________
Location of the device pond:_______________________
    Address or Intersection:_______________________
    Latititude, Longitude:_________________________
Date the device began operation: ___________________
Device dimensions. Depth (ft.): ____________________
    Area (ft. x ft.) ______________________________
Time since last rainfall (hr): ______________________
Quantity of last rainfall (in): _____________________
Rainfall Measurement Location: ____________________

Site Sketch (include inlets, outlets, north arrow, etc.)

Based on visual assessment of the site, answer the following questions and make photographic or video-graphic documentation:

1. Has visual inspection been conducted at this location before? □ Yes □ No □ **I don't know**
   1. a) If yes, enter date:______________________
   1. b) Based on previous visual inspections, have any corrective actions been taken?
      □ Yes □ No □ **I don't know** **(If yes, describe actions in comments box)**

2. Has it rained within the last 48 hours at this location? □ Yes □ No □ **I don't know**

3. Access
   3. a) Access to the underground sedimentation device is:
      □ Clear □ Partially obstructed □ Mostly obstructed □ Inaccessible
   3. b) If obstructed, the obstruction is (choose and provide comments) :
      □ temporary **and** □ no action needed **or** □ action needed
      □ permanent **and** □ before or during installation **or** □ new since installation
   3. c) Access to the upstream and downstream drainage is:
      □ Clear □ Partially obstructed □ Mostly obstructed □ Inaccessible
   3. d) If obstructed, the obstruction is (choose and provide comments) :
      □ temporary **and** □ no action needed **or** □ action needed
      □ permanent **and** □ before or during installation **or** □ new since installation

Comments

4. Inlet Structures
   4. a) How many inlet structures are present? □ 0 □ 1 □ 2 □ 3 □ 4 □ 5 □ > 5
   4. b) Are any of the inlet structures clogged? (If yes, mark location on site sketch above and fill in boxes below with items causing clogging (ie. debris, sediment, vegetation, etc.)

	Inlet #:	Inlet #:	Inlet #:	Inlet #:	Inlet #:
Partially					
Completely					
Not Applicable					

   4. c) Are any of the inlet structures askew or misaligned from the original design or otherwise in need of maintenance? (if yes, write in reason: frost heave, vandalism, unknown, etc.)

	Inlet #:	Inlet #:	Inlet #:	Inlet #:	Inlet #:
Reason					

5. Is a significant amount of water entering the underground device? □ Yes □ No □ **I don't know**
   5. a) If yes, what is the source?
      □ Recent rainfall/runoff event
      □ Leaking pipes or manholes
      □ Lawn irrigation
      □ Fire hydrant
      □ Other, specify____________________

6. Is there evidence of illicit storm sewer discharges?
   □ Yes □ No □ **I don't know (if yes, describe in comment box)**

7. Structure
   7. a) Are there excessive amounts of solids, debris, vegetation, or other objects that could be hindering performance or be re-suspended and exit the system during subsequent runoff events?
      □ Yes □ No □ **I don't know**
   7. b) Does the structure have any:
      □ Significant cracks
      □ Leaks
      □ Joint failures
      □ Structural instability
      □ Corrosion
      □ Other signs of damage or components requiring attention (describe in comment box)

Comments

8. Are any outlet structures clogged? □ No □ Partially □ Completely □ NA

8. a) If yes, specify the clogging material (i.e. debris, sediment, vegetation, etc.) in the box below.

	Outlet #:	Outlet #:	Outlet #:
Material			
Partial or Comp.			

8. b) Are any of the outlet structures askew or misaligned from the original design or otherwise in need of maintenance? (if yes, write in reason: frost heave, vandalism, unknown, etc.)

	Outlet #:	Outlet #:	Outlet #:
Reason			

9. Is there any evidence of any of the following downstream of the outlet structure?

□ Sediment deposition □ Erosion or channelization □ Other □ No

9. a) If sediment deposition is evident, what is the source?

- □ Erosion or channelization inside the filtration practice
- □ Erosion or channelization outside the filtration practice
- □ Construction site erosion
- □ Other, Specify____________________
- □ Unknown

10. Inspector's Recommendations. When is maintenance needed?

- □ Before the next rainfall
- □ Before the next rainy season
- □ Within a year or two
- □ No sign that any is required

Comments

12. Summarize the results of this inspection and write any other observations in the box below.

Summary and other observations

Optimizing Stormwater Treatment Practices:

A Handbook of Assessment and Maintenance

Field Data Sheet for Level 1 Assessment: Visual Inspection
Bioretention Practices (including Rain Gardens)

Inspector's Name(s):____________________
Date of Inspection:____________________
Location of the bioretention practice:____________________
Address or Intersection:____________________
Latitude, Longitude:____________________
Date the bioretention practice began operation: ____________________
Bioretention practice area (ft. x ft.): ____________________
Time since last rainfall (hr): ____________________
Quantity of last rainfall (in): ____________________
Rainfall Measurement Location: ____________________

Site Sketch (include inlets, outlets, north arrow, etc.)

Based on visual assessment of the site, answer the following questions and make photographic or video-graphic documentation:

1. Has visual inspection been conducted at this location before? □ Yes □ No □ **I don't know**
   1. a) If yes, enter date:____________________
   1. b) Based on previous visual inspections, have any corrective actions been taken?
      □ Yes □ No □ **I don't know** **(If yes, describe actions in comments box)**

2. Has it rained within the last 48 hours at this location? □ Yes □ No □ **I don't know**

3. Does this bioretention practice utilize pretreatment practices upstream?
   □ Yes □ No □ **I don't know (If yes, describe pretreatment practices in comment box)**

4. Access
   4. a) Access to the bioretention practice is:
      □ Clear □ Partially obstructed □ Mostly obstructed □ Inaccessible
   4. b) If obstructed, the obstruction is (choose and provide comments) :
      □ temporary **and** □ no action needed **or** □ action needed
      □ permanent **and** □ before or during installation **or** □ new since installation
   4. c) Access to the upstream and downstream drainage is:
      □ Clear □ Partially obstructed □ Mostly obstructed □ Inaccessible
   4. d) If obstructed, the obstruction is (choose and provide comments) :
      □ temporary **and** □ no action needed **or** □ action needed
      □ permanent **and** □ before or during installation **or** □ new since installation

Comments

5. Inlet Structures

5. a) How many inlet structures are present? □ 0 □ 1 □ 2 □ 3 □ 4 □ 5 □ > 5

5. b) Are any of the inlet structures clogged? (If yes, mark location on site sketch above and fill in boxes below with items causing clogging (i.e.,. debris, sediment, vegetation, etc.)

	Inlet #:	Inlet #:	Inlet #:	Inlet #:	Inlet #:
Partially					
Completely					
Not Applicable					

5. c) Are any of the inlet structures misaligned from the original design or otherwise in need of maintenance? (if yes, write in reason: frost heave, vandalism, unknown, etc.)

	Inlet #:	Inlet #:	Inlet #:	Inlet #:	Inlet #:
Reason					

6. Is there standing water in the bioretention practice? □ Yes □ No

6. a) If yes, does the water have:
- □ Surface sheen (from oils or gasoline)
- □ Murky color (from suspended solids)
- □ Green color (from algae or other biological activity)
- □ Other (describe In comment box)

7. Is there evidence of illicit storm sewer discharges?

□ Yes □ No □ **I don't know (if yes, describe in comment box)**

8. Does the bioretention practice smell like gasoline or oil? □ Yes □ No

9. What is the approximate percentage of vegetation coverage in the practice? ________ %

9. a) Does the current vegetation match the original design? □ Yes □ No □ Unknown

9. b) Is there the presence of:
- □ Weeds
- □ Wetland vegetation
- □ Invasive vegetation
- □ None of the above
- □ Other, specify________________________________

9. c) Does the vegetation appear to be healthy? □ Yes □ No (if no, describe in comment box)

9. d) Is the vegetation the appropriate size and density? □ Yes □ No (if no, describe in comment box)

Comments

10. Are there indications of any of the following in the bioretention practice? (If yes, mark on site sketch)
- □ Sediment deposition
- □ Erosion or channelization
- □ Excessive or undesirable vegetation (that needs mowing or removal)
- □ Litter or debris
- □ Other
- □ No

10. a) If sediment deposition is evident, what is the source?
- □ Erosion or channelization inside the infiltration practice
- □ Erosion or channelization outside the infiltration practice
- □ Construction site erosion
- □ Other
- □ Unknown

11. Are there indications of any of the following on the banks of the bioretention practice:
- □ Erosion or channelization
- □ Soil slides or bulges
- □ Excessive animal burrows
- □ Seeps and wet spots
- □ Poorly vegetated areas
- □ Trees on constructed slopes
- □ None of the above, the banks are in good condition
- □ Other, specify____________________

12. Are any overflow or bypass structures clogged? □ No □ Partially □ Completely □ NA

12. a) If yes, specify the clogging material (i.e. debris, sediment, vegetation, etc.) in the box below.

	Outlet #:	Outlet #:	Outlet #:
Material			
Partial or Comp.			

12. b) Are any of the overflow or bypass structures misaligned from the original design or otherwise in need of maintenance? (if yes, write in reason: frost heave, vandalism, unknown)

	Outlet #:	Outlet #:	Outlet #:
Reason			

13. Inspector's Recommendations. When is maintenance needed?
- □ Before the next rainfall
- □ Before the next rainy season
- □ Within a year or two
- □ No sign that any is required

Comments

14. Summarize the results of this inspection and write any other observations in the box below.

Summary and other observations

Optimizing Stormwater Treatment Practices:

A Handbook of Assessment and Maintenance

Field Data Sheet for Level 1 Assessment: Visual Inspection
Constructed Wetlands

Inspector's Name(s):____________________
Date of Inspection:____________________
Location of the constructed wetland:____________________
Address or Intersection:____________________
Latititude, Longitude:____________________
Date the constructed wetland began operation: ____________________
Area of the constructed wetland (ft. x ft.): ____________________
Time since last rainfall (hr): ____________________
Quantity of last rainfall (in): ____________________
Rainfall Measurement Location: ____________________

Site Sketch (include inlets, outlets, north arrow, etc.)

Based on visual assessment of the site, answer the following questions and make photographic or video-graphic documentation:

1. Has visual inspection been conducted at this location before? □ Yes □ No □ I don't know
   1. a) If yes, enter date:____________________
   1. b) Based on previous visual inspections, have any corrective actions been taken?
      □ Yes □ No □ I don't know **(If yes, describe actions in comments box)**

2. Has it rained within the last 48 hours at this location? □ Yes □ No □ I don't know

3. Does this constructed wetland untilize pretreatment practices upstream?
   □ Yes □ No □ **I don't know (If yes, describe pretreatment practices in comment box)**

4. Access
   4. a) Access to the contructed wetland is:
      □ Clear □ Partially obstructed □ Mostly obstructed □ Inaccessible
   4. b) If obstructed, the obstruction is (choose and provide comments) :
      □ temporary **and** □ no action needed **or** □ action needed
      □ permanent **and** □ before or during installation **or** □ new since installation
   4. c) Access to the upstream and downstream drainage is:
      □ Clear □ Partially obstructed □ Mostly obstructed □ Inaccessible
   4. d) If obstructed, the obstruction is (choose and provide comments) :
      □ temporary **and** □ no action needed **or** □ action needed
      □ permanent **and** □ before or during installation **or** □ new since installation

Comments

5. a) How many inlet structures are present? □ 0 □ 1 □ 2 □ 3 □ 4 □ 5 □ > 5

5. b) Are any of the inlet structures clogged? (If yes, mark location on site sketch above and fill in boxes below with items causing clogging (ie. debris, sediment, vegetation, etc.)

	Inlet #:	Inlet #:	Inlet #:	Inlet #:	Inlet #:
Partially					
Completely					
Not Applicable					

5. c) Are any of the inlet structures askew or misaligned from the original design or otherwise in need of maintenance? (if yes, write in reason: frost heave, vandalism, unknown, etc.)

	Inlet #:	Inlet #:	Inlet #:	Inlet #:	Inlet #:
Reason					

6. How many cells are in the wetland system? ____________

6. a) Is there standing water in the constructed wetland? □ Yes □ No

6. b) If yes, does the water in the wetland have:
- □ Surface sheen (from oils or gasoline)
- □ Murky color (from suspended solids)
- □ Green color (from algae or other biological activity)
- □ Other (describe In comment box)

7. Is there evidence of illicit storm sewer discharges?
□ Yes □ No □ **I don't know (if yes, describe in comment box)**

8. Does the constructed wetland smell like gasoline or oil? □ Yes □ No

9. What is the approximate percentage of vegetation coverage in the practice? ________ %

9. a) Does the current vegetation match the original design? □ Yes □ No □ Unknown

9. b) Is there the precense of:
- □ Weeds
- □ Invasive vegetation
- □ None of the above
- □ Other, specify____________________

9. c) Does the vegetation appear to be healthy? □ Yes □ No (if no, describe in comment box)

9. d) Is the vegetation the appropriate size and density? □ Yes □ No (if no, describe in comment box)

Comments

10. Are there indications of any of the following in the wetland? (If yes, mark on site sketch)
- □ Sediment deposition in excess of 50% of the sediment storage capacity
- □ Erosion or channelization
- □ Excessive or undesirable vegetation (that needs mowing or removal)
- □ Litter or debris
- □ Other
- □ No

10. a) If sediment deposition is evident, what is the source?
- □ Erosion or channelization inside the wet pond
- □ Erosion or channelization outside the wet pond
- □ Construction site erosion
- □ Other
- □ Unknown

11. Are there indications of any of the following on the banks of the filtration practice:
- □ Erosion or channelization
- □ Soil slides or bulges
- □ Excessive animal burrows
- □ Seeps and wet spots
- □ Poorly vegetated areas
- □ Trees on constructed slopes

12. Are any overflow or outlet structures clogged? □ No □ Partially □ Completely □ NA

12. a) If yes, specify the clogging material (i.e. debris, sediment, vegetation, etc.) in the box below.

	Outlet #:	Outlet #:	Outlet #:
Material			
Partial or Comp.			

12. b) Are any of the overflow or outlet structures askew or misaligned from the original design or otherwise in need of maintenance? (if yes, write in reason: frost heave, vandalism, unknown, etc.)

	Outlet #:	Outlet #:	Outlet #:
Reason			

13. Is there any evidence of any of the following downstream of the outlet structure?
□ Sediment deposition □ Erosion or channelization □ Other □ No

13. a) If sediment deposition is evident, what is the source?
- □ Erosion or channelization inside the filtration practice
- □ Erosion or channelization outside the filtration practice
- □ Construction site erosion
- □ Other, Specify____________________
- □ Unknown

Comments

14. Inspector's Recommendations. When is maintenance needed?
   - □ Before the next rainfall
   - □ Before the next rainy season
   - □ Within a year or two
   - □ No sign that any is required

Comments

15. Summarize the results of this inspection and write any other observations in the box below.

Summary and other observations

Optimizing Stormwater Treatment Practices:

A Handbook of Assessment and Maintenance

Field Data Sheet for Level 1 Assessment: Visual Inspection
Filter Strips and Swales

Inspector's Name(s):____________________
Date of Inspection:____________________
Location of the filter strip or swale:____________________
Address or Intersection:____________________
Latititude, Longitude:____________________
Date the filter strip or swale began operation: ____________________
Size of the practice. Depth and length, if swale (ft.): ____________________
Area (ft. x ft.), if filter strip ____________________
Time since last rainfall (hr): ____________________
Quantity of last rainfall (in): ____________________
Rainfall Measurement Location: ____________________

Site Sketch (include inlets, outlets, roads, north arrow, etc.)

Based on visual assessment of the site, answer the following questions and make photographic or video-graphic documentation:

1. Has visual inspection been conducted at this location before? □ Yes □ No □ I don't know
   1. a) If yes, enter date:____________________
   1. b) Based on previous visual inspections, have any corrective actions been taken?
      □ Yes □ No □ I don't know **(If yes, describe actions in comments box)**

2. Has it rained within the last 48 hours at this location? □ Yes □ No □ I don't know

3. Does this swale or filter strip untilize pretreatment practices upstream?
   □ Yes □ No □ **I don't know (If yes, describe pretreatment practices in comment box)**

4. Access
   4. a) Access to the swale or filter strip is:
      □ Clear □ Partially obstructed □ Mostly obstructed □ Inaccessible
   4. b) If obstructed, the obstruction is (choose and provide comments) :
      □ temporary **and** □ no action needed **or** □ action needed
      □ permanent **and** □ before or during installation **or** □ new since installation
   4. c) Access to the upstream and downstream drainage is:
      □ Clear □ Partially obstructed □ Mostly obstructed □ Inaccessible
   4. d) If obstructed, the obstruction is (choose and provide comments) :
      □ temporary **and** □ no action needed **or** □ action needed
      □ permanent **and** □ before or during installation **or** □ new since installation

Comments

5. a) How many inlet structures are present? □ 0 □ 1 □ 2 □ 3 □ 4 □ 5 □ > 5

5. b) Are any of the inlet structures clogged? (If yes, mark location on site sketch above and fill in boxes below with items causing clogging (ie. debris, sediment, vegetation, etc.)

	Inlet #:	Inlet #:	Inlet #:	Inlet #:	Inlet #:
Partially					
Completely					
Not Applicable					

5. c) Are any of the inlet structures askew or misaligned from the original design or otherwise in need of maintenance? (if yes, write in reason: frost heave, vandalism, unknown, etc.)

	Inlet #:	Inlet #:	Inlet #:	Inlet #:	Inlet #:
Reason					

6. Is there standing water in the swale or on the filter strip? □ Yes □ No
   6. a) If yes, does the water have:
      - □ Surface sheen (from oils or gasoline)
      - □ Murky color (from suspended solids)
      - □ Green color (from algae or other biological activity)
      - □ Other (describe In comment box)

7. Is there evidence of illicit storm sewer discharges?
   □ Yes □ No □ **I don't know (if yes, describe in comment box)**

8. What is the approximate percentage of vegetation coverage in the practice? ________ %
   8. a) Does the current vegetation match the original design? □ Yes □ No □ Unknown
   8. b) Is there the precense of:
      - □ Weeds
      - □ Invasive vegetation
      - □ None of the above
      - □ Other, specify________________________________
   8. c) Does the vegetation appear to be healthy? □ Yes □ No (if no, describe in comment box)
   8. d) Is the vegetation the appropriate size and density? □ Yes □ No (if no, describe in comment box)

9. Are there indications of any of the following within the filter strip or swale? (If yes, mark on site sketch)
   - □ Sediment deposition
   - □ Erosion or channelization
   - □ Excessive or undesirable vegetation (that needs mowing or removal)
   - □ Bare soil or lack of healthy vegetation significantly different from the original design
   - □ Litter or debris
   - □ Other
   - □ No

Comments

9. a) If sediment deposition is evident, what is the source?
   - □ Erosion or channelization inside the wet pond
   - □ Erosion or channelization outside the wet pond
   - □ Construction site erosion
   - □ Other
   - □ Unknown

10. Are there indications of any of the following on the banks of the swale:
   - □ Erosion or channelization
   - □ Soil slides or bulges
   - □ Excessive animal burrows
   - □ Seeps and wet spots
   - □ Poorly vegetated areas
   - □ Trees on constructed slopes
   - □ Not applicable

11. Are any overflow or outlet structures clogged? □ No □ Partially □ Completely □ NA

11. a) If yes, specify the clogging material (i.e. debris, sediment, vegetation, etc.) in the box below.

	Outlet #:	Outlet #:	Outlet #:
Material			
Partial or Comp.			

11. b) Are any of the overflow structures askew or misaligned from the original design or otherwise in need of maintenance? (if yes, write in reason: frost heave, vandalism, unknown, etc.)

	Outlet #:	Outlet #:	Outlet #:
Reason			

12. Is there any evidence of any of the following downstream of the outlet structure?

□ Sediment deposition □ Erosion or channelization □ Other □ No

12. a) If sediment deposition is evident, what is the source?
   - □ Erosion or channelization inside the filtration practice
   - □ Erosion or channelization outside the filtration practice
   - □ Construction site erosion
   - □ Other, Specify____________________
   - □ Unknown

13. Inspector's Recommendations. When is maintenance needed?
   - □ Before the next rainfall
   - □ Before the next rainy season
   - □ Within a year or two
   - □ No sign that any is required

Comments

12. Summarize the results of this inspection and write any other observations in the box below.

Summary and other observations

References

Addison H (1941) Hydraulic measurements: a manual for engineers. John Wiley & Sons, Inc., New York, NY

Agricultural and Resource Management Council of Australia and New Zealand and the Australian and New Zealand Environment and Conservation Council (ARMC-ANZ-ECC) (2000) Australian guidelines for urban stormwater management. Available at: http://www.environment.gov.au/water/publications/quality/pubs/urban-stormwater-management-paper10.pdf

Agricultural Experiment Stations (AES) (1988) Recommended chemical soil test procedures for the North-Central region. Agricultural experiment stations of Illinois, Indiana, Iowa, Kansas, Michigan, Minnesota, Missouri, Nebraska, North Dakota, Ohio, Pennsylvania, South Dakota and Wisconsin, and the US Department of Agriculture cooperating

Ahmed F, Gulliver JS (2011) User's manual for the MPD infiltrometer. St. Anthony Falls Laboratory, University of Minnesota, Minneapolis, MN

Ahmed F, Gulliver JS, Nieber JL (2011) A new technique to measure infiltration rate for assessing infiltration of BMPs. 12th International Conference on Urban Drainage, Porto Alegre, Brazil, 11–16 September 2011

Aldridge KT, Ganf GG (2003) Modification of sediment redox potential by three contrasting macrophytes: implications for phosphorus adsorption/desorption. Marine & Freshwater Research 54:87–94

American Public Health Association (APHA) (1998a) Section 4500-P Phosphorus. In: Clescerl LS et al (eds.) Standard methods for the examination of water and wastewater, 20th edn. American Public Health Association, American Water Works Association, and the Water Environment Federation, Washington, DC

American Public Health Association (APHA) (1998b) Standard methods for the examination of water and wastewater, 20th edn. American Public Health Association, American Water Works Association, and the Water Environment Federation, Washington, DC

American Society for Testing and Materials (ASTM) (2002) C33/C33M Standard specification for concrete aggregates. ASTM International, West Conshohocken, PA, www.astm.org

American Society for Testing and Materials (ASTM) (2005) D3385-09 Standard test method for infiltration rate of soils in field using double-ring infiltrometer. ASTM International, West Conshohocken, PA, www.astm.org

American Society for Testing and Materials (ASTM) (2007a) D1556-07 Standard test method for density and unit weight of soil in place by the sand-cone method. ASTM International, West Conshohocken, PA, www.astm.org

American Society for Testing and Materials (ASTM) (2007b) D3977-97 Standard test methods determining sediment concentrations in water samples. ASTM International, West Conshohocken, PA, www.astm.org

American Society for Testing and Materials (ASTM) (2008) D5640–95 Standard guide for selection of weirs and flumes for open-channel flow measurement of water. ASTM International, West Conshohocken, PA, www.astm.org

A.J. Erickson et al., *Optimizing Stormwater Treatment Practices: A Handbook of Assessment and Maintenance*, DOI 10.1007/978-1-4614-4624-8,

American Society for Testing and Materials (ASTM) (2009) C1701/C1701M Standard test method for infiltration rate of in place pervious concrete. ASTM International, West Conshohocken, PA, www.astm.org
American Society for Testing and Materials (ASTM) (2010a) D5126 / D5126M–90 Standard guide for comparison of field methods for determining hydraulic conductivity in vadose zone. ASTM International, West Conshohocken, PA, www.astm.org
American Society for Testing and Materials (ASTM) (2010b) D6640–01 Standard practice for collection and handling of soils obtained in core barrel samplers for environmental investigations. ASTM International, West Conshohocken, PA, www.astm.org
Ancion PY, Lear G, Lewis GD (2010) Three common metal contaminants of urban runoff (Zn, Cu & Pb) accumulate in freshwater biofilm and modify embedded bacterial communities. Environmental Pollution 158(8):2738–2745
Anderson DL, Siegrist RL, Otis RJ (1985) Technology assessment of intermittent sand filters. Report #832R85100, Municipal Environmental Research Laboratory, Office of Research and Development, US EPA, Washington, DC
Arai R, Tada K, Nakatani N, Okuno T, Ohta K (2009) Automatic measurement of dissolved inorganic nitrogen ions in coastal field using simplified flow injection method. Int J Offshore Polar Engineering 19(1):71–76
Arias CA, Del Bubba M, Brix H (2001) Phosphorus removal by sands for use as media in subsurface flow constructed reed beds. Water Research 35:1159–1168
Asian Development Bank (ADB) (2006) Proposed loan People's Republic of China: Wuhan wastewater and stormwater management project. Available at: http://www.adb.org/Documents/RRPs/PRC/37597-PRC-RRP.pdf
Asleson BC, Nestingen RS, Gulliver JS, Hozalski RM, Nieber JL (2009) Performance assessment of rain gardens. J Am Water Resources Association 45(4):1019–1031
Bagarello V, Iovino M, Elrick D (2004) A simplified falling-head technique for rapid determination of field-saturated hydraulic conductivity. Soil Science Society of America Journal 68(1):66–73
Barr Engineering (2011) SHSAM: sizing hydrodynamic separators and man-holes. Accessed 12 June, 2011 at http://www.barr.com/Environmental/SHSAM/SHSAMapp.asp
Barraud S, Gautier A, Bardin JP, Riou V (1999) The impact of intentional stormwater infiltration on soil and groundwater. Water Science and Technology 39(2):185–192
Barrett ME, Walsh PM, Malina JFJ, Charbaneau RJ (1998) Performance of vegetative controls for treating highway runoff. Journal of Environmental Engineering 124(11):1121–1128
Bedient PB, Huber WC (1992) Hydrology and floodplain analysis, 2nd edn. Addison-Wesley Publishing Company, Massachusetts, Reading, MA
Bell W, Stokes L, Gavan LJ, Nguyen TN (1995) Assessment of the pollutant removal efficiencies of Delaware sand filter BMPs. City of Alexandria Department of Transportation and Environmental Services, Alexandria Virginia
Black CA (1965) Methods of soil analysis, Part 2. Chemical and microbiological properties, Number 9 in the series AGRONOMY. American Society of Agronomy, Inc., Publisher, Madison, WI
Booth D, Hartley D, Jackson R (2002) Forest cover, impervious-surface area, and the mitigation of stormwater impacts. J Am Water Resources Association 38:835–845
Bos MG (ed) (1998) Discharge measurement structures. International Institute for land reclamation and improvement, Publication 20, Wageningen, the Netherlands
Brezonik PL, Stadelmann TH (2002) Analysis and predictive models of stormwater runoff volumes, loads, and pollutant concentration from watersheds in the Twin Cities metropolitan area, Minnesota, USA. Water Research 36:1743–1757
Brown E, Caraco D, Pitt R (2004) Illicit discharge detection and elimination: a guidance manual for program development and technical assessment. Center for Watershed Protection, Ellicott City, MD
Brutsaert W (2005) Hydrology: an introduction. Cambridge University Press, New York, NY

Bulc T, Slak AS (2003) Performance of constructed wetland for highway runoff treatment. Water Science and Technology 48:315–322

Camesano TA, Logan BE (1998) Influence of fluid velocity and cell concentration on the transport of motile and non-motile bacteria in porous media. Environ Sci Tech 32(34):1699

Carleton JN, Gizzard TJ, Godrej AN, Post HE, Lampe L, Kenel PP (2000) Performance of a constructed wetlands in treating urban stormwater runoff. Water Environment Research 72:295–304

Clark SE, Pitt R (2007) Influencing factors and a proposed evaluation methodology for predicting groundwater contamination potential from stormwater infiltration practices. Water Environment Research 79:29–36

Claytor RA, Schueler TR (1996) Design of stormwater filtering systems. Center for Watershed Protection for Chesapeake Research Consortium and US EPA, Solomons, MD

Clements W, Kiffney P (2002) Ecological effects of metals on benthic invertebrates. In: Simon TP (ed) Biological response signatures. CRC Press, Boca Raton, FL, pp 135–154

Gilson Company, Inc. (2003) AP-1B Asphalt Field Permeameter Operating Instructions. Gilson Company, Inc., Lewis Center, OH

Collins KA, Hunt WF, Hathaway JM (2010) Side-by-side comparison of nitrogen species removal for four types of permeable pavement and standard asphalt in eastern North Carolina. Journal of Hydrologic Engineering 15(6):512–521

Comings KJ, Booth DB, Horner RR (2000) Storm water pollutant removal by two wet ponds in Bellevue Washington. Journal of Environmental Engineering 126:321–330

Commonwealth of Australia (2002) Introduction to urban stormwater management in Australia. Available at: http://www.environment.gov.au/coasts/publications/stormwater/pubs/stormwater.pdf

Cooley LA Jr. (1999) Permeability of superpave mixtures: evaluation of field permeameters, NCAT Report No. 99–1, Auburn, AL

SoilMoisture Equipment Corp (1986) Guelph permeameter 2800ki operating instructions, Rev 8/86. SoilMoisture Equipment Corp, Santa Barbara, CA

Corsi SR, Graczyk DJ, Geis SW, Booth NL, Richards KD (2010) A fresh look at road salt: aquatic toxicity and water-quality impacts on local, regional, and national scales. Environ Sci Technology 44(19):7376–7382

Davis AP, Shokouhian M, Ni SB (2001) Loading estimates of lead, copper, cadmium, and zinc in urban runoff from specific sources. Chemosphere 44(5):997–1009

Davis AP, Shokouhian M, Sharma H, Minami C, Winogradoff D (2003) Water quality improvement through bioretention: lead, copper, and zinc removal. Water Environment Research 75(1):73–82

Deletic A, Fletcher TD (2006) Performance of grass filters used for stormwater treatment—a field and modeling study. Journal of hydrology 317(3–4):261–275

Dhamotharan S, Gulliver JS, Stefan HG (1981) Unsteady one-dimensional settling of suspended sediment. Water Resources Research 17:1125–1132

Dierkes C, Geiger WF (1999) Pollution retention capabilities of roadside soils. Water Science and Technology 39(2):201–208

Dingman SL (2002) Physical hydrology. Waveland Press, Long Grove, IL

Erickson AJ, Gulliver JS, Weiss PT (2007) Enhanced sand filtration for storm water phosphorus removal. Journal of Environmental Engineering 133(5):485–497

Erickson AJ, Gulliver JS, Kang JH, Weiss PT, Wilson CB (2010) Maintenance for stormwater treatment practices. J Contemp Water Research & Education 146:75–82

Erickson AJ, Gulliver JS, Weiss PT (2012) Capturing phosphates with iron enhanced sand filtration. Water Research 46(9):3032–3042

European Commission (2008). Water Information System for Europe (WISE) Waternote 8. Pollution: reducing dangerous chemicals and Europe's waters. Available at: http://ec.europa.eu/environment/water/participation/pdf/waternotes/water_note8_chemical_pollution.pdf

Fang F, Brezonik PL, Mulla DJ, Hatch LK (2002) Estimating runoff phosphorus loss from calcareous soils in the Minnesota River basin. Journal of Environmental Quality 31:1918–1929

Farahbakhshazad N, Morrison GM (2003) Phosphorus removal in a vertical upflow constructed wetland system. Water Sci Technol 48:43–50

Farrell AC, Scheckenberger RB (2003) An assessment of long-term monitoring data for constructed wetlands for urban highway runoff control. Water Quality Research Journal of Canada 38:283–315

Faulkner S, Richardson C (1991) Physical and chemical characteristics of freshwater wetlands soils. In: Hammer D (ed) Constructed wetlands for wastewater treatment. CRC Press, Boca Raton, Florida

Federal Interagency Sedimentation Project (FISP) (1941) Laboratory investigations of suspended sediment samplers. Iowa University Hydraulics Laboratory, Iowa City, IA

Ferguson BK (2005) Porous pavements, integrated studies in water management and land development. Taylor & Francis, London

Ferguson RI, Church M (2004) A simple universal equation for grain settling velocity. Journal of Sedimentary Research 74(6):155–160

Franzini JB, Finnemore EJ (1997) Fluid mechanics with engineering applications, 9th edn. McGraw Hill, New York, NY

Galli J (1992) Analysis of urban stormwater BMP performance and longevity in Prince George's County, Maryland. Metropolitan Washington Council of Governments, Washington, DC

Gettel M, Gulliver JS, Kayhanian M, DeGroot G, Br and J, Mohseni O, Erickson AJ (2011) Improving suspended sediment measurements by automatic samplers. Journal of Environmental Monitoring 13(10):2703–2709

Gosset WS (1908) The probable error of a mean. Biometrika 6(1):1–25

Gray JR, Glysson GD, Turcios LM, Schwarz GE (2000) Comparability of suspended-sediment concentration and total suspended solids data. US Geological Survey, Reston, VA

Gwinn WR, Parsons DA (1976) Discharge equations for HS, H, and HL flumes. ASCE Journal of the Hydraulics Division 102:73–88

Harper HH, Herr JL (1993) Treatment efficiencies of detention with filtration systems. SJRWMD Contract No. 90B103

Harrington WZ, Strohschein BL, Reedy D, Harrington JE, Schiller WR (1995) Pavement temperature and burns: streets of fire. Annals of Emergency Medicine 26(5):563–568

Herb WR, Janke B, Mohseni O, Stefan HG (2007a) Estimation of runoff temperatures and heat exports from different land and water surfaces. Project Report no. 488, St. Anthony Falls Laboratory, University of Minnesota, Minneapolis, MN. http://purl.umn.edu/113694

Herb WR, Mohseni O, Stefan HG (2007b) A model for mitigation of surface runoff temperatures by a wetland basin and a wetland complex. Project Report no. 496, St. Anthony Falls Laboratory, University of Minnesota, Minneapolis, MN. http://purl.umn.edu/115329

Herb WR, Janke B, Mohseni O, Stefan HG (2008) Thermal pollution of streams by runoff from paved surfaces. Hydrological Processes 22(7):987–999

Herb WR, Janke B, Mohseni O, Stefan HG (2009) Simulation of temperature mitigation by a stormwater detention pond. Journal of the American Resources Association 45(5):1164–1178

Herrera Environmental Consultants (1995) Lake Sammammish phase 2 restoration project, Lake Park storm water treatment facility, task 2: bench scale test results

Herschy RW (1995) Streamflow measurement. E & F N Spoon, London

Hillel D (1998) Environmental soil physics. Academic Press, Amsterdam

Howard A, Mohseni O, Gulliver JS, Stefan HG (2011) SAFL Baffle retrofit for suspended sediment removal in storm sewer sumps. Water Research 45(18):5895–5904

Hunt WF, Jarrett AR, Smith JT, Sharkey LJ (2006) Evaluating bioretention hydrology and nutrient removal at three field sites in North Carolina. J Irrigat Drain Eng-ASCE 132(6):600–608

Hunt WF, Lord WG (2006a) Urban waterways: bioretention performance, design, construction, and maintenance. North Carolina State University, North Carolina Cooperative Extension Service, Raleigh, NC

Hunt WF, Lord WG (2006b) Urban waterways: maintenance of stormwater wetlands and wet ponds. North Carolina State University, North Carolina Cooperative Extension Service, Raleigh, NC

Hussain CF, Brand J, Gulliver JS, Weiss PT (2006) Water quality performance of dry detention ponds with underdrains. Minnesota Department of Transportation Report 2006–43, December 2006. http://www.cts.umn.edu/Publications/ResearchReports/reportdetail.html?id=1120

Jang A, Seo Y, Bishop PL (2005) The removal of heavy metals in urban runoff by sorption on mulch. Environmental Pollution 133(1):117–127

Janke B, Mohseni O, Herb WR, Stefan HG (2009) Heating of rainfall runoff on residential and commercial roofs. Project Report no. 533, St. Anthony Falls Laboratory, University of Minnesota, Minneapolis, MN. http://purl.umn.edu/115559

Kadlec RH, Knight RL (1996) Treatment wetlands. Lewis Publishers, Boca Raton, FL

Kaushal SS, Groffman PM, Likens GE, Belt KT, Stack WP, Kelly VR, Band LE, Fisher GT (2005) Increased salinization of fresh water in the northeastern United States. Proceedings of the National Academy of Sciences 102:13517–13520

Kayhanian M, Suverkropp C, Ruby A, Tsay K (2007) Characterization and prediction of highway runoff constituent event mean concentration. J Environ Manage 85(2):279–295

Kayhanian M, Stransky C, Bay S, Lau SL, Stenstrom MK (2008) Toxicity of urban highway runoff with respect to storm duration. Science of the Sci Total Environment 389(2–3):386–406

Kennedy P (1999) The effects of road transport on freshwater and marine ecosystems. Prepared for the Ministry of Transport, Te Manatu Waka by Kingett Mitchell Ltd, Auckland, New Zealand

Klatt JG, Mallarino AP, Downing AJ, Kopaska JA, Wittry DJ (2003) Soil phosphorus, management practices and their relationship to phosphorus delivery in the Iowa Clear Lake watershed. Journal of Environmental Quality 32:2140–2149

Klute A (1986) Methods of soil analysis, Part I. Physical and mineralogical methods, 2nd edn. Soil Science Society of America, Inc. Publisher, Madison, WI

Koob TL (2002) Treatment of highway runoff using wet detention ponds: water quality, operation, and maintenance considerations. Washington State University, Pullman, WA

Kovacic DA, David MB, Gentry LE, Starks KM, Cooke RA (2000) Effectiveness of constructed wetlands in reducing nitrogen and phosphorus export from agricultural tile drainage. Journal of Environmental Quality 29:1262–1274

Laber J (2000) Constructed wetland system for storm water treatment. Journal of environmental science and health. Part A, Toxic/hazardous substances & environmental engineering 35:1279

Landers DH (1982) Effects of naturally senescing aquatic macrophytes on nutrient chemistry and chlorophyll a of surrounding waters. Limnol Oceanog 27:428–439

Landphair HC, McFalls JA, Thompson D (2000) Design methods, selection, and cost-effectiveness of stormwater quality structures. Texas Department of Transportation, Report 1837–1, Austin, TX

Lee GF, Jones-Lee A (2003) Synthesis and discussion of findings on the causes and factors influencing low Do in the San Joaquin River Deep Water Ship Channel near Stockon, CA: Including 2002 Data. Report Submitted to SJR low DO in the San TMDL Steering Committee and CALFED Bay-Delta Program, G. Fred Lee & Associates, El Macero, CA

Legret M, Nicollet M, Miloda P, Colandini V, Raimbault G (1999) Simulation of heavy metal pollution from stormwater infiltration through a porous pavement with reservoir structure. Water Science and Technology 39(2):119–125

Lin ZQ, Terry N (2003) Selenium removal by constructed wetlands: quantitative importance of biological volatilization in the treatment of selenium-laden agricultural drainage water. Environmental Science and Technology 37:606–615

Lindsey G, Roberts L, Page W (1991) Storm water management infiltration. Maryland Department of the Environment, Sediment and Storm Water Administration. Baltimore, MD. (http://www.p2pays.org/ref/17/16120.pdf)

MacBerthouex P, Brown LC (1996) Statistics for environmental engineers. Lewis Publishers, Washington, DC

Maehlum T, Jenssen PD, Warner WS (1995) Cold-climate constructed wetlands. Water Science and Technology 32:95

Maestre A, Pitt R (2005) The national stormwater quality database, version 1.1 A compilation and analysis of NPDES stormwater monitoring information. US EPA Office of Water, Washington, DC

Mallin MA, Ensign SH, Wheeler TL, Mayes DB (2002) Surface water quality: pollutant removal efficacy of three wet detention ponds. Journal of Environmental Quality 31:654–660

Mallin MA, Johnson VL, Ensign SH (2009) Comparative impacts of stormwater runoff on water quality of an urban, a suburban, and a rural stream. Environ Monit Assess 159(1–4):475–491

McCarthy EL (1934) Mariotte's bottle. Science 80:100

McDowell RW, Sharpley AA, Beegle DB, Weld JL (2001) Comparing phosphorus management strategies at a watershed scale. Journal of Soil and Water Conservation Conservat 56:306–315

Metcalf and Eddy, Inc. (1991) Wastewater engineering: treatment, disposal, and reuse. McGraw - Hill, New York, NY

Michigan Department of Environmental Quality (MDEQ) (2006) Part four: water quality standards. http://www.michigan.gov/documents/deq/wb-swas-rules-part4_254149_7.pdf

Mijangos-Montiel JL, Wakida FT, Temores-Pena J (2010) Stormwater quality from gas stations in Tijuana, Mexico. International Journal of Environmental Research 4(4):777–784

Minnesota Pollution Control Agency (MPCA) (2003) Minnesota rule 7050.0222. Specific standards of quality and purity for class 2 waters of the state; aquatic life and recreation. Minnesota Pollution Control Website: https://www.revisor.mn.gov/rules/?id=7050.0222

Moore DS, McCabe GP (2003) Introduction to the practice of statistics, 4th edn. W.H. Freeman and Company, New York, NY

Morrison GM, Revitt DM, Ellis JB, Balmer P, Svensson G (1983) Heavy metal partitioning between the dissolved and suspended solid phases of stormwater runoff from a residential area. Science of the Total Environment 33:237

Munoz-Carpena R, Regalado CM, Alvarez-Benedi J, Bartoli F (2002) Field evaluation of the new Philip-Dunne permeameter for measuring saturated hydraulic conductivity. Soil Science 167(1):9–24

National Academy of Sciences (NAS) (2000) Watershed management for potable water supply: assessing the New York City strategy. Commission on Geosciences, Environment and Resources, Washington, DC

Nelson P (2003) Index to EPA test methods, April 2003 revision. US Environmental Protection Agency, Region 1, Boston, MA. http://www.epa.gov/region1/info/testmethods/pdfs/testmeth.pdf

Nestingen R (2007) The comparison of infiltration devices and modification of the Philip-Dunne permeameter for the assessment of rain gardens. M.S. Thesis. University of Minnesota, Minneapolis, MN

New Jersey Department of Environmental Protection (NJDEP), Division of Watershed Management (2004) New Jersey stormwater best management practices manual, Trenton, NJ

New York State Attorney General's Office (NYSAGO) (2011) Reducing harmful phosphorus pollution in the New York City reservoirs through the Clean Water Act's "Total Maximum Daily Load" requirements: a case study of the New Croton Reservoir and recommendation to the EPA. May 27, 2011. (http://www.ag.ny.gov/bureaus/environmental/pdfs/phosphorus_report.pdf)

Novotny EF, Stefan HG (2010) Projections of chloride concentrations in urban lakes receiving road de-icing salt. Water Air and Soil Pollution 211(1–4):261–271

Novotny EV, Murphy D, Stefan HG (2008) Increase of urban lake salinity by road deicing salt. Science of the Science Total Environment 406(1–2):144

Novotny EV, Sander AR, Mohseni O, Stefan HG (2009) Chloride ion transport and mass balance in a metropolitan area using road salt. Water Resources Research 45:W12410. doi:10.1029/2009WR008141

Oberts G (1999) Return to Lake McCarrons: does the performance of wetlands hold up over time? Watershed Protection Technique 3:597–600

Parshall RL (1936) The Parshall measuring flume. Colorado State College Colorado Experiment Station, Fort Collins, CO

Paul J, Meyer J (2001) Streams in the urban landscape. Annual Review of Ecology and Systematics 32:333–365

Paus KH, Morgan J, Hozalski RM, Gulliver JS, Leiknes T (Unpublished) The influence of temperature and salt on toxic metal removal and retention in bioretention cells. University of Minnesota, Minneapolis, MN

Pepper IL, Gerba CP, Brusseau ML (1996) Pollution science. Academic Press, San Diego, CA

Philip JR (1993) Approximate analysis of falling-head lined borehole permeameter. Water Resources Research 29:3763–3768

Pitt R (2002) Receiving water impacts associated with urban runoff. In: D Hoffman, B Rattner, J Burton, BS, J Cairns, J, (eds.) Handbook of ecotoxicology, 2nd Edition. CRC Lewis, Boca Raton, FL

Pitt R, Clark S, Parmer K, Field R (1996) Groundwater contamination from stormwater infiltration. Ann Arbor Press, Inc., Chelsea, MI

Pitt R, Clark S, Field R (1999) Groundwater contamination potential from stormwater infiltration practices. Urban Water 1:217–236

Polta RC (2001) Fate and environmental impacts of sediments removed from stormwater ponds: a review. Environmental Services Division, Metropolitan Council, St. Paul, MN

Polta RC, Balogh S, Craft-Reardon A (2006) Characterization of stormwater pond sediments. Environmental Services Division, Metropolitan Council, St. Paul, MN

Pote DH, Danial TC, Nichols AN, Moore PA, Miller DM, Edwards DR (1999) Relationship between phosphorus levels in three ultisols and phosphorus concentrations in runoff. Journal of Environmental Quality 28:170–175

Rangsivek R, Jekel MR (2005) Removal of dissolved metals by zero-valent iron (ZVI): kinetics, equilibria, processes and implications for stormwater runoff treatment. Water Research 39(17):4153–4163

Rawls WJ, Brakensiek DL, Miller N (1983) Green-Ampt infiltration parameters from soils data. J Hydraul Div, ASCE 109(1):62–70

Rawls WJ, Gimenez D, Grossman R (1998) Use of soil texture, bulk density, and slope of the water retention curve to predict saturated hydraulic conductivity. Transactions ASAE 41(4):983–988

Reed GD (1981) Evaluation of automatic suspended solids sampling procedures. Res J Water Pollut Contr Fed 53:1481–1491

Regalado CM, Ritter A, Alvarez-Benedi J, Munoz-Carpena R (2005) Simplified method to estimate the Green-Ampt wetting front suction and soil sorptivity with the Philip-Dunne falling-head permeameter. Vadose Zone Journal 4(2):291–299

Reynolds WD, Elrick DE (1991) Determination of hydraulic conductivity using a tension infiltrometer. Soil Science Society of America Journal 55(3):633–639

Richards LA (ed) (1954) Diagnosis and improvement of saline and alkali soils. US Dept. of Agriculture handbook #60. Riverside, CA

Richardson JL, Vepraskas MJ (2001) Wetland soils: genesis, hydrology, landscapes, and classification. Lewis Publishers, Boca Raton, FL

Robbins RW, Glicker JL, Bloem DM, Niss BM (1991) Effective watershed management for surface water supplies. J Am Water Works Assoc 83:34–44

Robertson WD, Schiff SL, Ptacek CJ (1998) Review of phosphate mobility and persistence in 10 septic system plumes. Ground Water 36:1000–1010

Rouse H (1937) Modern conceptions of mechanics of fluid turbulence. Transactions ASCE 102:463–543

Sansalone JJ, Koran JM, Smithson JA, Buchberger SG (1998) Physical characteristics of urban roadway solids transported during rain events. Journal of Environmental Engineering 124(5):427–440

Saxton KE, Rawls W (2005) Soil water characteristics: hydraulics property calculator. USDA Agricultural Research Service and USDA-ARS, Hydrology and Remote Sensing Laboratory, Beltsville, MD. Accessed Jan 2012. http://hrsl.arsusda.gov/soilwater/index.htm

Schindler DW (1977) Evolution of phosphorus limitation in lakes: natural mechanisms compensate for deficiencies of nitrogen and carbon in eutrophied lakes. Science 195:260–262

Scholz M, Grabowiecki P (2007) Review of permeable pavement systems. Building and Environment 42(11):3830–3836

Schueler TR (1987) Controlling urban runoff: a practical manual for planning and designing urban best management practices. Metropolitan Washington Council of Governments, Washington, DC

Schueler T (1992) Design of wetland stormwater systems: guidelines for creating diverse and effective wetlands in the Mid-Atlantic region. Metropolitan Washington Council of Governments, Washington, DC

Schueler T (2000a) The importance of imperviousness. In: Schueler T, Holland HK (eds.) The practice of watershed protection. Center for Watershed Protection, Ellicott City, MD

Schueler T (2000b) Microbes in urban watersheds: concentrations, sources and pathways. In: Schueler T, Holland HK (eds.) The practice of watershed protection. Center for Watershed Protection, Ellicott City, MD

Schueler TR, Kumble PA, Heraty MA (1992) A current assessment of urban best management practices: techniques for reducing non-point source pollution in the coastal zone. Anacostia Restoration Team, Department of Environmental Programs, Metropolitan Washington Council of Governments, Washington, DC

Selbig WR, Bannerman R, Bowman G (2007) Improving the accuracy of sediment-associated constituent concentrations in whole storm water samples by wet-sieving. Journal of Environmental Quality 36(1):226–232

Shaw D, Schmidt R (2003) Plants for stormwater design: species selection for the upper midwest. Minnesota Pollution Control Agency, St. Paul, MN

Shilton A, Wilks T, Smyth J, Bickers P (2000) Tracer studies on a New Zealand waste stabilization pond and analysis of treatment efficiency. Water Science and Technology 42:343–348

Silvan N, Vasander H, Laine J (2004) Vegetation is the main factor in nutrient retention in a constructed wetland buffer. Plant Soil 258:179–187

Stefan HG, Cardoni J, Schiebe F, Cooper C (1983) A model of light penetration in a turbid lake. Water Resources Research 19:109–120

Stenstrom MK, Strecker EW (1993) Assessment of storm drain sources of contaminants to Santa Monica Bay, Volume II. UCLA ENG 93–63, Los Angeles, CA

Stenstrom MK, Silverman G, Bursztynsky TA (1982) Oil and grease in stormwater runoff. Association of Bay Area Governments, Berkley, CA. Accessed online at: (http://www.seas.ucla.edu/stenstro/r/r8)

Steuer J, Selbig W, Hornewer N, Prey J (1997) Sources of contamination in an urban basin in Marquette, Michigan, and an analysis of concentrations, loads, and data quality, Middleton, WI. Water-Resources Investigations Report 97–4242, US Geological Suvey, Madison, WI

Stokes GG (1851) Transactions of the Cambridge Philosophical Society 9 (Part II), 8

Sturm TW (2001) Open channel hydraulics, 1st edn. McGraw-Hill, Boston

Sullivan K, Martin DJ, Cardwell RD, Toll JE, Duke S (2000) An analysis of the effects of temperature on salmonids of the Pacific Northwest with implications for selecting temperature criteria. Sustainable Ecosystems Institute, Portland, OR. (http://www.sei.org/downloads/reports/salmon2000.pdf)

Sun X, Davis AP (2007) Heavy metal fates in laboratory bioretention systems. Chemosphere 66:1601–1609

Taylor JK (1987) Quality assurance of chemical measurements. Lewis Publishers, Chelsea, MI

Teledyne Isco Inc. (2006) Teledyne Isco - products - 4200 series flow meters. http://www.isco.com/products/products2.asp?PL=20230 Accessed on June 30, 2006

Thien SJ (1979) A flow diagram for teaching texture-by-feel analysis. J Agron Educ 8:54–55

Thomann RV, Mueller JA (1987) Principles of surface water quality modeling and control. Harper-Collins, New York, NY

Tornes L (2005) Effects of rain gardens on the quality of water in the Minneapolis–St. Paul metropolitan area of Minnesota, 2002–04. Report #2005-5189, US Geological Survey, Mounds View, MN. http://pubs.er.usgs.gov/usgspubs/sir/sir20055189

US Bureau of Reclamation (2001) Water measurement manual. US Environmental Protection Agency. Washington, D.C. http://www.usbr.gov/pmts/hydraulics_lab/pubs/wmm/

US EPA (1983) Results of the nationwide urban runoff program. WH-554, Water Planning Division, US Environmental Protection Agency, Washington, DC

US EPA (1992) Environmental impacts of stormwater discharges: a national profile. EPA 841-R-92-001, Office of Water, US Environmental Protection Agency, Washington, DC

US EPA (1997) Volunteer stream monitoring: a methods manual. EPA 841-B-97-003. http://www.epa.gov/volunteer/stream/index.html. US Environmental Protection Agency, Washington, DC

US EPA (1999a) Preliminary data summary of urban storm water best management practices. EPA-821-R-99-012. US Environmental Protection Agency, Washington, DC

US EPA (1999b) Storm water technology fact sheet: infiltration trench. EPA 832-F-99-019. US Environmental Protection Agency, Washington, DC

US EPA (2000a) Low Impact Development (LID): a literature review. EPA-841-B-00-005. US Environmental Protection Agency, Washington, DC

US EPA (2000b) National water quality inventory, 305(b) Report. EPA-841-R-02-001. US Environmental Protection Agency, Washington, DC

US EPA (2002) Urban stormwater BMP performance monitoring. EPA-821-B-02-001, http://epa.gov/waterscience/stormwater/monitor.htm. US Environmental Protection Agency, Washington, DC

US EPA (2004) EPA ground water & drinking water. Current drinking water standards. http://www.epa.gov/safewater/mcl.html. US Environmental Protection Agency, Washington, DC

US EPA (2011) National summary of state information. Accessed 27 May 2011 (http://iaspub.epa.gov/waters10/attains_nation_cy.control). US Environmental Protection Agency, Washington, DC

Unice KM, Logan BE (2000) The insignificant role of hydrodynamic dispersion on bacterial transport. Journal of Environmental Engineering 126(6):491

United Nations Millennium Project (UNMP) (2005) Investing in development: a practical plan to achieve the millennium development goals: overview. United Nations Development Program, New York, NY

Urban Drainage Flood Control District (1992) Best management practices. In: Urbonas B (ed) Urban storm drainage criteria manual, vol 3. Urban Drainage Flood Control District, Denver, CO

Wang L, Lyons J, Kanehl P (2003) Impacts of urban land cover on trout streams in Wisconsin and Minnesota. Trans Am Fish Soc 132:825–839

Warrick AW, Nielsen DR (1980) Spatial variability of soil physical properties in the field. In: Hillel D (ed) Applications of soil physics. Academic Press, New York, pp 319–344

Waschbusch RJ, Selbig WR, Bannerman RT (1999) Sources of phosphorus in stormwater and street dirt from two urban residential basins in Madison, Wisconsin, 1994–95. Water-Resources Investigations 99–4021, US Geological Survey, Madison, WI

Water Environment Federation (WEF) (1996) Wastewater sampling for process and quality control (Manual of practice). Water Environment Federation, Task Force on Wastewater Sampling for Process and Quality Control, Alexandria, VA

Watershed Management Institute (WMI) (1997) Operation, maintenance, and management of stormwater management systems. Prepared for: US EPA Office of Water, Washington, DC

Watts AW, Ballestero TP, Roseen RM, Houle JP (2010) Polycyclic aromatic hydrocarbons in stormwater runoff from sealcoated pavements. Environmental Science and Technology 44(23):8849–8854

Weiss PT, Gulliver JS, Erickson AJ (2005) The cost and effectiveness of stormwater management practices. http://purl.umn.edu/986. Minnesota Department of Transportation Report 2005–23, June 2005. http://www.cts.umn.edu/Publications/ResearchReports/reportdetail.html?id=1023

Weiss PT, Erickson AJ, Gulliver JS (2007) Cost and pollutant removal of storm-water treatment practices. Journal of Resources Planning and Management 133(3):218–229

Weiss PT, LeFevre G, Gulliver JS (2008) Contamination of soil and groundwater due to stormwater infiltration practices. SAFL Project Report No. 515, St. Anthony Falls Lab, Minneapolis, MN. http://purl.umn.edu/115341

Wenck (2003) Surface water pathogen study. Prepared for the Minnehaha Watershed District, Plymouth, MN

Wetzel RG (1975) Limnology. Saunders company, Philadelphia, PA

Wilson MA, Mohseni OM, Gulliver JS, Hozalski RM, Stefan HG (2009) Assessment of hydrodynamic separators for storm-water treatment. Journal of Hydraulic Engineering ASCE 135(5):383–392

Winandy JE, Barnes HM, Falk RH (2004) Summer temperatures of roof assemblies using western red cedar, wood-thermoplastic composite, or fiberglass shingles. Forest Products Journal 54(11):27–33

Winer R (2000) National pollutant removal performance database for stormwater treatment practices, 2nd edn. Center for Watershed Protection, US EPA Office of Science and Technology, Ellicott City, MD

Wisconsin Department of Natural Resources (WDNR) (2004) Guidelines for designating fish & aquatic life uses for Wisconsin surface waters. http://www.dnr.state.wi.us/org/water/wm/wqs/wbud/UDG_FINAL_2004.pdf Madison, Wisconsin. WI

Wossink A, Hunt B (2003) The economics of structural stormwater BMPs in North Carolina. University of North Carolina Water Resources Research Institute, Report 2003–344

Wright J, Bergsrud F (1991) Irrigation scheduling: checkbook method. University of Minnesota Extension Service Report Number AG-FO-1322-C. Department of Agriculture Engineering, University of Minnesota, St. Paul, MN

Wu JS, Holman RE, Dorney JR (1996) Systematic evaluation of pollutant removal by urban wet detention ponds. Journal of Environmental Engineering 122:983–988

Yu SL, Kuo JT, Fassman EA, Pan H (2001) Field test of grassed-swale performance in removing runoff pollution. Journal of Water Resources Planning and Management 127(3):168–171

Zang GL, Burghardt W, Lu Y, Gong ZT (2001) Phosphorus-enriched soils of urban and suburban Nanjing and their effect on groundwater phosphorus. Journal of Plant Nutrition and Soil Science 164:295–301

Zhang RD (1997) Determination of soil sorptivity and hydraulic conductivity from the disk infiltrometer. Soil Science Society of America Journal 61(4):1024–1030

Zvomuya F, Gupta SC, Rosen CJ (2005) Phosphorus leaching in sandy outwash soils following potato-processing wastewater application. Journal of Environmental Quality 34:1277–1285

Index

A.J. Erickson et al., *Optimizing Stormwater Treatment Practices: A Handbook of Assessment and Maintenance,* DOI 10.1007/978-1-4614-4624-8,

MIX
Papier aus verantwortungsvollen Quellen
Paper from responsible sources
FSC® C105338

If you have any concerns about our products,
you can contact us on
ProductSafety@springernature.com

In case Publisher is established outside the EU,
the EU authorized representative is:
Springer Nature Customer Service Center GmbH
Europaplatz 3, 69115 Heidelberg, Germany

Printed by Libri Plureos GmbH
in Hamburg, Germany